ADVANCES IN STRUCTURAL RELIABILITY

ON RELIABILITY AND RISK ANALYSIS

A series devoted to the publication of courses and educational seminars given at the Joint Research Centre, Ispra Establishment, as part of its education and training program. Published for the Commission of the European Communities, Directorate-General Telecommunications, Information Industry and Innovation.

The publisher will accept continuation orders for this series which may be cancelled at any time and which provide for automatic billing and shipping of each title in the series upon publication. Please write for details.

ADVANCES IN STRUCTURAL RELIABILITY

*Proceedings of the Advanced Seminar on Structural Reliability,
held at the Joint Research Centre, Ispra, Italy, 4–8 June 1984*

Edited by

ALFREDO C. LUCIA

*Commission of the European Communities,
Joint Research Centre, Ispra Establishment,
Ispra, Italy*

D. REIDEL PUBLISHING COMPANY

A MEMBER OF THE KLUWER  ACADEMIC PUBLISHERS GROUP

DORDRECHT / BOSTON / LANCASTER / TOKYO

Library of Congress Cataloging-in-Publication Data

Advanced Seminar on "Structural Reliability"
   (1984 : Joint Research Centre, Ispra, Italy)
   Advances in structural reliability.
   (Ispra courses on reliability and risk analysis)
   Includes index.
   1. Structural stability—Congresses. 2. Reliability (Engineering)—Con-
gresses. 3. Safety factor in engineering—Congresses. I. Lucia, Alfredo C.
II. Title. III. Series.
TA656.A38   1984   624.1'7   86–31506
ISBN 90–277–2429–6

Commission of the European Communities   Joint Research Centre Ispra (Varese), Italy

Publication arrangements by
Commission of the European Communities
Directorate-General Telecommunications, Information Industry and Innovation, Luxembourg

EUR 9831
© 1987 ECSC, EEC, EAEC, Brussels and Luxembourg

LEGAL NOTICE
Neither the Commission of the European Communities nor any person acting on behalf of the
Commission is responsible for the use which might be made of the following information.

Published by D. Reidel Publishing Company
P.O. Box 17, 3300 AA Dordrecht, Holland

Sold and distributed in the U.S.A. and Canada
by Kluwer Academic Publishers,
101 Philip Drive, Assinippi Park, Norwell, MA 02061, U.S.A.

In all other countries, sold and distributed
by Kluwer Academic Publishers Group,
P.O. Box 322, 3300 AH Dordrecht, Holland

All Rights Reserved
No part of the material protected by this copyright notice may be reproduced or utilized
in any form or by any means, electronic or mechanical, including photocopying,
recording or by any information storage and retrieval system,
without written permission from the copyright owner.

Printed in The Netherlands

TABLE OF CONTENTS

FOREWORD

Reliability is a rapidly expanding field of research.
Engineers, mathematicians, physicists and others are making important contributions, accounting for its multidisciplinary nature.
The papers collected in this book (lectures given in the Advanced Seminar on Structural Reliability, held in Ispra on June '84) deals with recent developments of some aspects of theory as well as application of structural reliability.
Several subjects are dealt with, ranging from load combination, treatment of uncertainties, damage modelling and prediction to application to nuclear pressure vessels, industrial tanks, structural systems, etc.
Two papers deal with fuzzy sets theory which seems to constitute a coherent frame to handle imprecise, vague information for which formal-logics could be inadequate.
The problem of combining and using pieces of information coming from different sources and of different "quality", is a central one in structural reliability whose existence (as a scientific discipline) is essentially consequence of the existence of the uncertainty (or more generally "imperfection") of our knowledge.
A primary concern of reliability is consequently to rank information and knowledge according to some figure of confidence, to represent them analytically and to use the all of them in a rational frame.
To a large measure, the development and advancies of the reliability theory owe their success to the merging of tools and techniques belonging to information theory, fracture mechanics, non destructive inspection, probability theory, signal analysis, logics: which contributes to the challenging character of reliability.

GAUSSIAN SAFETY MARGINS

Ove Ditlevsen
Department of Structural Engineering
Technical University of Denmark
Build. 118 DK 2800 Lyngby

ABSTRACT. This review paper describes the "method of Gaussian safety margins" as a method by which a quite general type of reliability problem may be approximated by a reliability problem of series system type in terms of a set of linear safety margins in a Gaussian distribution space. This method is the refined result of contributions published by several people during the last 10 to 15 years.

INTRODUCTION

Structural reliability theory is herein considered as a branch of applied probability theory. The main issue of this theory is to define an event called "failure" and to set up a "probability space" that contains that event. This modeling part of structural reliability theory is based on the theory of mechanics of structures including theories of failure. Further it is based on statistical information about the uncertainty of the relevant parameters or, perhaps, knowledge about the inherent stochastic nature of the structural materials or the applied loads. Finally, but not least important, it is guided by the pragmatic consideration that it should be possible by some manageable mathematical algorithm to calculate the probability of failure. The result may be needed as a part of the input to a decision procedure based on utility theory. More directly it may be needed for comparison with standardized maximal values fixed by some code authority. In passing it should be emphasized that this comparison may cause troubles in case the failure probability is calculated by a reliability analysis model to which the standardized values are not adapted a priori.

----

*   This is a revised reprint of a part of a paper presented at
     Fourth International Conference on
     Applications of Statistics and Probability in Soil and Structural
     Engineering.
     Universita di Firenze (Italy) 1983, Pitagora Editrice.

A. C. Lucia (ed.), Advances in Structural Reliability, 1–15.
© *1987 by ECSC, EEC, EAEC, Brussels and Luxembourg.*

Standardized failure probability values are inseparable from a given class of reliability analysis models. Usually they are chosen by calibration to simple structures of different safety classification. Of course, these test structures are each for their purpose considered to be of both safe and economical design. If the actual reliability analysis is carried out by a non-standardized mathematical model, an allowable failure probability value for that model should be chosen such that it leads approximately to the dimensions of the simple test structures. Even though different reliability analysis models per definition will be inconsistent with each other, this principle ensures that they are almost consistent at least on the class of test structures. To the writer's knowledge no existing national codes or international code recomdations give an explicit specification of such a class of test structures. This is an unfortunate situation for the practical implementation of new developments in the structural reliability theory. At the very least any structural safety code should contain a paragraph saying that other than the authorized reliability analysis models with their adjoined decision rules may be approved by the authorities if they for common, simple structures by and large lead to the same structural dimensions as the authorized methods.

This paper deals with the problem of calculating failure probabilities in a manageable way for a fairly large class of structural reliability problems. The class is defined by the following conditions:

1. The failure event can be represented as a subset of the n-dimensional real space for some finite n.
2. The uncertainty measures of the considered model can be represented as a probability measure on this n-dimensional space.
3. There is a probability measure preserving mapping from a subset Ω of this space onto the n-dimensional space, R^n, such that the induced probability measure on this image space is standardized normal (zero mean vector, unit covariance matrix).
4. The complement of the failure event relative to the subset Ω (of probability 1) by the same mapping has a simply connected subset S of R^n as image.
5. The boundary ∂S of S is piecewise differentiable, i.e. it consists of a finite number of pieces each of which has a normal unit vector that varies continuously across the corresponding smooth part of ∂S.
6. The origin O of the image space is a point in S. The smallest Euclidian distance

$$\delta = \min_{\bar{x} \in \partial S} \sqrt{x_1^2 + \ldots + x_n^2} \tag{1}$$

from the origin to the boundary ∂S is large in the sense that $\Phi(-\delta)$ is close to zero. Here $\Phi(\cdot)$ is the standardized normal distribution function.

7. There is at most a finite number of local minima of the distance $\sqrt{\bar{x}'\bar{x}}$ from O to $\bar{x} \in \partial S$. Each such point of local minimal distance

from 0 has a neighbourhood in which ∂S is differentiable.

The following terminology is used herein: The space postulated in point 1 is called the physical formulation space while the subset Ω is called the set of possible observations. The image space of Ω by the probability measure preserving mapping is called the standardized Gaussian formulation space or the transformed formulation space. The subset S is called the safe set while its complement F is called the failure set. The boundary $\partial S = \partial F$ is called the failure surface. The same terminology is used for those subsets of the physical formulation space that maps into S, F, and ∂S respectively. The points of ∂S at which $\underline{x}'\underline{x}$ has local minimum are called the locally most likely failure points while the point (points) of strict minimum is (are) called the globally most likely failure point (points).

The method of failure probability calculation discussed herein has just recently been fully clarified in its fundamental principles through several papers of different authors, eg. [4,5,6,7,8,9,11,15,17,18,21,22, 26,29,30]. An appropriate name is the "method of Gaussian safety margins". It is an approximation method resulting from two completely different types of approximations made after each other. The first approximation is only needed if the safe set S is not polyhedral and convex. It is as follows: First locate some or all of the locally most likely failure points. Let there be q of these. Obviously they are among the orthogonal projections of the origin 0 on the failure surface. A system of q hyperplanes is next defined such that each hyperplane of the system has its origin projection point coinciding with a locally most likely failure point. This set of hyperplanes defines a convex polyhedral set, S', with q faces. The complement of this convex polyhedral set, F', has a probability which clearly is less than $\Phi(-\beta_1) + \ldots + \Phi(-\beta_q)$ in which $\beta_1, \ldots, \beta_q$ are the distances from 0 to the q origin projection points, respectively. The first approximation consists in simply changing the safe set S to S', that is, to approximate the failure probability by the probability of F'. Recently Breitung [4,5] has shown that

$$\frac{\Phi^{-1}(P(S))}{\Phi^{-1}(P(S'))} \to 1 \text{ as } \delta \to \infty \tag{2}$$

where the limit passage takes place through similarity expansions of the pair (S,S'). Thus a change of S to S' allows for an asymptotically correct result as $\delta \to \infty$ for the generalized reliability index [8]

$$\beta_G = \Phi^{-1}(P(S)) \tag{3}$$

Remark 1: A certain correction of S' is needed if also the calculation of the failure probability, P(F), is required to be asymptotically correct as $\delta \to \infty$. Breitung [4,5] has shown that

$$\frac{1-P(S)}{1-P(S')} \to \alpha \text{ as } \delta \to \infty \tag{4}$$

in which

$$\alpha = [(1-\delta\kappa_1)(1-\delta\kappa_2)\cdot\ldots\cdot(1-\delta\kappa_{n-1})]^{-\frac{1}{2}} \qquad (5)$$

and $\kappa_1, \ldots, \kappa_{n-1}$ are the principal curvatures of the failure surface ∂S at the globally most likely failure point. Note that the numbers $\delta\kappa_1,\ldots,\delta\kappa_{n-1}$ are independent of δ. If there are more than one of these points, each point contributes with an α value as given in (4) and the right side of (3) is changed to the sum of these α values. An asymptotically correct calculation of P(F) is therefore obtained by parallel shifts of all the boundary hyperplanes of S' that have the distance δ from the origin. For each such hyperplane the corrected distance is

$$\delta' = -\Phi^{-1}(\alpha\Phi(-\delta)) \qquad (6)$$

Obviously a more complete correction is obtained by parallel shift of all the boundary hyperplanes such that the ith hyperplane from having the distance β_i gets the distance β_i' from the origin. The corrected distance β_i' is given by (6) after setting $\delta = \beta_i$ and α to the value defined by (5) at the corresponding locally most likely failure point.

The theory of the rest of the paper is independent of this correction of the convex polyhedral set S' to a new convex polyhedral set S". If it is made, the only influence of the correction is that $\beta_1, \ldots, \beta_q$ should be changed to $\beta_1', \ldots, \beta_q'$ in all the relevant formulas. End of Rem.1.

The second approximation concerns the calculation of the probability P(F') = 1 - P(S') for the convex polyhedral set S' assuming that the probability P(S') is close to 1. Generally an exact calculation is impracticable. However, simple upper and lower bounds on the probability of the union of several events are known [9,12,20] and their narrowness for heigh reliability type applications have in recent years been pointed out in several papers. These now well-known bounds are simple because they are expressed solely in terms of the probabilities of the single events that make up the union event and the probabilities of the intersections between any two of these events.

The calculation of the bounds causes no practical problems if the number of faces q of the polyhedral set is moderate. However, several types of practically important structural systems need not be very large before the number q becomes so large that also the calculation of the bounds becomes impracticable. Even more severely, it becomes a serious problem to identify the most important faces of the polyhedral set, i.e. those faces that have distances to the origin equal or close to the minimal distance. Several attempts are reported of developing search procedures to identify these faces. More directly, actually, their concern is to identify the "most dominant failure modes" of a structural system [1,23,24,25,28]. As it will be demonstrated in the following the search for a failure mode is, in fact, equivalent with the search for a part of the failure surface ∂S that contains a locally most likely failure point

and which subsequently is approximated by a face of the polyhedral set
S'.

Unfortunately, due to the difficulty of the problem, these known
search procedures are mainly based on intuitive arguments. No rigorous
proofs of their validity are given. At best some of the methods, when
applied to elastic-ideal plastic frame or truss structures, identify
plastic mechanisms believed to be among the most dominant. However, the
convex polyhedral set defined by a genuine subset of the set of possible
mechanisms will have a probability which is larger than the probability
on the polyhedral set defined by all plastic mechanisms. The difference
may be small as compared to the failure probability, of course. However,
according to Remark 1 it may be quite important that the identified subset of mechanisms contains those mechanisms that define faces in minimal
or close to minimal distance from the origin 0. It seems to be difficult
to prove that this is achieved by the known search procedures. The problem is serious because the obtained lower bound on the failure probability is less relevant for engineering decision than is an upper bound.

Augusti and Baratta [2,3] has pointed out that reliability analysis
of plastic structures should be based on the lower bound theorem of plasticity theory (static theorem of admissible stress fields) rather than
the upper bound theorem (kinematic theorem of mechanisms). Application
of this lower bound theorem leads to upper bounds on the failure probability. However, such analysis is in general quite difficult and often
the best bounds known in deterministic theory are not nearly as close to
the exact result as are the best bounds obtained from the kinematic
theorem. An analysis reported in [16] takes starting point in the lower
bound theorem. Upper bounds on the failure probability are obtained by
systematic use of the method of Gaussian safety margins. Comparisons
with specific examples of known narrow bounds on the exact results confirm that the upper bounds are not particularly close to the exact results. It is interesting, however, that the analysis besides the upper
bound reveals a strategy of search for the most dominant plastic mechanisms. It seems quite successful for frame structures even with a very
large number of possible mechanisms.

The following is a short description of the general technique of
formulating the convex polyhedral set S' in terms of a set of Gaussian
safety margins. It includes the determination of the locally most likely
failure points. The technique is well-known from the literature and it
includes the principle of normal tail approximation [11] which is the
basis for the celebrated Rackwitz-Fiessler algorithm [26]. What is worth
pointing out, however, is that almost all the calculations can formally
be made within the simple first and second moment algebra (uncertainty
algebra [10]). The generic elements of the uncertainty algebra are
Gaussian safety margins. This will become clear in the following.

CONVEX POLYHEDRAL APPROXIMATION TO SAFE SET

Gaussian Safety Margins Forming Series System

Let the coordinates of the transformed formulation space be $(x_1, \ldots, x_n)$
(column matrix $\bar{x}$) and let the equation of the hyperplane of the ith face

of S' be $\bar{a}_i' \bar{x} + b_i = 0$ (prime = transposed) in which $\bar{a}_i$ is the column matrix of constant coefficients $(a_{i1}, \ldots, a_{in})$ and b_i is a positive constant. Since 0 is an internal point of S', the vector $-\bar{a}_i$ is directed out of S' orthogonally to the face. Let $(X_1, \ldots, X_n)$ (column matrix $\bar{X}$) be the standardized Gaussian random vector (zero mean vector, unit covariance matrix) of the transformed formulation space. We obviously have an outcome of $\bar{X}$ in S' if and only if $\bar{a}_i' \bar{X} + b_i \geq 0$ for all $i \in \{1, \ldots, q\}$ (assuming for convenience that the boundary $\partial S'$ is a subset of S'). Thus the event $\bar{X} \in S'$ is equivalent to the event that

$$M_i = \bar{a}_i' \bar{X} + b_i \geq 0 \tag{7}$$

for all $i \in \{1, \ldots, q\}$. The random variables $M_1, \ldots, M_q$ are called Gaussian safety margins. The reliability index β_i of M_i is defined by

$$\beta_i = \frac{E[M_i]}{D[M_i]} = \frac{b_i}{\sqrt{\bar{a}_i' \bar{a}_i}} \tag{8}$$

which is identified as the distance from the origin 0 to the ith hyperplane. It is obvious that the set of Gaussian safety margins $\bar{M}$ as defined by (7) is invariant to an arbitrary inhomogeneous linear mapping of the transformed n-dimensional formulation space onto a new n-dimensional space. We have

$$\bar{M} = \bar{\bar{A}}' \bar{X} + \bar{b} = \bar{\bar{A}}_1' \bar{X}_1 + \bar{b}_1 \tag{9}$$

in which $\bar{X} = \bar{\bar{K}} \bar{X}_1 + \bar{k}$, where $\bar{\bar{K}}$ and $\bar{k}$ are matrices defining the inverse mapping. In the new space the image of S' is a polyhedral set defined by all safety margins being nonnegative. Thus the simplified reliability problem of calculating the probability of S' may as well be considered in any transformed formulation space having a nonsingular normal probability measure. We get the reliability index of the ith safety margin by

$$\beta_i = \frac{E[M_i]}{D[M_i]} = \frac{\bar{a}_i' E[\bar{X}] + b_i}{\sqrt{\bar{a}_i' \bar{\bar{C}}_{\bar{X}} \bar{a}_i}} \tag{10}$$

Of course, this reliability index is only interpretable as the Euclidian distance from the origin to the ith hyperplane in those particular spaces where the mean vector $E[\bar{X}]$ is zero and the covariance matrix $\bar{\bar{C}}_{\bar{X}}$ is the unit matrix $\bar{\bar{I}}$, that is, in the standardized normal spaces. The reliability indices $\beta_1, \ldots, \beta_q$ are invariant and so are all the correlation coefficients $\rho[M_i, M_j]$, $i,j \in \{1, \ldots, q\}$. Furthermore these numbers are invariant to multiplication of the safety margins by arbitrary positive constants. Note that this is not the case for the means $E[M_i]$ and the covariances $Cov[M_i,M_j]$. However, if β_i and $\rho_{ij} = \rho[M_i,M_j]$ are known for all $i,j \in \{1,\ldots,q\}$ then the probability of S' is the same as the probability that a set of Gaussian random variables of mean vector $(\beta_1,\ldots,\beta_q)$ and covariance matrix $\bar{\bar{P}} = \{\rho_{ij}\}$ gets an outcome with nonnegative values in all coordinates. This probability is close to 1 and it may therefore be quite accurately evaluated by exact bounding technique [9,12] or it may be approximated by the second order Taylor expansion with respect to the correlation coefficients with expansion from equicorrelation [13]. A method of Hohenbichler [19] is also available.

Regression Method for Determination of Most Likely Failure Points

Let $M = \bar{a}' \, \bar{X} + b$ be a Gaussian safety margin and let us consider the linear regression $\hat{E}[\bar{X}|M]$ of $\bar{X}$ on M. It may be characterized as an inhomogeneous linear expression $\bar{\alpha} + \bar{\beta} \, M$ which is uncorrelated with the residual $\bar{X} - (\bar{\alpha}+\bar{\beta} \, M)$ and has mean $E[\bar{X}]$, [10]. The last condition implies that $\bar{\alpha} + \bar{\beta} \, M$ may be written as $E[\bar{X}] + \bar{\beta}(M-E[M])$ while the first condition requires that

$$Cov[\bar{X}-\bar{\beta} \, M, \bar{\beta}' \, M] = (Cov[\bar{X},M] - \bar{\beta} \, Var[M]) \, \bar{\beta}' \qquad (11)$$

is the zero matrix. Thus $\bar{\beta} = Cov[\bar{X},M]/Var[M]$, and the linear regression of $\bar{X}$ on M may be written

$$\hat{E}[\bar{X}|M] = E[\bar{X}] + \frac{Cov[\bar{X},M]}{Var[M]} \, (M-E[M]) \qquad (12)$$

In particular, if $E[\bar{X}] = \bar{o}$, $Cov[\bar{X},\bar{X}'] = \bar{\bar{I}}$ and $M = \bar{a}' \, \bar{X} + b$ we get

$$\hat{E}[\bar{X}|M=0] = \frac{-\bar{a}}{\sqrt{\bar{a}'\bar{a}}} \, \beta \qquad (13)$$

where $\beta = E[M]/D[M]$ is the reliability index of the safety margin. It is seen that $\hat{E}[\bar{X}|M=0]$ is the vector to the origin projection point on the hyperplane $\bar{a}' \, \bar{x} + b = 0$. Substitution of $\bar{X} = \bar{\bar{K}} \, \bar{X}_1 + \bar{k}$ into the left side of (13) gives

$$\hat{E}[\bar{X}|M=0] = \hat{E}[\bar{\bar{K}} \, \bar{X}_1 + \bar{k}|M=0] = \bar{\bar{K}} \, \hat{E}[\bar{X}_1|M=0] + \bar{k} \qquad (14)$$

using that $\hat{E}$ is a linear functional. This shows that $\hat{E}[\bar{X}_1|M=0]$ is the image point of the origin projection point $\hat{E}[\bar{X}|M=0]$ by the considered inhomogeneous linear mapping. As such it is called the most likely failure point on the hyperplane $\bar{a}'(\bar{\bar{K}} \, \bar{x}_1+\bar{k}) + b = 0$ in the image space. It is obvious that a locally most likely failure point $\bar{x}_o$ on the failure surface in the standardized normal space is also a locally most likely failure point on the tangent hyperplane to the failure surface at $\bar{x}_o$. Thus

$$\bar{x}_o = \hat{E}[\bar{x}|M=0] \tag{15}$$

in which M is the Gaussian safety margin corresponding to the tangent hyperplane. Clearly (15) is both a necessary and sufficient condition for x_o being a locally most likely failure point. Since a tangent hyperplane maps onto a tangent hyperplane by an inhomogeneous linear mapping, it follows that the condition (15) is valid also in the image space. A simple iterative procedure for the determination of a locally most likely failure point is as follows. Let $\bar{x}_1$ be a point on the failure surface and let M_1 be the Gaussian safety margin corresponding to the tangent hyperplane at $\bar{x}_1$. Calculate the linear regression $\hat{E}[\bar{x}|M_1]$. If $\bar{x}_1 = \hat{E}[\bar{x}|M_1=0]$, then $\bar{x}_1$ is a locally most likely failure point and M_1 is the Gaussian safety margin corresponding to a face of the convex, polyhedral set S'. Otherwise, substitute $\hat{E}[\bar{x}|M_1]$ into the equation of the failure surface. This gives an equation solely in M_1. If it has solutions, the solution closest to zero, $M_1 = u_1$, defines a point $\bar{x}_2 = \hat{E}[\bar{x}|M_1=u_1]$ on the failure surface. Take this point, $\bar{x}_2$, as starting point like $\bar{x}_1$ and construct a third point $\bar{x}_3$ on the failure surface. Repeat this procedure over and over. In this way a sequence $\bar{x}_1, \bar{x}_2, \ldots, \bar{x}_m, \ldots$ is defined. If it is convergent, then the limit point $\bar{x}_o$ is a point on the failure surface which satisfies the condition (15). The proof of this is given in [11]. Thus $\bar{x}_o$ is a locally most likely failure point. The corresponding limit M_o of the sequence of Gaussian safety margins $M_1, M_2, \ldots, M_m, \ldots$ defines a face of the convex, polyhedral set S'. It is seen that the calculations are based on uncertainty algebraic formulas. Of course, modifications may be introduced that speed up convergence or that turn divergence of the described procedure into convergence.

Principle of Normal Tail Approximation

Let $T : \Omega \sim R^n$ be the mapping from the subset Ω of the physical formulation space (x space) onto the standardized Gaussian formulation space ($\bar{y}$ space) and assume that T has an inverse mapping $T^{-1} : R^n \sim \Omega$ which is differentiable with continuous partial derivatives everywhere. Further assume that the image surface $T^{-1}(\partial S)$ of the failure surface ∂S is given directly by an equation, $G(\bar{x}) = 0$, in which the function G is negative if and only if x corresponds to a point of the failure event. Assume that it has continuous partial derivatives $G_{,i}(\bar{x})$, $i \in \{1,\ldots,n\}$, such that the vector $(G_{,1}(\bar{x}),\ldots,G_{,n}(\bar{x}))$ vanishes nowhere on the surface.

Then the following fundamental theorem is valid (proof in [11,14]).

THEOREM. Principle of normal tail approximation: Let $\bar{x}$ be a point on the failure surface $T^{-1}(\partial S)$, that is, a point for which $G(\bar{x}) = 0$. By the transformation T the point $\bar{x}$ maps into a locally most likely failure point $\bar{y}$ on ∂S if and only if it is the most likely failure point on the tangent hyperplane to the surface $T^{-1}(\partial S)$ at $\bar{x}$ given that the probability measure in the physical formulation space has formally been changed to a Gaussian probability measure with mean vector $\bar{\mu}$ defined by

$$\bar{\mu} = \bar{x} - \bar{\bar{A}}\,\bar{y} \tag{16}$$

$$\bar{y} = T(\bar{x}) \tag{17}$$

and covariance matrix $\bar{\bar{A}}\bar{\bar{A}}'$. The matrix $\bar{\bar{A}}$ is defined by

$$A = \{a_{ij}\} = \left\{\frac{\partial x_i}{\partial y_j}(\bar{y})\right\} \tag{18}$$

in which a_{ij} is the partial derivative at $\bar{y}$ of the ith coordinate function of T^{-1} with respect to the jth axis.

The distance from the origin O to $\bar{y}$ is equal to the reliability index of the Gaussian safety margin corresponding to the tangent hyperplane at $\bar{x}$ and the normal probability measure defined by (16) and (18).

This theorem shows that only a slight modification is necessary of the iteration principle of the previous section in order to use the function G and the transformation T directly to determine the Gaussian safety margins corresponding to the faces of the convex, polyhedral set S':

Choose $\bar{x}_1$ such that $G(\bar{x}_1) = 0$. Determine $\bar{y}_1 = T(\bar{x}_1)$ and the matrix $\bar{\bar{A}}$. Calculate $\bar{\mu}$ and $\bar{\bar{A}}\,\bar{\bar{A}}'$. Define $\bar{X}$ to have normal distribution with mean vector $E[\bar{X}] = \bar{\mu}$ and covariance matrix $\mathrm{Cov}[\bar{X},\bar{X}'] = \bar{\bar{A}}\,\bar{\bar{A}}'$. Let M_1 be the Gaussian safety margin

$$M_1 = \sum_{i=1}^{n} G_{,i}(\bar{x}_1)(X_i - x_{1i}) \tag{19}$$

in which x_{1i} is the ith coordinate of $\bar{x}_1$. Determine the linear regression $E[\bar{X}|M_1]$. If $\bar{x}_1 = E[\bar{X}|M_1=0]$, then the theorem in combination with the condition (15) shows that $\bar{y}_1 = T(\bar{x}_1)$ is a locally most likely failure point. Otherwise, solve the equation

$$G(\hat{E}[\bar{X}|M_1]) = 0 \tag{20}$$

with respect to M_1. Among the solutions (if there are any) take that closest to zero, $M_1 = u_1$, and define $\bar{x}_2 = E[\bar{X}|M_1=u_1]$. Start all over again with $\bar{x}_2$ in place of $\bar{x}_1$. Repeat over and over to construct a sequence $\bar{x}_1$, $\bar{x}_2$, ..., $\bar{x}_m$, ... of points on $T^{-1}(\partial S)$. If it is convergent with limit point $\bar{x}$, then $\bar{y} = T(\bar{x})$ is a locally most likely failure point. Proof of this property is given in [11].

The reliability index of the corresponding Gaussian safety margin may be calculated as $\beta = \sqrt{\bar{y}'\bar{y}}$, or as $\beta = E[M]/D[M]$, in which M is the limit of the sequence M_1, M_2, ..., M_m, ... of Gaussian safety margins of the form as (19). In fact, M is given by (19) with $\bar{x}$ substituted in place of $\bar{x}_1$. The correlation coefficient, ρ_{12}, between the two Gaussian safety margins corresponding to the tangent hyperplanes at two different origin projection points $\bar{y}_1$, $\bar{y}_2$ both determined by this algorith may be calculated directly from $\bar{y}_1$, $\bar{y}_2$ as

$$\rho_{12} = \frac{\bar{y}_1'\bar{y}_2}{\sqrt{\bar{y}_1'\bar{y}_1 \ \bar{y}_2'\bar{y}_2}} \tag{21}$$

or it may be calculated as the correlation coefficient between the two limit Gaussian safety margins

$$M_1 = \sum_{i=1}^{n} G_{,i}(\bar{x}_1)(X_{1i}-x_{1i})$$

$$M_2 = \sum_{i=1}^{n} G_{,i}(\bar{x}_2)(X_{2i}-x_{2i}) \tag{22}$$

in which $\bar{x}_1 = T^{-1}(\bar{y}_1)$, $\bar{x}_2 = T^{-1}(\bar{y}_2)$ and

$$\text{Cov}[\bar{X}_1,\bar{X}_1'] = \bar{\bar{A}}_1\bar{\bar{A}}_1'$$

$$\text{Cov}[\bar{X}_2,\bar{X}_2'] = \bar{\bar{A}}_2\bar{\bar{A}}_2' \tag{23}$$

$$\text{Cov}[\bar{X}_1,\bar{X}_2'] = \bar{\bar{A}}_1\bar{\bar{A}}_2'$$

where $\bar{\bar{A}}_1$, $\bar{\bar{A}}_2$ are determined at $\bar{x}_1$ and $\bar{x}_2$ respectively.

Case of Mutually Independent Basic Variables

The simplest case of a transformation T appears when the random variables X_1, ..., X_n of the physical formulation space are mutually independent with density functions $f_1, \ldots, f_n$ and corresponding distribution functions $F_1, \ldots, F_n$ respectively. Then T may simply be defined by the equations

$$\Phi(y_i) = F_i(x_i) \quad , \quad i = 1, \ldots, n \tag{24}$$

uniquely defining y_i as function solely of x_i. It follows from (18) that $a_{ij} = 0$ for $i \neq j$. Denoting a_{ii} by σ_i and the ith element of $\bar{\mu}$ by μ_i it is easily seen that μ_i and σ_i are determined by the equations

$$\Phi\left(\frac{x_i - \mu_i}{\sigma_i}\right) = F_i(x_i) \tag{25}$$

$$\frac{1}{\sigma_i} \Phi\left(\frac{x_i - \mu_i}{\sigma_i}\right) = f_i(x_i) \tag{26}$$

in which $\bar{x} = (x_1, \ldots, x_n)$ is the point at which the tangent hyperplane to the surface $T^{-1}(\partial S)$ is considered.

These equations (25) and (26) are the origin of the name "principle of normal tail approximation". They simply determine the mean μ_i and standard deviation σ_i of that normal distribution that has the same distribution function value and density function value at the approximation point x_i as the given distribution of X_i. This principle is the basis for the well-known Rackwitz-Fiessler algorithm [26]. It has been generalized by Hohenbichler and Rackwitz [18] to be applicable also when the variables are mutually dependent simply by changing the right side of (24) to the conditional distribution function $F_i(x_i | x_1, \ldots, x_{i-1})$, $i = 1, \ldots, n$. This special transformation is called a Rosenblatt transformation [27] by these authors.

It is emphasized that in spite of the word "approximation" in the name this principle gives an exact determination of the most likely failure points in the transformed formulation space. It may be generalized to a principle of defect and/or truncated normal tail approximation [15]. This generalized principle aims at decreasing the curvature of the failure surface at any specific locally most likely failure point in the transformed formulation space defined by the principle. Thereby a more accurate polyhedral approximation to the safe set may be obtained. This, in fact, is supported by the asymptotic results of Breitung [4,5].

REDUCTION OF PROBLEM DIMENSION

Dimension Reduction of Physical Formulation Space

Even though the simple second moment algebra may be used to calculate

upper and lower bounds on the probability of the convex, polyhedral set
S', this calculation becomes impracticable if the number of faces of S'
becomes large.

In some cases it is possible to simplify the problem by reducing
the dimension of the physical formulation space. Typically this possibility may be considered when a subset $x_1, \ldots, x_m$, say, of the basic variables $x_1, \ldots, x_n$ enters the equation of the failure surface solely in terms of $r < m$ functions $z_1 = g_1(x_1, \ldots, x_m), \ldots, z_r = g_r(x_1, \ldots, x_m)$ of $x_1, \ldots, x_m$. In principle we may then reduce the dimension from n to $n - m + r$ using $z_1, \ldots, z_r, x_{n-m+1}, \ldots, x_n$ as a new set of basic variables. Naturally this reduction of dimension only pays off if it is simpler to calculate the joint distribution of the corresponding random variables $Z_1, \ldots, Z_r, X_{n-m+1}, \ldots, X_n$ than to work directly with the original n-dimensional formulation space. Often, however, structural reliability analysis applications have cases where $Z_1, \ldots, Z_r$ are particularly simple functions of $X_1, \ldots, X_m$ such as linear functions or functions of the form as $\max\{X_1, \ldots, X_m\}$ or $\min\{X_1, \ldots, X_m\}$.

Reduction of Set of Most Likely Failure Points

A much more difficult problem is that the locally most likely failure
points may be so numerous that it becomes impracticable to find all of
them. Even more seriously, it seems difficult to give rigorous principles that may help to identify the most important of these points without knowing all of them. It is obvious that the convex, polyhedral set
constructed from any genuine subset of the set of most likely failure
points will have a larger probability than the probability of S'. Of
course, in applications this is not to the side of safe design. However,
if most of the reliability indices corresponding to all the locally most
likely failure points are somewhat larger than the smallest reliability
index, the difference will not be large provided those few hyperplanes
that have reliability indices among the small ones are taken into account. It is important in this respect that the golbally most likely
failure point can be identified. As previously mentioned, the problem is
particularly relevant for structural systems that may fail in a large
number of different modes. Each failure mode defines a piece of the
failure surface ∂S in the transformed formulation space. If the considered failure mode is not entirely prevented to occur by other more dominating failure modes, it will contain at least one locally most likely
failure point and thus contribute to the construction of S' with at
least one face.

A manageable reliability bounding theory for ideal-plastic frame
structures is given in Ref. 16. Specifically the theory is about frame
structures that are subjected to random loads at given points and which
may develop yield hinges solely at points of a given set of points. Between these points the beam elements are infinitely rigid and at the

points relative rotation may occur only if the bending moments reach the respective yield moments. These are random but independent of the relative rotations in accordance with the assumption of ideal plasticity.

REFERENCES

1. Ang, A.H.-S. and H.-F. Ma, 'On the Reliability of Structural Systems', Proceedings of ICOSSAR'81: Structural Safety and Reliability, T. Moan and M. Shinozuka, eds., (Elsevier, Amsterdam, 1981) 103-134.

2. Augusti, G. and A. Baratta, 'Limit Analysis of Structures with Stochastic Strength Variations', J. Struct. Mech. 1 (1) (1972) 43-62.

3. Augusti, G. and A. Baratta, 'Probabilistic Limit Analysis and Design of Structures and Earthwork: The Static Approach.'Proceedings of the 2nd International Conference on Application of Statistics and Probability in Soil and Structural Engineering, Aachen, (1975) (Deutsche Gesellschaft für Erd- und Grundbau e.V., Essen, F.R. Germany) 7-26.

4. Breitung, K., 'An Asymptotic Formula for the Failure Probability, DIALOG 6-82: Euromech 155 (1982)', Danish Engineering Academy, Lyngby, Denmark.

5. Breitung, K., 'Asymptotic Approximations for Multinormal Integrals', J. Engrg. Mech. ASCE 110 (EM3) (1984) 357-366.

6. Cornell, C.A., 'Bounds on the Reliability of Structural Systems', J. Struct. Div. ASCE 93 (ST1) (1967) 117-200.

7. Cornell, C.A., 'A Probability - based Structural Code', ACI J. 66 (12) (1969) 974-985.

8. Ditlevsen, O., 'Generalized Second Moment Reliability Index', J. Struct. Mech. 7 (4) (1979) 435-451.

9. Ditlevsen, O., 'Narrow Reliability Bounds for Structural Systems', J. Struct. Mech. 7 (4) (1979) 453-472.

10. Ditlevsen, O., Uncertainty Modeling, (McGraw-Hill, New York, 1981).

11. Ditlevsen, O., 'Principle of Normal Tail Approximation', J. Engrg. Mech. ASCE 108 (EM6) (1981) 1191-1208.

12. Ditlevsen, O., 'System Reliability Bounding by Conditioning', J. Engrg. Mech. ASCE 108 (EM5) (1982) 708-718.

13. Ditlevsen, O., 'Taylor Expansion of Series System Reliability', J. Engrg. Mech. ASCE 110 (EM2) (1984) 293-307.

14. Ditlevsen, O., 'Basic Reliability Concepts', in Reliability Theory and its Application in Structural and Soil Mechanics . P. Thoft-Christensen, ed., NATO ASI Series E (1983), Martinus Nijhoff Publishers, The Hague.

15. Ditlevsen, O., 'Defect and/or Truncated Normal Tail Approximation', Structural Safety 2 (1) (1984) 65-70.

16. Ditlevsen, O. and P. Bjerager, 'Reliability of Highly Redundant Plastic Structures', J. Engrg. Mech. ASCE 110 (EM5) (1984) 671-693.

17. Hasofer, A.M. and N.C. Lind, 'An Exact and Invariant First-order Reliability Format', J. Engrg. Mech. ASCE 100 (EM1) (1974) 111-121.

18. Hohenbichler, M. and R. Rackwitz, 'Non-normal Dependent Vectors in Structural Safety', J. Engrg. Mech. ASCE 107 (EM6) (1981) 1227-1238.

19. Hohenbichler, M. and R. Rackwitz, 'First-Order Concepts in System Reliability', Structural Safety 1 (3) (1983) 177-188.

20. Kounias, E.G., 'Bounds for the Probability of a Union, with Applications'. The Annals of Mathematical Statistics 39 (6) (1968) 2154-2158.

21. Madsen, H.O., 'Some Experience with the Rackwitz-Fiessler Algorithm for the Calculation of Structural Reliability under Combined Loading', DIALOG-77, Danish Engineering Academy, Lyngby, Denmark (1977) 73-98.

22. Moses, F. and D.E. Kinser, 'Analysis of Structural Reliability', J. Struct. Div., ASCE 93 (ST5) (1967) 147-164.

23. Moses, F., 'System Reliability Developments in Structural Engineering', Structural Safety 1 (1) (1982) 3-13.

24. Murotsu, Y., H. Okada, M. Yonezawa, and K. Taguchi, 'Reliability Assessment of Redundant Structures', Proceedings of ICOSSAR'81: Structural Safety and Reliability, T. Moan and M. Shinozuka, eds., (Elsevier, Amsterdam, 1981) 315-329.

25. Murotsu, Y., H. Okada, M. Yonezawa, M. Grimmelt, and K. Taguchi, 'Automatic Generation of Stochastically Dominant Modes of Structural Failure in Frame Structure', Bulletin of University of Osaka Prefecture, Series A, Vol. 30, 2 (1981) 85-101.

26. Rackwitz, R., and B. Fiessler, 'Structural Reliability under Combined Random Load Sequences', Comput. Struct. 9 (1978) 489-494.

27. Rosenblatt, M., 'Remarks on a Multivariate Transformation', Annals of Math. Stat., Vol. 23, 470-472.

28. Thoft-Christensen, P. and J.D. Sørensen, 'Calculation of Failure
    Probabilities of Ductile Structures by the β-Unzipping Method'. Re-
    port No. 8208 (1982), Institute of Building Technology and Struc-
    tural Engineering, Aalborg Universitetscenter, Aalborg, Denmark.

29. Vanmarcke, E.H., 'Matrix Formulation of Reliability Analysis and
    Reliability-Based Design', Computers and Structures 3 (1973)
    757-770.

30. Veneziano, D., 'Contributions to Second Moment Reliability', Res.
    Rept. R74-33, Dept. of Civil Eng., Massachusetts Institute of Tech-
    nology (1974).

LOAD COMBINATION

F. Casciati
Dept. of Struct. Mechanics
University of Pavia
Via Luino 12
I 27100 Pavia

ABSTRACT. Four classes of problems are considered in order to provide
a framework for the systematization of the multi-fold developments
recently pursued in the field of the load combination for structural
reliability purposes. The mutual relations between the different trends
of the current research on load combination are analysed by providing an
accurate definition of the class in which the single research can be
categorized.
   The class with more practical applications groups the research
efforts whose common objective is the achievement of the goal of
probabilistic design within the deterministic scheme of current design
criteria. Some results obtained by the author and his coworkers in this
field are sketched and discussed.

1. INTRODUCTION

Let $\mathbb{F}^n$ be the space of the n loads $Q_i (i=1,\ldots,n)$ acting upon a
structure. $\mathbb{F}^n$ is a subspace of the Euclidean space $\mathbb{R}^m$ of all the m
design variables $X_j (j =1,\ldots,m)$ which characterize the structural
problem being investigated. Given an ultimate limit state the objective
of the load combination problem is to combine the stochastic model of
the actions with the conditional probability of failure (CPF) associated
with any point of $\mathbb{F}^n$ in order to assess the global reliability of the
structure. For deterministic structures the boundary $\partial\mathbb{D}$ of the safety
domain $\mathbb{D}$ substitutes for the CPF.
If the multivariate load condition includes time-varying stochastic
loads, some circumstances makes difficult the assessment of the
reliability.
   1) Each varying-in time load is generally modelled independently
of the other possible concurrent actions. It follows that the actual
input acting on the structural system is rarely know as a stochastic-
-mechanics problem requires, i.e. by the knowledge of the properties of
the single processes and by the relevant cross - properties.
   2) The concept of "more unfavourable level" is strictly related,
for a single action, with either the maximum or the minimum value of

17

A. C. Lucia (ed.), Advances in Structural Reliability, 17–28.
© *1987 by ECSC, EEC, EAEC, Brussels and Luxembourg.*

its intensity. By contrast the more unfavourable load combination
depends on the shape of the boundary of the safety domain and, hence,
it is different from limit-state to limit-state, i.e. the problem is
highly structural dependent.

3) Even if special assumptions are made in order to remove the
previous two difficulties, the available stochastic approaches are
often inadequate. This is the case of structures assumed to behave
in a linear elastic way (linear boundary of the safety domain) for
which the total load can be expressed by a linear function of the
load intensities (linear input combination).

State-of-the-art papers on load combination have been presented
at some recent international conferences (|1|, |2| and |3| among
others) and many recent books on structural reliability or stochastic
mechanics devote one or more sections to the problem|4||5||6|. The
objective of this paper is to put in right light the different trends
of the current research on load combination and to clarify the
relations between them.

Technical problems in which the theoretical difficulties are
predominant on the operative aspect give rise in fact to two parallel
research fields: the first branch tackles the problem in its more
general formulation and attempts to solve it in a general way or under
suitable assumptions; the second branch neglects the central problem
and attempts to achieve technical results by introducing simplified
models or bounding techniques. This general trend caused a
diversification of interests also in the researchers working in the
load combination area and the current research efforts can be grouped
into four categories:

1) approaches which pursue the solution of the problem in its
general formulation;

2) approaches which pursue the rigorous solution of a simplified
problem of common practical interest;

3) proposal of approximate models enabling to pursue general
results;

4) translation of the previous results into common rules of
practice.

The first two categories tackle the problem in its general
formulation; classes 3) and 4) are attempts to achieve immediate
technical results.

2. APPROACHES TO THE PROBLEM

The projection on $\mathbb{F}^n$ of the boundary of the safety domain $\mathbb{D}^*$ in $\mathbb{R}^m$
is well described by the conditional probability of failure (CPF) $|6|$

$$P_\Gamma(q_1,q_2,\ldots,q_n) = \text{Prob (failure occurs } |Q_1 = q_1,$$
$$Q_2 = q_2,\ldots,Q_n = q_n) \tag{1}$$

where Q_i means the i-th load and $(q_1,q_2,\ldots,q_n)$ is any point of $\mathbb{F}^n$.
The CPF is an extension to structures with random properties of

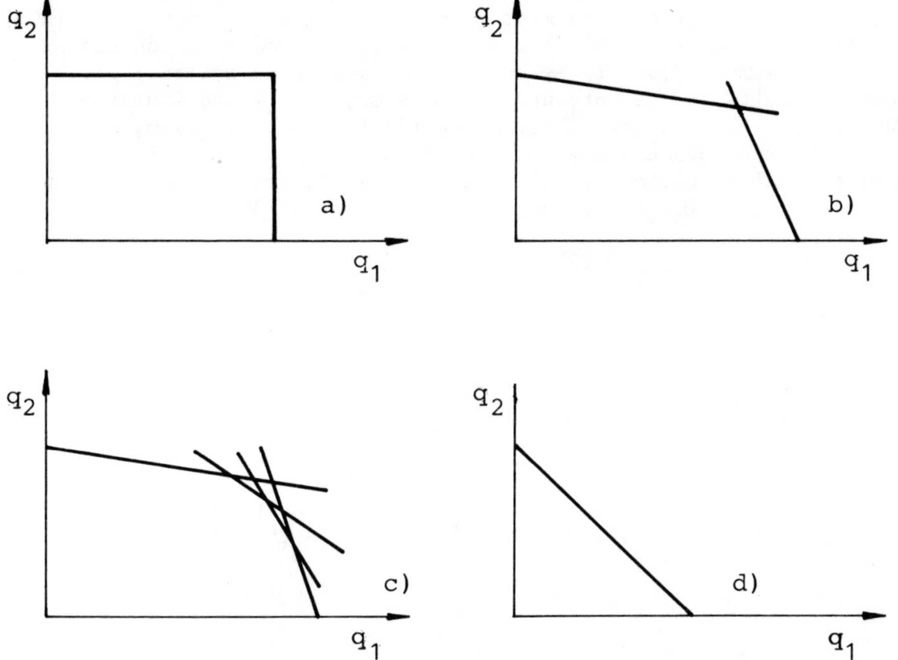

Figure 1. Possible boundaries $\partial\mathbb{D}$ of the safety domain $\mathbb{D}$ in $\mathbb{F}^n$:
a) the mode of failure under one load is always activated by an ∞
value of the other; b) the mode of failure under one load is reached
for a limit value which depends on the magnitude of the other;
c) intermediate modes of failure are also present; d) both the loads
activate the same mode of failure (linear safety domain) (from $|31|$).

the concept of boundary $\partial\mathbb{D}$ of the safety domain $\mathbb{D}$ in $\mathbb{F}^n$ (Figure 1):
for deterministic structures, in fact, $\partial\mathbb{D}$ separates the points of $\mathbb{F}^n$
where $P_\Gamma=0$ (safe points $\in \mathbb{D}$) from the points where $P_\Gamma=1$ (unsafe points
$\notin \mathbb{D}$). Due to this remark, the following considerations are developed
for a deterministic system. The aptitude of the results to be
generalized to random structures is discussed as a successive step.

2.1. The rigorous approach

It is well known that the safety of a deterministic structure under
one stochastic load can be evaluated by solving a barrier crossing
problem, the barrier being the load carrying capacity of the structure
(Fig. 2a and c). Therefore, the load combination is a special aspect
of the multivariate extension of such a problem: the objective of the
analysis is the assessment of the probability of the first excursion
from the safety domain of the vector-valued random process $Q(t)$.
However, even if the hypothesis of deterministic structure is accepted,

the problem is solved in the univariate scheme for very particular
cases $|7|$. The function $Q(t)$ must be once or twice differentiable in
order to make use of classical results (Rice's formula) which can be
extended to some random processes with discontinuous realization.
Stronger results can be obtained for Poisson, Markov and Gaussian
random processes. For the multivariate first-passage problem, Gauss
and Gauss-Markov processes have to be considered in order to make
use of the multivariate generalization of Rice's result for the mean
out-crossing rate (mean rate of exit events out of $\mathbb{D}$) ν_D:

$$P_{fail}(t) = P_{fail}(0) + \int_0^t \nu_D(\tau) \, d\tau \tag{2}$$

with

$$\nu_D(\tau) = \int_{\partial\mathbb{D}} E \left[\max \{0, \sum_{i=1}^n \alpha_i \, \dot{Q}_i \} \right.$$
$$\left. |\underline{Q}(\tau) = \underline{q}\right] \, p_{\underline{Q}}(\underline{q}) \, d(\partial\mathbb{D}) \tag{3}$$

In Eqs.(2) and (3) $P_{fail}(t)$ means the probability of failure at
time t; $E[\ ]$ denotes expectation; $p_Q(q)$ is the joint prob. density of $\underline{Q}(t)$;
a superimposed dot means derivative in time; $\underline{\alpha}$ is the unit vector
normal to $\partial\mathbb{D}$ at $\underline{q} \in \partial\mathbb{D}$.

Mean crossing rates of isotropic processes out of square regions
and out of spheres in $\mathbb{F}^n$ are presented in Ref. $|8|$, where the safe
regions are time invariat and convex, while Gaussian vector processes
are considered with Q and $\dot{Q}$ for any fixed t mutually independent.
Further results have been obtained by using a similar approach in
Refs. $|9|$ and $|10|$. Significant improvements have been presented in
$|11|$: i) non-Gaussian load vectors are mapped into Gaussian vectors;
ii) safe convex polyhedrons are considered and this permits the author
to bound the failure probability from above an below; iii) no assumption
about the time derivate $\dot{Q}(t)$ is needed and $\dot{Q}(t)$ may not even exist.
Research under progress is attempting to implement the bounding
approach presented in $|11|$ in a general purpose computer code; it is
also investigating the possibility of extending the results to non
deterministic structures.

2.2. Rigorous approach to a simplified problem

A basic procedure for structural analysis is founded on the assumptions
of linear structural behaviour and linear combination of the load
intensity variables or their (linear or non linear) functions. Then,
the load combination problem can be formulated as follows: given n
loads Q_i, i = 1,...,n, which are modelled by stationary independent
stochastic processes, derive the stochastic properties of the process

$$Q(t) = \sum_{i=1}^n Q_i(t) \tag{4}$$

or

$$R(t) = \sum_{i=1}^n c_i \, L_i(t) \tag{5}$$

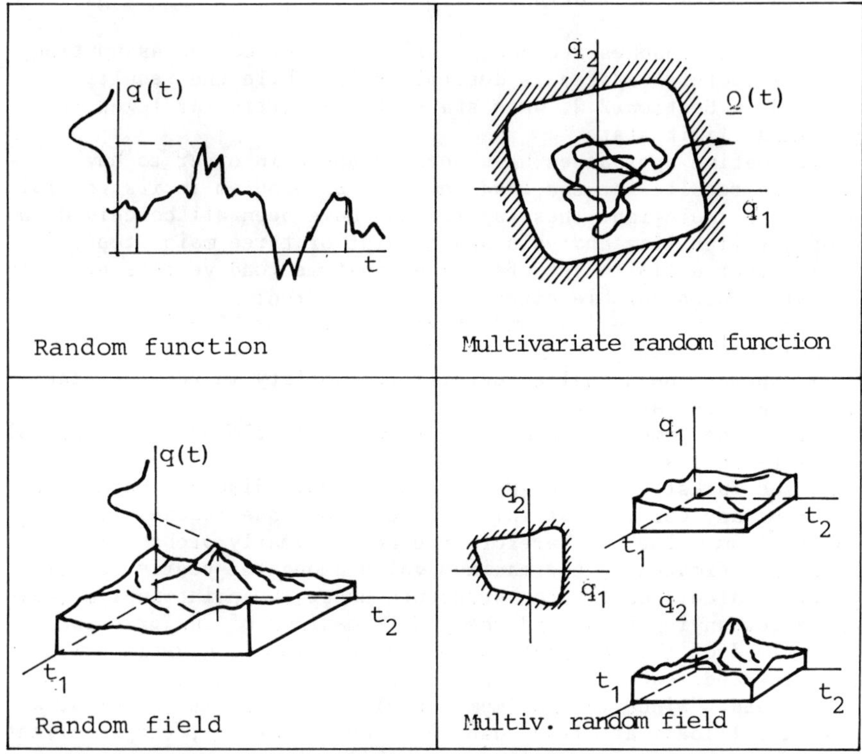

Figure 2. First-excursion problem for univariate and multivariate random process and field (from |7|).

where $L_i(t)$ is a scalar function of the load process Q_i and c_i is a structural influence coefficient. The knowledge of the stochastic properties of the sum $Q(t)$ of the individual process or of the total load effect $R(t)$, in fact, enables to assess the probability that $Q(t)$ or $R(t)$ crosses with positive slope (up-crosses) the relevant given barrier during a given reference period.

The probability distribution of the extreme value of the sum process was given as a multiple integral in |12| and by its Laplace transform in |13|. Results of practical interest in this field have been reached by using the "point crossing method" |14|. Unfortunately, the relevant solving formula becomes impractical when three a more processes are present in the load effect. The adoption of "translation processes", obtained from Gaussian processes by non-linear transformations, has been therefore proposed in order to make more efficient the solution procedure |15|.

2.3. Load combination models

The previous approaches are characterized by the common assumption
that the structural system is deterministic, while the results
summarized in Section 2.2. were stated for a particular (even if
very common) limit state.
       Alternative approaches have been proposed in order to have
approximate results for the load combination problem in its general
mechanical formulation. These approaches have been all conceived in
view of a reliability analysis and consist of three main steps:
       1) select a finite set of costant-in-time load vectors by which
all possible unfavourable situations are covered;
       2) solve the time independent reliability problem associated
with each load vector;
       3) express the actual structural reliability in terms of the
values calculated at step 2).
       Three load combination models are widely employed in structural
reliability analysis.
       1) The Turkstra's rule |16| considers "only discrete points in
time during any one year at which one of the loads reaches its
annual maximum value" . Therefore the actual yearly probability of
failure is estimated by the maximum value among the yearly failure
properties calculated for the structure under the point-in-time values
of the accompanying loads and the yearly maximum of the principal
variable load of the combination. Of course, any time variant load
must be rotated, each load taking the position of the principal
variable load. "Since the maximum annual total load may occur at a
time when all loads are less than their annual maxima, the idealization
will underestimate yearly failure probability".
       2) The Ferry-Borges-Castanheta model |17| is sketched in Fig.3.
Let n be the number of loads which are modelled as independent Poisson
square wave processes of arrival rate λ_i; the finite set of load
combinations is stated as follows:
a) the loads are ordered for decreasing arrival rates λ_i, i=1,...,n;
b) in the first load combination the maximum value of the n-th load
is associated with the point in time values of the other n-1 actions;
c) in the second load combination the maximum value of the (n-1)-th
load is associated with the point-in-time values of the first (n-2)
actions and with the maximum value in $1/\lambda_2$ of the n-th load;
d) etc. ...
The actual structural reliability is the lower among the values
calculated in the n load-combination above defined. The model is
very accurate for long-duration actions, but it cannot be applied
to the combination of short duration loads, where the probability
of simultaneous occurrence plays a basic rôle (see Fig.4).
       3) The load-coincidence model, proposed by Wen |18| and adopted
by several researchers |19||20||21||22| overcomes this limitation of
the Ferry-Borges-Castanheta model. Here the load combination for
which the reliability analysis is conducted are: the principal load
associated with a zero value of the other actions; the principal
load defined by the simultaneous occurrence of two actions; the

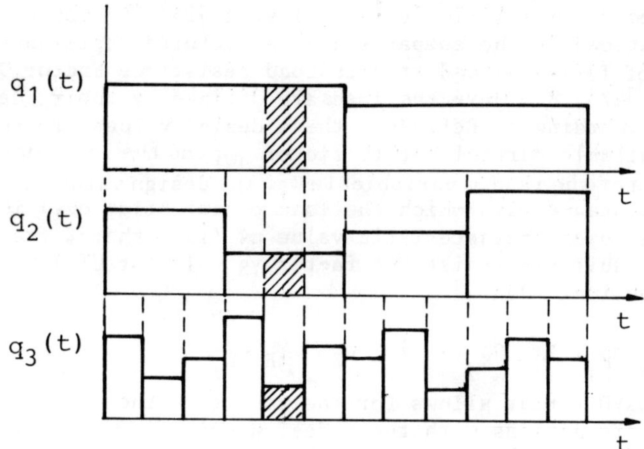

Figure 3.   Sketch of the Ferry-Borges Castanheta model

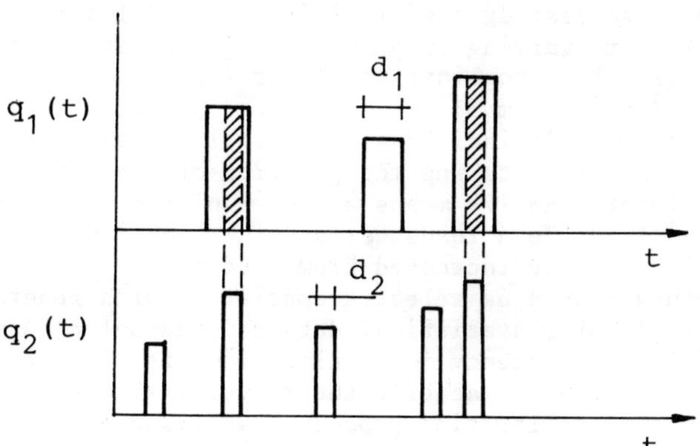

Figure 4.   Sketch of the "load coincidence" model

principal load defined by the simultaneous occurrence of
three actions. The actual probability of failure is then overestimated
by the sum of all the calculated probabilities of failure. The
probabilities of failure under four or more simultaneous loads are
generally negligible.

2.4. Derivation of rules of common practice

"For practical reasons, the goal of probabilistic design has to be
achieved within the deterministic framework of current design criteria"
|23||24|. All theoretical result must therefore be translated into

suggestions for a Load and Resistance Factor Design (LRFD) format.

Present semiprobabilistic design criteria $|25||26|$ conceive the safety verifications as the comparison of a factored resistance with a combination of factored-load effects(Load Resistance Factor Design (LRFD) format) $|27||28|$. Here the loads are defined by their design values (q_{di}). According to Ref. $|26|$, these design values are the product of a suitable partial safety factor γ_{Qi} and the 95% fractile q_{Ki} of the distribution of the random variabiable Q_i. The design value u_D of the generalized resistance with which the load effect $U(\gamma_{Qi} q_{Ki})$ must be compared is its lower characteristic value u_K (i.e. the 5% fractile) multiplied by a suitable resistance factor $\gamma_R \leq 1$: let $U(\ )$ be the limit state function, then

$$U(\gamma_{Q_1} q_{K_1}, \gamma_{Q_2} q_{K_2}, \ldots) \leq u_D = \gamma_R u_K \qquad (6)$$

The same LRFD format allows for the fact that the simultaneous occurrence of more actions with their design value is an unlikely event by substituting for Eq.(6)

$$U(\gamma_D d_K, \gamma_{Q_0} q_{K_0}, \gamma_{Q_1} \psi_{o_1} q_{K_1}, \ldots, \gamma_{Q_{n-1}} \psi_{o(n-1)} q_{K_{n-1}}) \leq u_D \qquad (7)$$

where γ_D, γ_{Q_i} = load factors; d_K = characteristic dead load; q_K = characteristic time-varying load over a given reference period (f.i. 50 years); ψ_{oi} = load combination factor ($\psi_{oi} \leq 1$); n = number of time varying loads. The symbol q_{K_0} denote the principal variable load of the combination; any individual time-variant load must be rotated in Eq.(7), each load taking the position of the principal variable load. Therefore, Eq.(7) means a system of inequalities as q_{K_0} must be rotated into the n loads q_{K_i} and the number of time-variant loads in combination must be increased from 1 to n.

The load factors should be selected consistent with general desired reliability levels, statistical data and the selected safety checking format $|27|$. Consistency is attained in Refs. $|29|$ and $|30|$ by designing the structures to acheive the target limit state probability p_{f_0}. The resulting calibration procedure is discussed with some detail in the next section.

3. LOAD COMBINATION BY PARTIAL SAFETY FACTORS

In Refs. $|29|$ and $|25|$ reliability calculations are conducted in terms of the safety index β, to be compared with a target value β_o associated with p_{f_0}. The assessment of β requires to individuate a checking point x^* in $\mathbb{R}^m$ and the safety factor γ_{X_j} of any variable X_j is defined by the relation

$$\gamma_{X_j} = \frac{x_j^*}{x_{K_j}} \qquad (8)$$

where x_{K_j} is the characteristic value of X_j. If X_j coincides with a load Q_i, then $\gamma_{X_j} = \psi_{oi}\gamma_{Qi}$. The set of safety factors defined by Eq.(8)

TABLE I. Probabilistic definition of the actions: L=live load; E=earthquake excitation; S=snow action; W=wind (from |31|)

Type of load	L	E	W	S
Type of CDF	Ext I (max)	Ext I or II (max)	Ext I (max)	Ext I (max)
C.o.v. of the max over 50 years	25%	70 or 35%	18.6%	20%
Mean arrival rate [per year]	12	2	24	6
Mean duration of each occurrence [days]	2	7×10^{-4}	1	20

TABLE II. Influence of ψ_{oi} on the probability of failure in 50 years for the standard CPF contour surfaces (Fig.5): γ_{Qi}=1.5; coef. of variation (c.o.v.) of the carrying capacity 10% (all the values of probability have to be multiplied by the factor 10^{-3}).

Load combination	$\psi_{oi} =$		
	1	0.2	0.05
E''_{35} W S	6.72	6.83	7.62
E^I_{70} W S	4.13	4.24	5.00
L W S	1.16	1.26	1.95

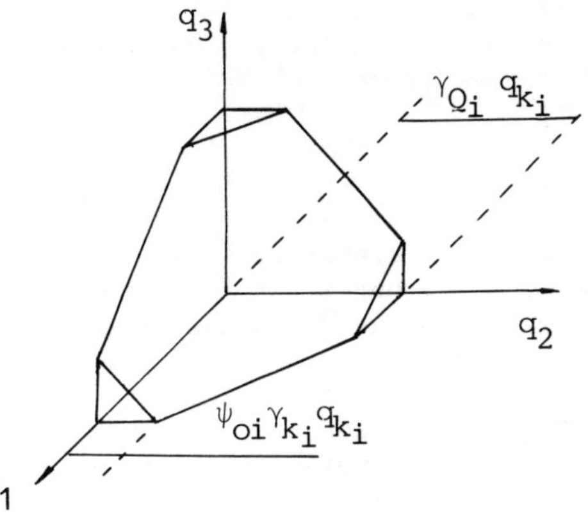

Figure 5. Shape of the standard contour surface introduced to idealize the CPF in Ref.|31|.

changes from design situation to design situation. By contrast, the
set of load factors of a LRFD format is constant for several design
situations.It follows that the associated reliabilities deviate from
the target value $(1-pf_o)$. The optimal set of load factors will be
therefore selected which minimizes the extent of this deviation.
This can be made using either the Turkstra rule $|25|$ or the Ferry-Borges
Castanheta model $|29|$.

A deep criticism of the approach summarized in Eq.(8) is developed
in Ref.$|32|$. Attention is only focussed here on the fact that despite
the extent of the deviation is minimized, the selected set of safety
factors provides also unconservative designs(i.e. $p_{fail} > p_{fo}$). In
order to reduce (not to eliminate) this effect, the safety factors
have to be selected for design situations quite similar and hence
common unified safety factors cannot be pursued.

In order to achieve unified values of the safety factors for
general purpose codes, a procedure founded on the concept of marginal
level of safety has been proposed in Ref.$|31|$ and justified from a
theoretical point of view in Ref.$|32|$. The target reliability associated
with a single permanent or long-duration load is assumed to be much
higher than the global target reliability $1-p_{fo}$. The load safety
factors γ_{Qi} are calculated independently of the actual number of
loads in combination, while the load combination factors ψ_o of the
short-duration actions are derived from probabilistic calculations
exploiting the marginal target reliability. In order to make possible
these results one does not pursue the values of γ_{Qi} and ψ_{oi} associated
with a single design situation; all the calculations are referred
to a standard case defined by an appropriate piece-wise-linear shape
of the contour surfaces of the CPF in $\mathbb{F}^n$.

Then, if one checks the safety of any structure in the vertices
of this standard CPF, the general convexity of the safety domains
(for further details see Ref.$|31|$), leads to the conclusion that
the reliability of the structure is greater than the target one.

In addition to the advantages above specified(load factors
independent of the design situation and of the number of loads in
combination), the standard CPF can also be selected in such a way
that the load factors are also independent of the mutual ratios
between the loads $|31|$ (Figure 5). For the data of Table I, Table II
shows the influence of ψ_{oi} (see Figure 5) on the global probability
of failure in 50 years for the standard case.

4. CONCLUSIONS

The combination of the loads acting upon a structure is a multi-fold
field of investigation which leads to either the discussion of very
sophisticated mathematical problems or to the definition of the design
values of the actions to be introduced in deterministic safety
verifications. The research effort of the last decade have provided
exact and approximate results that can be used at different levels
in the design process. When a nuclear power plant structural component
or an off-shore structure has to be analysed, the probability of first

excursion of the stochastic load vector from the relevant safety domain
can be assessed and for this purpose the simpler approaches holding in
a linear context can be adopted. By contrast, load and resistance
factored design (LRFD) formats can be useful in common professional
practice; the evaluation of the corresponding safety factors can be
pursued by the code-maker making use of appropriate calibration
procedures. In particular the safety factors can be calibrated by
techniques founded on the first order reliability methods (FORM) when
very limited set of design situations have to be considered, while
special procedures based on the introduction of conservative models
of the structures can be adopted for general purpose codes.

AKNOWLEDGEMENTS

The present work makes use of several results achieved by Prof.
Lucia Faravelli of the University of Pavia, to whom the author is
thankful for many useful discussions.Founds have been provided by
the Italian National Research Council (CNR).

5. REFERENCES

|1| Ditlevsen O, and Madsen H.O., "Probabilistic Modelling of Man-Made
    Load Process and Their Individual and Combined Effects", in
    Structural Safety and Reliability, Proc. of ICOSSAR '81 , Eds.
    T.Moan and M.Shinozuka (Elsevier, 1981), pp. 103-104
|2| Shinozuka M., "Stochastic Characterization of Loads and Load
    Combinations", in *Sructural Safety and Reliability*, Proc. of
    ICOSSAR '81, Eds. T.Moan and M.Shinozuka (Elsevier, 1981), pp.
    57-76
|3| Tichy M., "The Science of Structural Actions", in Proc.of 4th
    ICASP, Eds. G.Augusti et al. (Pitagora, 1983), pp. 295-321
|4| Ang A.H-S. and Tang W.H., *"Probability Concepts in Engineering
    Planning and Design"*, J.Wiley & Sons, Vol.$\underline{2}$, 1983
|5| Thoft-Cristensen P. and Baker M.J., *Structural Reliability Theory
    and its Applications"*, Springer-Verlag,1982
|6| Augusti G., Baratta A. and Casciati F., *"Probabilistic Methods
    in Structural Engineering"*, Chapman & Hall, 1984
|7| Veneziano D.,*"Random Processes for Engineering Application"*,
    Draft Copyright, MIT, 1978
|8| Veneziano D., Grigoriu M. and Cornell C.A., "Vector Model for
    System Reliability", *J. Engrg.Mech.*, ASCE $\underline{103}$, (EM3), 1977, pp.
    441-460
|9| Larrabee R.D. and Cornell C.A. "Upcoming Rate Solution for Load
    Combinations", *J.Struct.Div.*, ASCE $\underline{105}$, (ST1), 1979, pp.125-132
|10| Breitung K. and Rackwitz R., "Nonlinear Combination of Load
    Processes", *J.Struct.Mech.*, $\underline{10}$(2), 1982, pp.145-166
|11| Ditlevsen O., "Gaussian Outcrossing from Safe Convex Polyhedrons",
    J.Eng.Mech.Div., ASCE $\underline{109}$(EM1), 1983, pp.127-148.
|12| Hasofer A.M., "Time Dependent Maximum of Floor Live Load",

Technical Note, *J.Engrg.Mech.*, ASCE 100 (EM5), 1974, pp.1086-1091

|13| Gaver D.P. and Jacobs P.A., "On Combination of Random Loads", SIAM, *J. of Appl.Math.*, 1981

|14| Larrabee R.D. and Cornell C.A., "Combination of Various Load Processes", *J. of the Struct.Div.*, ASCE $\underline{107}$(ST1), 1981, pp.223-239

|15| Grigoriu M., "Load Combination Analysis by Translation Processes", *DIALOG 6-82*, Proc. EUROMECH 155, Lyngby, 1982, pp.165-183

|16| Turkstra C.J., *"Theory and Structural Design Decision"*, Solid Mechanics Study N.2, Univ.Waterloo, Waterloo, Ontario, Canada, 1972

|17| Ferry-Borges J. and Castanheta M., *"Structural Safety"*, Laboratorio Nacional de Engenharia Civil, Lisbon, 1972

|18| Wen Y.K., "Statistical Combination of Loads", *J.Struct.Div.*, ASCE $\underline{103}$ (ST5), 1977, pp.1079-1093

|19| Schwarz R.F. and Schuëller G.I., "Reliability of Structures under Combined Loading", Proc. ASCE Spec.Conf. on Prob.Mech. and Struct. Rel., Tucson, 1979, pp.112-120

|20| Der Kiureghian A., "Reliability Analysis under Stochastic Loads", *J.Struct.Div.*, ASCE $\underline{106}$ (ST2), 1980, pp.411-429

|21| Casciati F. and Faravelli L., "Calibration of the Load Enhancement Factors", *Engineering Structures*, $\underline{2}$(2), 1980, pp.113-122

|22| Krée P., "Utilisation des Probabilités et des Statistiques pour l'Evaluation de la Securité des Ouvrages", in *La Securité des Constructions*, Eyrolles, Paris, 1976

|23| Cornell C.A., Ravindra M.K. and Chou C.K., "Development of Load Combinations for Design of Nuclear Components: a Probabilistic Methodology". Trans. of the 6th SMiRT, Paris, France, 1981, J6/2

|24| Schwartz M.W., "Derivation of Load Factors for Design of Nuclear Power Plants Components", Trans. of the 6th SMiRT, Paris, France, 1981, J6/1

|25| Ellingwood B. et al., "Development of a Probability Based Load Criterion for American National Standard A58", NBS, Special Publ. N.$\underline{577}$, June 1980

|26| Common Unified Rules for Different Type of Construction and Material, *Bulletin d'Information*, N.124E, CEB, Paris, April 1978

|27| Galambos T.V. et al. "Probability Based Load Criteria: Assessment of Current Design Practice", *J.Struct.Div.*, ASCE $\underline{108}$(ST5), 1982, pp.959-977

|28| Ellingwood B. et al., "Probability Based Load Criteria: Load Factors and Load Combinations", *J.Struct.Div.*,ASCE $\underline{108}$ (ST5), 1982, pp.978-997

|29| CIRIA, "Rationalisation of Safety and serviceability Factors in Structural Codes", Report 63, London, 1977

|30| Faravelli L., "A Proposal on Load Combination for Level I Formats", Engineering Structures $\underline{4}$(3), 1982, pp.197-206

|31| Casciati F. and Faravelli L., "Load Combination by Partial Safety Factors", Nuclear Engineering and Design, $\underline{75}$, 1982, pp.439-452

|32| Faravelli L.,"Derivation of Rational Safety Factors for LRFD Formats", in *Stochastic Methods in Structural Mechanics*, F.Casciati and L.Faravelli Eds., Pavia, 1983

THE RELIABILITY OF STRUCTURES IN THE PRESENCE OF IMPERFECT INFORMATION

John Munro
Civil Engineering Department
Imperial College of Science and Technology
London   SW7 2BU
England

ABSTRACT.   A distinction is drawn between randomness and fuzziness and
it is concluded that civil engineering decision-making frequently
involves imprecise data. The safety of a simple joist subject to a
single load is assessed from the possibility that the load will exceed
the strength. This element analysis is extended to frames by use of
fuzzy programming. First the problem of optimal plastic design is
considered in the face of imprecise loading. The ideas are then extended
to limit analysis where the yield criterion is stated imprecisely. The
corresponding programs provide the load factor for any given support.
By varying this support the possibility distribution for the frame load
factor is derived. This measure of structural strength is combined
with the possibility of the actual load to assess the safety. Finally
a general approach is developed for decision analysis where the data
may be known deterministically, statistically or imprecisely. A
formalism based on a hierarchy of fuzzy programs is outlined.

1.   INTRODUCTION

The earlier developments of civil engineering science have been
principally concerned with those branches - structural engineering,
hydraulic engineering, geotechnical engineering - which are strongly
based on the corresponding branches of mechanics. Thus the integrated
study of mechanics constitutes a solid foundation and unifying
influence to those diverse engineering principles. However, in recent
years the main growth has been in those areas of civil engineering
which are concerned with the environment. These include transportation
engineering, water resources engineering and public health engineering
and their common feature is their involvement in planning decisions and
in environmental impact analyses.

Quantifiable decision-making requires the creation of mathematical
models. The simplest models are those which infer complete certitude.
They lead to relatively simple problems of optimisation and the greatly
expanding literature on mathematical programming is evidence of the
progress being made in this direction. However for most civil engineer-

A. C. Lucia (ed.), *Advances in Structural Reliability*, 29–48.
© 1987 by ECSC, EEC, EAEC, Brussels and Luxembourg.

ing problems, the assumption of complete certitude is inappropriate.
It will be convenient herein to divide incertitude into two types.
First, natural *randomness* which causes *uncertainty* and secondly
imprecision which causes *fuzziness*. The more familiar randomness can be
handled mathematically through the well-established theory of probability and it will be argued herein that the less familiar concepts of
imprecision should be handled by the newly developing theory of possibility. Many decision-making problems in civil engineering contain both
types of incertitude and it is important to distinguish them and to
create appropriate mathematical models.

The mathematical tools appropriate to decision-making in environmental studies have found further applications within the mechanics-
based branches of civil engineering. Thus mathematical programming has
been successfully exploited in a wide range of studies in structural
mechanics [15,16] and in geomechanics [20]. The study of structural
reliability [13] has generally been based on probability theory and the
random natures of the loading and of the structural resistance have
been combined to assess the probability of structural failure. However
these calculated probabilities have not been found to be in agreement
with the actual frequencies of structural malfunction. This discrepancy has been attributed to the omission of any consideration of gross
errors. The occurrence of such gross errors is not a random phenomenon.
However it is fuzzily related to such matters as the degree and com-
petence of site control and these can generally be assessed and stated
in the form of professional opinions. Such opinions are of considerable
importance even if expressed imprecisely. The handling of such
imprecision within a mathematical model is vital to the development of
a rational measure of structural safety.

2. FUZZINESS

The ideas of set theory have been extensively exploited in laying the
foundations of probability theory. The concept of the membership (or
non-membership) of a candidate with respect to a set is also of
fundamental importance in the foundation of Boolean logic. The idea of
some intermediate class ("maybe") between the binary states of "yes"
and "no" was introduced by Lukasiewicz and the Polish school of
logicians. The extension of this intermediate state to a continuous
spectrum of degrees of support in the range 0 to 1 was achieved by
Zadeh [33] in his theory of fuzzy sets. He later developed a relationship between linguistic variables [34] and fuzzy sets and this opened
out new possibilities in, for example, rendering expert opinions in
computer-compatible form. The first important application of fuzzy set
theory to problems of structural reliability was due to Blockley [3].
The present lecture is essentially a summary of the material presented
in references [5,19,22,24,25 and 26].

It is well-known that gross errors are often more important in the
assessment of the likelihood of structural failure than the randomness
of strength and loading. Thus it might be sensible to first consider

the relatively simple case of the reliability of a structural element
where the prior knowledge is contained in expert opinions regarding
strengths and loadings. It is common in timber technology for a first
estimate of strength to be obtained from a visual inspection of the
joists. This assessment is essentially imprecise. Consider strength
to be divided into four stress grades - the first being the strongest
and the fourth being the weakest. Now the fuzzy set representations
[23] of some typical linguistic variables are given whereby only the
supports are listed

$$\underline{strong} = \{1, 0.7, 0.1, 0\}$$
$$\underline{very\ strong} = \{1, 0.49, 0.01, 0\}$$
$$\underline{quite\ strong} = \{1, 0.84, 0.32, 0\} \qquad (2.1)$$
$$\underline{weak} = \{0, 0.1, 0.7, 1\}$$
$$\underline{not\ very\ strong} = \{0, 0.51, 0.99, 1\}$$

It may be the expert's opinion that the visual inspection tends to
over-estimate the strength and he may express this view in the
following way.

"If the visual inspection indicates <u>very strong</u> then the actual
state is <u>quite strong</u> or if the visual inspection indicates <u>quite
strong</u> then the actual state is <u>weak</u> or if the visual inspection
indicates <u>quite weak</u> then the actual state is <u>very weak</u> or if the
visual inspection indicates <u>weak</u> then the actual state is <u>very very
weak</u>."

Then

$$R' \equiv [(\underline{very\ strong}) \times (\underline{quite\ strong})]$$
$$R'' \equiv [(\underline{quite\ strong}) \times (\underline{weak})]$$
$$R''' \equiv [(\underline{quite\ weak}) \times (\underline{very\ weak})] \qquad (2.2)$$
$$R'''' \equiv [(\underline{weak}) \times (\underline{very\ very\ weak})]$$

where the symbol $\times$ indicates the cartesian product [34] of fuzzy sets.
The fuzzy relationship (R_{VA}) between the visual inspection and the
actual state in the expert's opinion is given by

$$R_{VA} \equiv R' \cup R'' \cup R''' \cup R'''' \qquad (2.3)$$

where the symbol $\cup$ represents the union of fuzzy sets [33,34]. For
this case [24] the relationship is

$$R_{VA} = \begin{bmatrix} 1 & 0.84 & 0.70 & 1 \\ 0.49 & 0.49 & 0.70 & 0.84 \\ 0.01 & 0.10 & 0.49 & 0.84 \\ 0 & 0.01 & 0.49 & 1 \end{bmatrix} \qquad (2.4)$$

and this encodes the expert view. If a visual inspection indicates a
fuzzy set $\tilde{V}$ then the corresponding actual state is the fuzzy set $\tilde{A}$
where

$$\tilde{V} \circ R_{VA} \equiv \tilde{A} \qquad\qquad (2.5)$$

and the dot symbol indicates fuzzy composition [23,33,34]. Thus if $\tilde{V}$
represents the linguistic assessment not strong then

$$\tilde{V} \equiv \{0,\ 0.3,\ 0.9,\ 1\}$$

and

$$\tilde{A} \equiv \{0.3,\ 0.3,\ 0.49,\ 1\} \qquad\qquad (2.6)$$

The supports for each stress-grade indicated by $\tilde{A}$ are termed the
"possibilities" that the joist will belong to that stress-grade.
In a similar way, the subjective views of a concrete engineer can
be sought with respect to the general influence on the ultimate moment
of a cross-section of such matters as the concrete mixing and placing,
and the fixing of the reinforcement. Fuzzy relations can be assembled
and the fuzzy set representation of the ultimate moment can be deduced
for any linguistic statements regarding the standard of site control.
This fuzzy measure of the ultimate moment for an element - say a
simply-supported beam in bending - can readily be expressed in terms
of the value of the load parameter to cause failure of the element.
This fuzzy measure of resistance ($\tilde{R}$) can then be compared with the
corresponding measure of the applied loading ($\tilde{S}$) and the possibility
distribution for ($\tilde{S} - \tilde{R}$) can be obtained. Suppose the states of $\tilde{R}$ and
$\tilde{S}$ are discretised to $\{0,1,2,3,4,5,6...\}$ then the value of 4 for $S - R$
can be made up of ($S = 6$, $R = 2$), ($S = 5$, $R = 1$) and ($S = 4$, $R = 0$).
The lesser of the supports for each pair is assessed and the maximum
of such minimal supports is the corresponding support for ($S - R$) to
take the value of 4. By considering all such values the possibility
distribution for ($S - R$) is obtained and the safety of the element is
assessed by comparing the support for positive values of ($S - R$)
corresponding to failure with the negative values which correspond to
non-failure. Suitable reliability levels can be specified depending
on the type of structure and the perceived consequence of failure.
The above analysis is for structural elements and the corresponding analysis for frames requires the evaluation of the fuzzy set for
frame resistance ($\tilde{R}$) from the varying fuzzy sets for the plastic
moments of resistance of the critical sections. This is achieved
through the fuzzy programs for plastic analysis.

3.   FUZZY PROGRAMMING

Mathematical Programming (MP) is one of the most powerful and widely
used tools of operations research (OR) and systems engineering. It is
the science and art of constrained optimisation of functions. The constraints are generally inequations but for the special case where the
constraints are exclusively equations then MP reduces to classical

(i.e. Lagrangean) optimisation. Similarly when the objective and constraint functions are all linear then the mathematical program (MP)
reduces to a linear program (LP). MP has been successfully applied to
several branches of civil engineering including transportation engineering, water resources engineering and structural engineering. In
particular MP has been shown to be a convenient basis for the plastic
analysis and synthesis of structural systems [6,15,17,28]. It is of
historic interest that MP and limit analysis were anticipated simultaneously in an 1823 memoire to the French Academy of Sciences by
Fourier but it was about one and a half centuries later before these
ideas were fully developed. Interestingly in the late nineteen-
thirties, significant advances were made in the USSR by Kantorovitch
with respect to LP and by Gvozdev with respect to limit analysis.
However, these developments were not fully exploited nor were they
immediately brought to the attention of the international scientific
community. In the nineteen-fifties and later, MP and engineering
plasticity flourished and the general symbiosis between the two subjects was fostered by this continued growth and by the increasing
availability of large computing facilities.

The LP of plastic limit analysis for the safe (or static) method
has constraints of static admissibility and the stipulations are
obtained from the yield condition. Thus for a simple flexural frame,
the plastic moments of resistance of the critical sections are the
stipulations of the program and are assumed to be known with complete
precision. For a reinforced concrete frame, where the sections are
singly-reinforced and under-reinforced, this moment depends on the
strengths of the steel reinforcement and of the concrete, the steel
area and the dimensions of the concrete section. These parameters are
generally known with varying degrees of precision. The depth of the
reinforcement will depend on the quality of the steel-fixing of the
contractor. Thus subjective views regarding such matters as site
control can be converted to fuzzy supports for a range of possible
values of plastic moments of resistance as has been previously
indicated. The deterministic LP with crisp stipulations has thus been
converted to a program whose stipulations are fuzzy.

Another important OR tool is Bayesian decision analysis. In this
subject, a fundamental problem is concerned with the evaluation of the
least-biased (or minimally-prejudiced) prior probabilities in the face
of sparse prior statistical knowledge. It has been shown that this
problem leads to a non-linear program (NLP) of a rather special form
[9,11,16,30]. The stipulations of this program are known mean (or
expected) values of functions of the random variable whose probability
distribution is being sought. The objective function to be maximised
is the Shannon entropy function [30] which measures the total uncertainty associated with the probabilities. The prior probabilities
may then be updated to posterior probabilities through Bayes' rule when
new statistical knowledge becomes available. Thus the means of the
functions of the random variable which were used to encode the prior
statistical knowledge are changed by the Bayesian updating and hence
the prior values were merely estimates. It has been pointed out [16,
22] that it is often unreasonable to imply complete prior confidence

with respect to these means and that some subjective (and therefore imprecise) measure of confidence should be incorporated into the specification of the prior knowledge.

Thus a wide range of problems are concerned with MPs which have fuzzy stipulations. When MPs have this form of imprecision entering into their objective or constraint functions, they are dubbed "Fuzzy Programs" and this section will concentrate on such programs where the fuzziness is confined to the stipulation of linear constraints.

First consider a deterministic LP with two objectives.

$$
\begin{array}{c}
\boxed{\begin{array}{l}
\text{Min } z_1 = \underset{\sim}{c}_1^T \underset{\sim}{x} \\[2ex]
\text{Min } z_2 = \underset{\sim}{c}_2^T \underset{\sim}{x} \\[2ex]
\underset{\sim}{A} \underset{\sim}{x} \geqslant \underset{\sim}{b} \\[2ex]
\underset{\sim}{x} \geqslant \underset{\sim}{0}
\end{array}}
\end{array}
\qquad (3.1)
$$

Multiple objectives [29] cause particular difficulty and it is convenient to convert the objectives to goals.

$$
\begin{aligned}
z_1 &\leqslant z_1' \\[2ex]
z_2 &\leqslant z_2'
\end{aligned}
\qquad (3.2)
$$

where z_1' and z_2' are crisp levels of aspiration with respect to the two goals. The MP (3.1) now consists of a system of inequalities.

$$
\begin{aligned}
\underset{\sim}{E}\,\underset{\sim}{x} &\geqslant \underset{\sim}{h} \\[2ex]
\underset{\sim}{x} &\geqslant \underset{\sim}{0}
\end{aligned}
\qquad (3.3)
$$

where

$$
\underset{\sim}{E} \equiv
\begin{bmatrix}
-\underset{\sim}{c}_1^T \\ \hline
-\underset{\sim}{c}_2^T \\ \hline
\underset{\sim}{A}
\end{bmatrix}
\qquad \text{and} \qquad
\underset{\sim}{h} \equiv
\begin{bmatrix}
-z_1' \\ \hline
-z_2' \\ \hline
\underset{\sim}{b}
\end{bmatrix}
$$

The ith inequality

$$
\sum_j e_{ij}\, x_j \geqslant h_i
\qquad (3.4)
$$

will now be softened in the sense that it is no longer mandatory to satisfy (3.4) but it is still necessary to satisfy the modified inequality

$$\sum_{ij} e_{ij} x_j \geqslant h_i - s_i \qquad\qquad (3.5)$$

where s_i is the (non-negative) measure of the softness that has been introduced into the constraint. Satisfaction of the original inequality (3.4) is preferred but its contravention is tolerated with diminishing acceptability until the slack of the modified inequality (3.5) is fully taken up. This view is indicated on figure 1 and is represented mathematically by the form of support (μ_i) given below.

$$\begin{aligned}
&\text{(i)} \quad \mu_i = 1 && \text{if } \sum_j e_{ij} x_j \geqslant h_i \\
&\text{(ii)} \quad \mu_i s_i = \sum_j e_{ij} x_j - h'_i && \text{if } h'_i \leqslant \sum_j e_{ij} x_j \leqslant h_i && (3.6) \\
&\text{(iii)} \quad \mu_i = 0 && \text{if } \sum_j e_{ij} x_j \leqslant h'_i
\end{aligned}$$

where $h'_i = h_i - s_i$.

Thus the fuzziness regarding the goals and constraints has now been combined into a single set of fuzzy constraints. Using the Bellman-Zadeh criterion [2] for decision-making in a fuzzy environment the support for the decision for any set of decision variables ($\underset{\sim}{x}$) is given by the minimum of the supports as one spans across the ensemble of fuzzy constraints. A new variable (γ) is now defined to be that minimal support. The maximising decision is determined by that value of the vector $\underset{\sim}{x}$ which maximises γ. Thus this form of fuzzy program can be converted to the following non-fuzzy LP.

$$\boxed{\begin{aligned}
&\text{Max } \gamma \\
&s_i \gamma \leqslant \sum_j e_{ij} x_j - h'_i \\
&x_j \geqslant 0 \qquad 0 \leqslant \gamma \leqslant 1 \\
&i = 1,2 \ldots \qquad j = 1,2 \ldots
\end{aligned}} \qquad (3.7)$$

The above LP can be solved by the simplex algorithm [7] and the optimal value of γ gives a measure of the acceptability of the optimal decision in the face of the prescribed imprecision.

This fuzzy programming approach has been modified [25] for the case where a single uncontroversial (and non-fuzzy) objective exists. For this case, the original objective function can be optimised and an aspiration level can be set as a goal for the level of acceptability. The fuzzy constraints can be transformed to non-fuzzy form as previously described. Therefore using this modified form of fuzzy programming, the following non-fuzzy LP is obtained.

$$\boxed{\begin{aligned}
\text{Min } z &= \sum_j c_j\, x_j \\[4pt]
s_i\, \gamma &\leq \sum_j a_{ij}\, x_j - b_i' \\[4pt]
\gamma &\geq \gamma_L \\[4pt]
x_j &\geq 0
\end{aligned}} \qquad (3.8)$$

where, as before, s_i is the softness introduced into the ith constraint and b_i' is given by

$$b_i' = b_i - \overset{.}{s}_i$$

and γ_L is the minimal value of support to ensure a sufficiently acceptable decision. Numerical examples of both techniques have been given
elsewhere [5,25].

4. OPTIMAL PLASTIC DESIGN AND ANALYSIS

In this section the optimal plastic design of a flexural frame
will first be considered. The simplest class of problem is where

    (i)   the geometry and topology are fixed *a priori*
    (ii)   a single loading state is considered
    (iii) the constitutive relations are those of perfect
           plasticity
    (iv)   the plasticity is controlled by a single stress-
           resultant

and   (v)   a single linearised design criterion is applied.

With the above restriction, the mesh primal LP takes the following form
[18,28]

$$\boxed{\begin{aligned}
\text{Min } z &= [\underset{\sim}{\ell}^T \mid \cdot\,]
\begin{bmatrix} \underset{\sim}{d} \\ \hline \underset{\sim}{p} \end{bmatrix} \\[10pt]
[\underset{\sim}{J} \mid -\underset{\sim}{N}^T\underset{\sim}{B}]
\begin{bmatrix} \underset{\sim}{d} \\ \hline \underset{\sim}{p} \end{bmatrix}
&\geq \underset{\sim}{N}^T \underset{\sim o}{m} \\[10pt]
\underset{\sim}{d} &\geq \underset{\sim}{0}
\end{aligned}} \qquad (4.1)$$

where $\underset{\sim}{d}$ are the design variables and $\underset{\sim}{\ell}$ are the lengths of the members
fixed by the corresponding design variables. The indeterminacies (or
more generally, the mesh actions) are denoted by $\underset{\sim}{p}$ and $\underset{\sim}{B}$ is the mesh
static matrix. The matrix $\underset{\sim}{J}$ describes the incidence of the design
variables with respect to the critical sections and $\underset{\sim o}{m}$ is the vector of

the particular solution stress-resultants at the critical sections. The
The normality matrix $\underset{\sim}{N}$ is defined by

$$\underset{\sim}{N} \equiv [\underset{\sim}{I} \mid -\underset{\sim}{I}] \tag{4.2}$$

where $\underset{\sim}{I}$ is the identity matrix.

If the mesh actions $\underset{\sim}{p}$ are expanded (in terms of non-negative components

$$\underset{\sim}{p} = \underset{\sim}{p}^{+} - \underset{\sim}{p}^{-} \tag{4.3}$$

$$\underset{\sim}{p}^{+} \geqslant \underset{\sim}{0} \qquad \underset{\sim}{p} \geqslant \underset{\sim}{0}$$

then the problem of plastic limit design is transformed to the form of
LP (3.1) but with a single constraint.

$$\underset{\sim}{A} \equiv [\underset{\sim}{J} \mid -\underset{\sim}{N}^{T}\underset{\sim}{B} \mid \underset{\sim}{N}^{T}\underset{\sim}{B}]$$

$$\underset{\sim}{c}^{T} \equiv [\underset{\sim}{\ell}^{T} \mid \cdot \mid \cdot \,]$$

$$\underset{\sim}{x} \equiv \begin{bmatrix} \underset{\sim}{d} \\ \hline \underset{\sim}{p}^{+} \\ \hline \underset{\sim}{p}^{-} \end{bmatrix} \tag{4.4}$$

$$\underset{\sim}{b} \equiv \underset{\sim}{N}^{T}\underset{\sim}{m}_{o}$$

If the objective is replaced by a goal then the constraint set (3.3) is
recovered and if all the stipulations of this set are fuzzified as
indicated in (3.6) then the non-fuzzy version of the fuzzy program is
given by (3.7). The solution of this program gives a (deterministic)
design which has optimal supports in the face of the fuzzy goal and of
the fuzziness of $\underset{\sim}{m}_{o}$. This latter vector is obtained from the load and
geometry and generally the fuzziness will stem from an imprecise knowledge of the loading. Thus the γ factor is a form of reliability
index associated with the design.

Alternatively one can use the modified fuzzy programming technique
[25] in which the original objective function is minimised and a lower
bound is set to γ. This corresponds to seeking the minimal value of
the linearised weight function subject to an acceptable value of the
reliability index. This latter procedure appears to have advantages
over the former for the structural design problem. The applications of
fuzzy linear programming to plastic limit design have been developed
more fully elsewhere [25].

In a similar way the problem of plastic limit analysis can be
formulated as a primal-dual LP pair [17,28] and for the static program
the stipulations are the plastic moments of resistance. In the usual
form of analysis these moments are assumed to be known with precision;
that is to say they are crisp numbers. It has been indicated earlier

that a fuzzy view of such matters as site control of the concrete pro-
duction and of steel-fixing can be transformed into a fuzzy number for
the plastic moment of resistance of a section of a reinforced concrete
member. These fuzzy representations may vary for different sections
depending on such matters as the difficulty of placing concrete and the
congestion of the reinforcement. Thus the fuzzy programming formulation [26] of the plastic limit analysis problem will convert the
fuzzy strengths of the sections to some overall measure of the fuzzy
strength of the frame as expressed through a loading parameter.

   If the second method of fuzzy programming [25] is adopted then the
load factor corresponding to any value of overall support (γ_L) can be
deduced and if this support is varied then the possibility distribution
for the frame load factor (or resistance) can be derived. The fuzzy
reliability analysis then continues as has been previously described for
an element.

5. FUZZY STATISTICS

The important problem of evaluating the prior probabilities for a
Bayesian decision analysis can be tackled by the Jaynes-Shannon formalism [9,30] and this involves solving a rather special form of NLP.
If X is a random variable and $\{\bar{g}_i\}$ is a set of functions of the random
variable then the prior statistical knowledge consists of the mean
values $\{g_i\}$ of the functions. The discrete state values $\{x_j\}$ are known
and the corresponding probabilities $\{p_j\}$ are to be determined. The
Shannon entropy (H) is the measure of total uncertainty and the
variables $\{p_j\}$ are determined from the following program.

$$
\boxed{
\begin{aligned}
&\text{Max } H = - \sum_j p_j \ln p_j \\[4pt]
&\sum_j p_j = 1 \\[4pt]
&\sum_j g_i (x_j) p_j = \bar{g}_i \\[4pt]
&i = 1, 2, \ldots
\end{aligned}
}
\tag{5.1}
$$

   Most civil engineering decision-problems involve both uncertainty
due to randomness and fuzziness due to imprecision. Two classes of
problems can be identified for the present purpose. An example of the
first is associated with the ready-mixed-concrete industry. The con-
troller makes decisions regarding the mix to be produced in a plant
during a given time period and this involves making estimates regarding
future orders for the various standard mixes. The way in which the
available statistics for previous similar periods may be analysed
objectively has been presented elsewhere [27] and the resulting probabilities have been incorporated in a Bayesian decision analysis. However, other influences such as weather and traffic conditions can affect
the order states but these are not known statistically. Expert opinions
may be sought regarding the general influence of, for example, weather

conditions on the incoming orders and these will usually be expressed
verbally. By using linguistic variables, fuzzy relationships can be
built up to express this link and the corresponding matrices form the
basis of a generalised Bayes' rule. Thus the statistical aspects of
the problem can be analysed through the maximum-entropy formalism and
the additional fuzzy information (such as the current weather conditions) can be incorporated through the generalised Bayes' rule to
yield the fuzzy probabilities. It should be noted that the entropy can
increase or decrease during this process depending on whether the fuzzy
considerations tend to contradict or confirm the statistical considerations. The fuzzy probabilities can be utilised in a Bayesian decision
analysis as before. An illustration of this process for a different
problem has been presented elsewhere [24] and will be discussed further
in section 6.

In the second class of problem the expert opinion is concerned with
the relevance of the available statistics. For example, test results of
the 28-day strength of a particular concrete may be available but, due
to the incorrect use of an additive, an expert concrete technologist may
express doubt regarding the relevance of these test results with respect
to the prediction of the long-term strength of the placed concrete. For
this class of problem the introduction of fuzziness cannot reduce and,
in general, will increase the total incertitude. It has been pointed
out [16,21,22] that the program (5.1) requires complete prior confidence with respect to these means even though a Bayesian updating of
the priors will almost inevitably follow. It appeared to be advant-
ageous to introduce some subjective doubt regarding the specification
of the prior statistical knowledge. The first step was to convert the
stipulations $\bar{g}_i$ from crisp numbers to ranges of values lying between
specified lower and upper bounds.

$$\bar{g}_i^L \leqslant \bar{g}_i \leqslant \bar{g}_i^U \tag{5.2}$$

The MP (5.1) with the bounds (5.2) can now be recognised as a non-
linear inexact program [31]. The cited report [16] went on to introduce fuzzy supports for various stipulation states within the specified
bounds. More recently [22] the modified fuzzy programming method outlined earlier has been applied directly to this problem since all the
statistical constraints are linear with respect to the probabilities.
Thus the fundamental problem is reduced to the following form:

$$
\boxed{
\begin{array}{l}
\text{Max } H = -\sum_j p_j \ln p_j \\[2ex]
\sum_j p_j = 1 \\[2ex]
\sum_j p_j g_i(x_j) = \tilde{g}_i \\[2ex]
i = 1,2,\ldots
\end{array}
}
\tag{5.3}
$$

The evaluation of the prior probabilities in the face of imprecisely known statistical parameters is achieved through the fuzzy-equality-constrained maximisation of entropy.

The general method can be illustrated with respect to a relatively simple problem where the statistical knowledge is confined to the mean (m) of the random variable. If this parameter is known precisely then the maximum entropy formalism leads to the exponential distribution. However if the mean is a fuzzy number ($\tilde{m}$) then the corresponding constraint can be represented as

$$\sum_j x_j p_j = \tilde{m} \tag{5.4}$$

and can be read as the left-hand-side being approximately equal to m.

One possible form of support is indicated in figure 2 and is given mathematically by:

$$\begin{aligned}
\mu_1 &= 0 & &\text{if } \sum_j x_j p_j \leqslant m - \epsilon \\
\epsilon\mu_1 &= \sum_j x_j p_j - m + \epsilon & &\text{if } m - \epsilon \leqslant \sum_j x_j p_j \leqslant m \\
\epsilon\mu_1 &= m + \epsilon - \sum_j x_j p_j & &\text{if } m \leqslant \sum_j x_j p_j \leqslant m + \epsilon
\end{aligned} \tag{5.5}$$

The modified fuzzy programming technique [22,25] reduces the problem to the following convex program with linear constraints.

$$\boxed{\begin{aligned}
\text{Min } z &= \sum_j p_j \ln p_j \\
\sum_j p_j &= 1 \\
\epsilon\gamma &\leqslant \sum_j x_j p_j - m + \epsilon \\
\epsilon\gamma &\leqslant m + \epsilon - \sum_j x_j p_j \\
\gamma &\geqslant \gamma_L \\
0 &\leqslant p_j \leqslant 1
\end{aligned}} \tag{5.6}$$

6. TOWARDS A GENERAL THEORY OF STRUCTURAL RELIABILITY

The plastic moments of resistance of the critical sections of a frame may be known deterministically, statistically or imprecisely. For the first case the collapse load factor of the frame can be deduced from linear programming. In the second case the probability distribution of the collapse load factor can be obtained from stochastic programming whilst in the third case the possibility distribution for the collapse load factor can be calculated using fuzzy programming.

It has become usual to present structural reliability as an entirely statistical problem whereas it has been shown how this subject may be developed for a purely fuzzy situation. In reality the available information may be mixed and the fundamental problem is to derive a

general theory of structural reliability incorporating both randomness
and fuzziness. In the present discussion it will be assumed that the
prior knowledge is primary statistical but that additional fuzzy information may alos be available. Two types of fuzzy information will be
distinguished herein. In the first type the prior probabilities of
strength (of, for example, a reinforced concrete beam) have been calculated through the maximum-entropy program in which the stipulations
are the relevant expected values obtained from tests on the concrete
and steel. Additional (and fuzzy) information is concerned with an
expert's view regarding other influences on strength such as site
control of the placing of concrete and the fixing of reinforcement.
This view can be incorporated into a fuzzy relationship as discussed in
section 2 and a generalised Bayes' rule [24] can be devised to up-date
the probabilities.

Thus if the maximum-entropy prior probabilities for four strength-
grades based on prior tests are

$$p_1 = 0.2 \qquad p_2 = 0.4 \qquad p_3 = 0.3 \qquad p_4 = 0.1$$

and if the fuzzy relation R, which links the tests results with the
actual *in situ* strength, corresponds to the expert opinion that "if the
test result is <u>very strong</u> then the actual strength is <u>quite strong</u>
ELSE it is <u>weak</u>" then, using the fuzzy sets of section 2, R is given by

$$R = \begin{bmatrix} 1 & 0.84 & 0.32 & 0 \\ 0.49 & 0.49 & 0.51 & 0.51 \\ 0.01 & 0.10 & 0.70 & 0.99 \\ 0 & 0.01 & 0.70 & 1 \end{bmatrix}$$

The posterior probabilities (p_j') [24] are given by

$$p_j' = \frac{\sum_i r_{ij}\, p_i}{\sum_k \sum_j r_{jk}\, p_j} \qquad\qquad (6.1)$$

where r_{ij} are the elements of the matrix R. Thus

$$p_1' = 0.204 \qquad p_2' = 0.207 \qquad p_3' = 0.281 \qquad p_4' = 0.308$$

It will be seen that the additional fuzzy information has both
flattened the distribution and pushed the mode to a lower strength-
grade. It will be seen that when the additional information is purely
statistical then the sample likelihoods replace the fuzzy relations and
Bayes' rule is recovered.

The second type of fuzzy information is concerned with expert views
on the relevance of the prior statistics [22]. It has been previously
shown that the evaluation of prior probabilities in the face of crisp
prior statistical knowledge is achieved through constrained maximisation
of the information entropy. Attempts have been made to modify the
information entropy to produce a measure of total incertitude [4,12,19]
and to use this measure to deduce fuzzy probabilities. Whilst these

techniques have made some progress, they suffer from certain fundamental
disadvantages [21,22]. They are corrections to the information entropy
which are somewhat arbitrarily defined so that the probabilities will be
altered to fuzzy probabilities in a more-or-less rational way. The
beauty of the original Jaynes-Shannon formalism [9,30] was that a very
wide class of problems corresponded to a single unifying extremal
principle and recently this principle has been extended and generalised
[22] to the case where the prior statistics are stated fuzzily. The
probabilities are deduced through the solution of a fuzzy non-linear
program [22]. The general considerations for the evaluation of prior
probabilities are displayed in figure 3. A somewhat similar consideration can be given to the loadings and the prior probabilities of
strength and load can be combined to deduce the probability of failure
in the usual way. At any stage these prior probabilities can be updated
to the corresponding posterior probabilities when new statistical or
fuzzy information becomes available. This updating can be achieved
using Bayes' rule for the new statistical information and using the
generalised Bayes' rule when the new information is fuzzy.

7. CLOSURE

Imprecision should enter into realistic mathematical modelling of many
civil engineering problems. This form of incertitude may be handled by
methods based on fuzzy sets. The problem of structural reliability is
concerned both with randomness and imprecision. A purely fuzzy problem
can be handled by expressing expert opinions linguistically and con-
structing the corresponding fuzzy relations. The fuzzy set representations of the plastic moments of resistance of the critical sections of
a frame become the stipulations of the linear program of plastic limit
analysis. Solution of this fuzzy program yields the fuzzy set for the
collapse load parameter. This latter set can be linked with the
corresponding loading set to produce the fuzzy set which measures
reliability.
     The natural randomness of loading and of material strengths can be
processed objectively through the maximum-entropy formalism. However
most civil engineering decision problems - and, in particular,
structural reliability - have both random and fuzzy aspects. A general
theory must be capable of including both types of information.
     If expert opinions are available with respect to the validity of
the prior statistics then the prior probabilities are obtained from the
solution of a fuzzy non-linear program. These priors can be updated to
posterior probabilities by Bayes' rule when new statistical information
becomes available. When the new information is fuzzy then fuzzy
relationships may be employed in a form of generalised Bayes' rule to
generate the appropriate posteriors. At each stage the probability of
failure can be deduced from the probabilities of load and resistance.

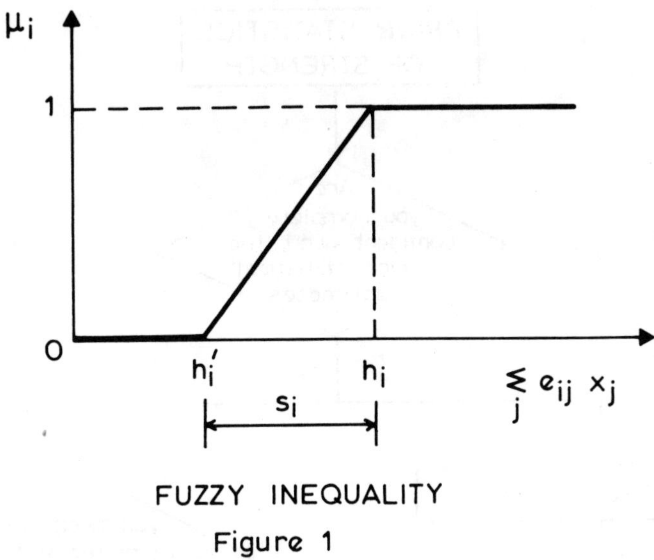

FUZZY INEQUALITY

Figure 1

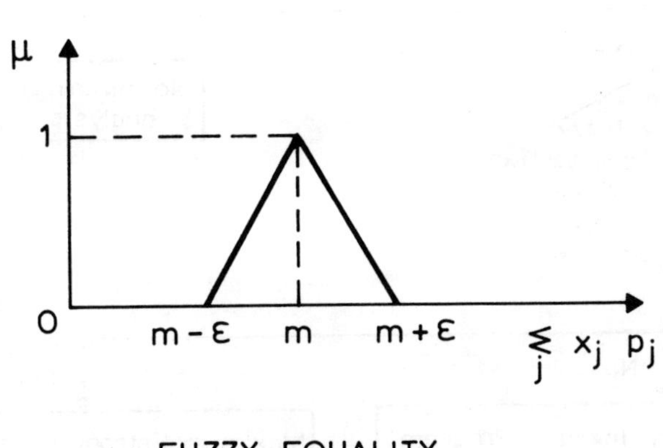

FUZZY EQUALITY

Figure 2

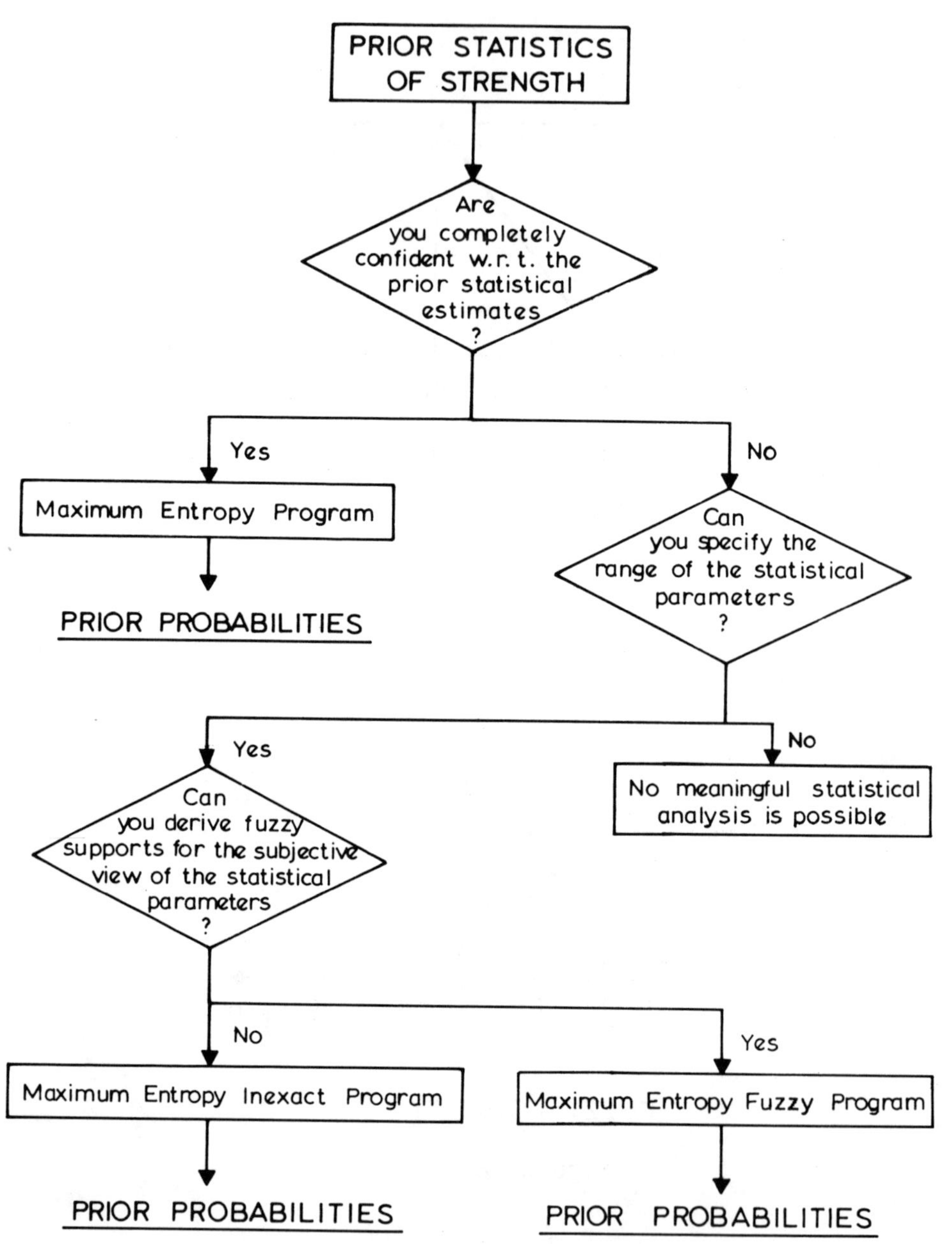

PRIOR STATISTICS OF STRENGTH
Are you completely confident w.r.t. the prior statistical estimates ?
Yes
No
Maximum Entropy Program
Can you specify the range of the statistical parameters ?
PRIOR PROBABILITIES
Yes
No
Can you derive fuzzy supports for the subjective view of the statistical parameters ?
No meaningful statistical analysis is possible
No
Yes
Maximum Entropy Inexact Program
Maximum Entropy Fuzzy Program
PRIOR PROBABILITIES
PRIOR PROBABILITIES

Figure 3

REFERENCES

[1]   BACKMAN, B.F., BROWN, C.B., JOWITT, P.W. and MUNRO, J. (1983)
        'Statistical Mechanics of Granular Materials' in Advances in
        the Mechanics and the Flow of Grnular Materials, Vol. 1
        (edited by M. Shahinpoor) Trans. Tech. Publications, Germany.

[2]   BELLMAN, R.E. and ZADEH, L.A. (1970)
        'Decision-Making in a Fuzzy Environment'
        Management Science, 17, 4, B.141-164.

[3]   BLOCKLEY, D.I. (1975)
        'Predicting the Likelihood of Structural Accidents'
        Procs. Inst. Civ. Engrs., Part 2, 59, 659.

[4]   BROWN, C.B. (1980)
        'Entropy Constructed Probabilities' Journal of the Engineering
        Mechanics Division, ASCE, 106, EM4, August, 633-640.

[5]   CHUANG, P-H and MUNRO, J. (1983)
        'Linear Programming with Imprecise Data'
        Civil Engineering Systems, 1, September, 37-41.

[6]   COHN, M.Z. and MAIER, G. (editors) (1979)
        Engineering Plasticity by Mathematical Proramming,
        Pergamon Press, New York.

[7]   DANTZIG, G.B. (1963)
        Linear Programming and Extensions. Princeton University
        Press, Princeton.

[8]   DeLUCA, A. and TERMINI, S. (1972)
        'A Definition of a Non-Probabilistic Entropy in the Setting of
        Fuzzy Sets Theory', Information and Control, 20, 301-312.

[9]   JAYNES, E.T. (1968)
        'Prior Probabilities', IEEE Trans. Systems, Science and
        Cybernetics. SSC-4, 3, 227.

[10]   JOWITT, P.W. (1979)
        'Bayesian Estimates of Material Properties from Limited Test
        Data', Engineering Structures, 1, 4, 170-178.

[11]   JOWITT, P.W. and MUNRO, J. (1975)
        'The Influence of Void Distribution and Entropy on the
        Engineering Properties of Granular Media'
        ICASP-2. Proceedings of the Second International Conference
        on the Applications of Statistics and Probability in Soil and
        Structural Engineering, Aachen.

[12]   KAM, T-Y and BROWN, C.B.   (1983)
       'Updating Parameters with Fuzzy Entropies', Journal of
       Engineering Mechanics, **109**, December, 1334-1343.

[13]   THOFT-CHRISTENSEN, P. and BAKER, M.J.   (1982)
       Structural Reliability Theory and its Applications.
       Springer-Verlag.

[14]   LOUCKS, D.P., STEDINGER, J.R. and HAITH, D.A.   (1981)
       Water Resource Planning and Analysis. Prentice-Hall,
       Englewood Cliffs, N.J.

[15]   MAIER, G. and MUNRO, J.   (1982)
       'Mathematical Programming Applications to Engineering Plastic
       Analysis'. Applied Mechanics Reviews, **35**, 12, 1631-1643.

[16]   MUNRO, J.   (1975)
       'Statistical Decision Analysis in Timber Engineering and
       Prototype Testing', Building Research Establishment -
       Princes Risborough Laboratory Report.

[17]   MUNRO, J.   (1979)
       'The Application of Mathematical Programming to Structural
       Problems', IASS World Congress, Madrid. Procs., **3**, 5.189-
       5.207.

[18]   MUNRO, J.   (1979)
       'Optimal Plastic Design of Frames', Chapter 7 of reference
       [6].

[19]   MUNRO, J.   (1979)
       'Uncertainty and Fuzziness in Engineering Decision-Making',
       Procs. First Canadian Seminar on Systems Theory for the Civil
       Engineer, Calgary, 113-133.

[20]   MUNRO, J.   (1982)
       'Plastic Analysis in Geomechanics by Mathematical Programming'
       in Numerical Methods in Geomechanics J.B. Martins (editor)
       Reidel.

[21]   MUNRO, J.   (1984)
       Contribution to the Discussion of reference [12] above.

[22]   MUNRO, J.   (1984)
       'Fuzzy Programming and Imprecise Statistics'
       Second Canadian Seminar on Systems Theory for the Civil
       Engineer, Calgary. May. (To appear in Civil Engineering
       Systems)

[23]   MUNRO, J.   (1984)
       'Systems Engineering' Chapter 13 of Mathematical Methods in
       Engineering, G.A.O.Davies (editor), John Wiley, Chichester.

[24] MUNRO, J. and BROWN, C.B.   (1983)
          'The Safety of Structures in the Face of Uncertainty and
          Imprecision' ICASP-4. Procs. Fourth International Conference
          on Applications of Statistics and Probability in Soil and
          Structural Engineering, Florence. 695-711.

[25] MUNRO, J. and CHUANG, P-H.   (1984)
          'Optimal Plastic Design with Imprecise Data'
          To be published in Journal of Engineering Mechanics, ASCE.

[26] MUNRO, J. and CHUANG, P-H.   (1984)
          'Optimal Plastic Limit Analysis with Imprecise Data'
          Euromech Colloquium Mathematical Programming for the Plastic
          Analysis of Structures, London, 1984.

[27] MUNRO, J. and JOWITT, P.W.   (1978)
          'Decision Analysis in the Ready-Mixed-Concrete Industry'
          Procs. Institution of Civil Engineers, Part 2, **65**, 41-52.

[28] MUNRO, J. and SMITH, D.L.   (1972)
          'Linear Programming Duality in Plastic Analysis and Synthesis'
          Procs. International Symposium on Computer-Aided Structural
          Design. Warwick, **1**, A1.22-A1.54.

[29] OSYCZKA, A.   (1984)
          Multicriterion Optimization in Engineering. Ellis Horwood,
          Chichester.

[30] SHANNON, C.E.   (1984)
          'The Mathematical Theory of Communication' Bell System
          Technical Journal, **27**, 279-428 and 623-656.

[31] SOYSTER, A.L.   (1973)
          'Convex Programming with Set-Inclusive Constraints and
          Applications to Inexact Linear Programming' Operations
          Research, **21**, 1154-1157.

[32] WILLIAMS, A.C.   (1965)
          'On Stochastic Linear Programming' SIAM Journal Applied
          Mathematics, **13**, 917-940.

[33] ZADEH, L.A.   (1965)
          'Fuzzy Sets' Information and Control, **8**, 338-353.

[34] ZADEH, L.A.   (1973)
          'Outline of a New Approach to the Analysis of Complex Systems
          and Decision Processes' IEEE Trans. Systems, Man and
          Cybernetics, SMC **3**, 28-44.

[35]   ZIMMERMAN, H-J.   (1976)
        'Description and Optimisation of Fuzzy Systems'
        International Journal of General Systems, **2**, 209-215.

[36]   ZIMMERMAN, H-J.   (1978)
        'Fuzzy Programming and LP with Several Objective Functions'
        <u>Internatinal Journal of Fuzzy Sets and Systems</u>, **1**, No. 1,
        45-55.

STRUCTURAL SAFETY INFERENCE BY THE USE OF FUZZY SET THEORY

D I Blockley
Department of Civil Engineering
University of Bristol
BRISTOL
BS8 1TR
UK

ABSTRACT. The paper is in three parts. The first discusses the ideas
of fuzziness and probability and the mathematical manipulations involved.
The second outlines a computer system called fuzzy relational inference
language (FRIL). The third part outlines two major applications of FRIL
to mine safety and to nuclear safety. The work on these examples is
still in its early stages.

1. THE CONCEPTS

1.1 Fuzziness and Probability

The distinction between fuzziness and probability can be drawn in two
ways, conceptually and mathematically. Fuzziness is vagueness and
imprecision of definition. For example in everyday language we speak
of a "tall man" and, without any information, an idea is communicated.
Someone with a scientific attitude may say that, in our search for
truth, this level of imprecision is unsatisfactory and that he will not
be satisfied until the man's height is measured to the nearest mm .
An engineer, on the other hand would first examine the problem to be
solved. If he had to buy a casual shirt for the man from a shop where
the shirts were labelled short, medium, tall, then the problem can be
solved immediately. However, for other problems, where more information
is required, he will judge whether or not the penalty (cost) to be paid
for acquiring it is worthwhile. If it is not, or if indeed the
information is simply unavailable, unattainable or not measureable then
he must make a decision based on his personal judgement. The measure
of imprecision suggested by Zadeh is the fuzzy membership function and
is a generalisation of the the indicator or characteristic function of set
theory. This function can be interpreted in various ways but the
interpretation favoured here is that it represents a restriction on the
set of possible values that the variable can take. It is a measure of
semantic definition, it is not a measure of randomness or chance or
belief about a possible outcome, which are the various well known
interpretations of probability. The membership function is therefore

49

A. C. Lucia (ed.), Advances in Structural Reliability, 49–72.
© *1987 by ECSC, EEC, EAEC, Brussels and Luxembourg.*

a fuzzy possibility restriction and much of the mathematics is aimed
towards combining these measures to provide the least restrictive set
of possibiliites. Probability, is of course, a relative measure because
all possible outcomes must be known and the probabilities distributed
so that they sum to one. It is quite possible to define probability
measures of fuzzy events (e.g. the probability that a tall man will come
into my office whilst I am writing this paper) or even fuzzy probabilities
(e.g. this event is very probable).

The membership function (χ) is a mapping from the sample space (S)
to the range (0,1) whereas the normal interpretation of probability (p)
is as a mapping from the power set (E) or the set of all subsets of S
to the range of (0,1).

$$\text{Thus for example if} \qquad S = (a_1, a_2, a_3)$$

$$E = \{(a_1), (a_2), (a_3), (a_1\ a_2), (a_1\ a_3),$$

$$(a_2\ a_3), S, \emptyset\}$$

$$\text{and} \quad \chi: S \rightarrow (0,1) \qquad p: E \rightarrow (0, 1)$$

A fuzzy relation between two sets is a set of points defined on the
cartesian product. The membership values of each of these points may
again be interpreted as restrictions on the possible strength of the
relationship to allow for imprecision of definition. This is a
generalisation to a many to many mapping of the normal functional
expression $y = g(x)$ which is a one to one or many to one mapping.

1.2 Mathematical Operations

Gaines (1,2) has shown that the mathematical operations associated with
both probability theory and fuzzy set theory are really special cases of
a more general result. The mathematical axioms chosen should be
appropriate for the problem being addressed. The problem has been outlined by Blockley et al (3) and can be illustrated by the use of a
voting model.

Consider a population, each member of which can respond to certain
questions with a binary yes or no reply. For some question A let us
assume that the proportion voting yes, p(A), is p and for some question
B the proportion voting yes, p(B), is q. As p and q are proportions
they are measures on the interval (0,1) and do not necessarily sum to
one. Fig. 1 shows how p and q could be distributed over the power set
of this sample space A, B where $\bar{A}$ represents not A, and U_A is the truth
of A. Let us denote the way in which the votes are distributed over
the power set as P_1, P_2, P_3, P_4 in Fig. 1.

Clearly $p = p_2 + p_3 = p(A)$

$$q = p_3 + p_4 = p(B) \qquad (1)$$

$$p_1 + p_2 + p_3 + p_4 = 1$$

and the individual values of p_i cannot be determined without additional
information. We require to find values for compound propositions such
as $p(A$ and $B)$, $P(A$ or $B)$, $P(A \supset B)$.

However as $p(A$ and $B) = p_3$; $p(A$ or $B) = p_2 + p_3 + p_4$ then we know
already that $p(A$ and $B) \leqslant$ Min $(p,q) \leqslant$ Max $(p,q) \leqslant p(A$ or $B)$ $\qquad (2)$

To fix values for conjunction and disjunction additional assumptions must be made.

Three alternative assumptions are a) mutual dependence,
b) statistical independence, c) mutual exclusion and maximum perversity.

a) Mutual Dependence

Under this assumption, any person who votes true for the proposition
with the lower votes must also vote true for the other.

Thus $p(A$ and $B) = p_3 = $ MIN $(p,q) = c$

$$p(A \text{ or } B) = p_2 + p_3 + p_4 = (p - c) + c + (q - c)$$

$$= p + q - \text{MIN } (p,q)$$

$$= \text{MAX } (p,q)$$

$$p(A \supset B) = p_1 + p_3 + p_4 = 1 - (p - c)$$

$$= 1 - p + \text{MIN } (p,q) = \text{MIN } (1, 1 - p + q)$$

These rules are those of Lukasievicz multivated logic. Fig. 2 shows
how the votes would be distributed under this assumption.

b) Statistical Independence

Here the distribution of votes is random. Thus of the p who voted true
for A, the expected value of the proportion who vote true for B is q.

$$P(A \text{ and } B) = P_3 = pq$$

$$p(A \text{ or } B) = P_2 + P_3 + P_4 = p + q - pq$$

$$p(A \supset B) = 1 - p + pq$$

These are the results of probability theory, and the distribution of votes is shown in Fig. 3. It is interesting to note that this same result is obtained if the p_i are determined by maximising the entropy. Thus if the problem is formulated as follows

$$\text{Maximise } H = - \sum_{i=1}^{4} P_i \, \ell n P_i$$

$$\text{subject to } \sum_{i=1}^{4} P_i = 1, \; P_2 + P_3 = p, \; P_3 + P_4 = q$$

$$\text{then } P_1 = (1 - p)(1 - q), \; P_2 = p(1 - q), \; P_3 = pq, \; P_4 = q(1-p)$$

showing that this is the unbiased estimate of the p_i, when no information is given about the relationship between A and B.

c) Mutual Exclusion and Maximum Perversity

Under this heading there are three possibilities

    i)   All voting is clear. If a person votes true for one he must vote false for the other.

    ii)   If a person votes true for one he must vote false for the other but if he votes false for one he can vote true or false for the other.

    iii)   If a person votes false for one he must vote true for the other but if he votes true for one he can vote true or false for the other.

These case on the truth table for A and B are

A	B	Case
T	T	(iii)
T	F	(i), (ii), (iii)
F	T	(i), (ii), (iii)
F	F	(ii)

Fig. 4 shows that distribution of voters in these cases and the results
are as follows:-

	(i)	(ii)	(iii)
p(A and B)	0	0	p+q-1
p(A or B)	p+q=1	p+q (p<1-q)	1 (p>1-q)
p(A⊃B)	1-p=q	1-p	q

and combining these we get

$$p(A \text{ and } B) = \text{MAX } (0, p + q - 1)$$

$$p(A \text{ or } B) = \text{MIN } (1, p + q)$$

$$p(A \supset B) = \text{MAX } (1 - p, q)$$

Finally let us consider the range of possible values under equations (1).

$$p(A \text{ and } B) = \{\text{MAX } (0, p + q - 1) \text{ to MIN } (p,q) \}$$

$$p(A \text{ or } B) = \{\text{MAX } (p, q) \text{ to MIN } (1, p + q)\}$$

$$p(A \supset B) = \{\text{MAX } (q, 1-p) \text{ to MIN } (1, 1 - p + q)\}$$

The bounds on these values come from the conditions of mutual dependence
(a) and maximum perversity (c). Figs. (5, 6) show these bounds applied
to the conjunction and disjunction of fuzzy sets X_1, X_2. The votes
p, q here are interpreted as fuzzy membership function values for X_1,
X_2 respectively.

1.3 Connectives for a Multi-Valued Logic

Gaines (1,2) has discussed in some detail the relationships between the
various multi-valued logics. He assumed the axioms of a "mathematical
lattice", which is a partially ordered set in which any two elements
have a greatest lower bound and a least upper bound. He then discussed
the characteristics of a continuous, order preserving valuation or
function p on the lattice, to the interval (0,1). He showed that the
inequality (2) can be derived with these minimal assumptions but that
to isolate particular values it is necessary to make assumptions about
implication and negation. He defined a standard uncertainty logic which
underlies both Rescher's probability logic and fuzzy logic. He
demonstrated that the essential difference between the two is that in
probability it is assumed that the law of the excluded middle applies,
whereas in fuzzy logic this is not the case. Thus his standard

uncertainty logic together with $p(x \vee \bar{x}) = 1$ results in Rescher's
probability logic but with $p(x \supset y) = 1$ OR $p(y \supset x) = 1$ we get Lukasiewicz'
logic or fuzzy logic.

Thus the use of min and max operations in fuzzy logic is not
sufficient to discriminate the logic from that of probability theory.
Our association of addition and multiplication as natural operations
upon probabilities comes from our frequent interest in statistically
independent events not from the logic of probability itself.

The use of conditional probability is clearly an attempt to isolate
a value in the range of possibilities. Thus referring back to the
voting model, if the population were also asked to vote on $p(A/B)$ with
a proportion r voting this way then –

$$p(A \cap B) = p(A/B)\ p(B) = rq$$

Likewise the population could vote on $p(B/A)$ with a result s

$$\text{then } p(A \cap B) = p(B/A)\ p(A) = sp$$

The justification of the multiplication here by (for example) De Finetti
(4) is based on a particular view of the way in which uncertainty is to
be modelled. De Finetti's results were extended by Lindley (5) who
showed that if a person in considering and event E about which he is
uncertain describes that uncertainty by a number x and there is a score
function of x, then a known transform of x must be a probability. Thus
if the voting group were all to behave rationally then $rq = sp$. However
in complex problems with many hidden dependencies between those factors
discussed and those hidden or unknown, it seems unlikely that one would
find this criteria satisfied. Of course it could be argued that this
lack of rationality must be isolated and judgements about how to vote
changed. Sometimes that maybe practicable but often it will not.
Clearly if the problem being addressed is such that it is realistic to
work with conditional probabilities then one should do so because it
provides specific numbers on the range of possible inferences for AND
and OR. However, it is necessary to be aware of the mathematical
restrictions of doing so, chiefly a compliance to the law of the
excluded middle.

If one assumes a max min logic then as shown by Bellman and Giertz
($_6$) it not only complies with other very reasonable axioms but also is
strongly truth functional (1,2). Fig. 5 shows that the minimum rule
for conjunction provides a least restrictive bound. The maximum rule
for disjunction follows this assumption. To take the least restrictive
bounds for both disjunctions and conjunction would always be valid but
too pessimistic for practical inference. One could carry the bounds
through a calculation always working with intervals but that maybe
computationally inefficient. Max, min rules represent the smallest
changes to votes or to the functions. The important idea, however, is
that by loosening the mathematical formalism one may make valid inferences
without having to describe complex dependencies as dictated by the
mathematics but rather by the number of logic statements of inference
which are included in the model.

2. FRIL

2.1 Fuzzy Relational Inference Language (FRIL)

FRIL is a computer specification language. It enables the setting up
of a data base and a knowledge base within the computer which can be
interrogated by the user sitting at a terminal. The logic to be used
within the system can be specified by the user. There are four units.
Firstly tables of facts or data known as virtual relations; secondly,
rules of inference or heuristics known as virtual relations; thirdly,
set theoretic relations such as >, =, <; fourthly queries.

a) Base Relations:- Each relation has a name (which can be a concept
or a proposition) and a series of attributes. Each row of the relation
or table is a set of valuations of those attributes. Associated with
that row is a measure χ on scale (0,1) which is the "truth" or
dependability with which it is believed that that set of valuations
belongs to the relation. For example;

tall-man	height	χ	heavy man	weight	χ
	1.76 m	0.1		76 kg	0.2
	1.78 m	0.5		82 kg	0.4
	1.80 m	0.9		88 kg	0.8
	1.82 m	1.0		94 kg	1.0

b) Virtual Relations (rules):- These are relations connected by
logical operators (if, and, or in this paper). New (virtual) relations
c an be formed by combining base relations. Again these can be combined
in other rules forming a hierarchical tree-like structures of rules
e.g. large-man (h,w) if tall-man (h) and heavy-man (w) would form a
table of pairs like (1.76 m, 76 kg) with χ = 0.1 and (1.80 m, 82 kg)
with χ = 0.4. The χ values are obtained by taking the minimum value
of χ for the elements separately. The full table expressed in matrix
form is:

large-man	76	82	88	94	kg
1.76 m	.1	.1	.1	.1	
1.78 m	.2	.4	.5	.5	
1.80 m	.2	.4	.8	.9	
1.82 m	.2	.4	.8	1.0	

The attributes need not be measures such as metres or kg but could be
names e.g. John, Miriam.

2.2 Examples Using FRIL

a) Given four designs D1, D2, D3, D4 each have a safety factor (SF) value and a cost ($£/m^2$). The base relations are as follows:

QUALITY	Name	S.F.	Cost	χ	SAFE	S.F.	χ	ECONOMIC	Cost	χ
	D1	1.3	7	0.5		1.1	0.1		5	1
	D2	1.5	6	0.8		1.3	0.3		6	0.9
	D3	1.5	5	1		1.5	0.8		7	0.5
	D4	1.7	9	0.6		1.7	1		8	0.3
						2.0	1		9	0.1

Question: Which designs are unsafe (not safe)?

WHICH {x, (NOT SAFE (y) AND QUALITY (x,y, -)) }

First we calculate

NOT SAFE	S.F.	$\bar{\chi}$
	1	1
	1.1	0.9
	1.3	0.7
	1.5	0.2

by $\chi_{\text{NOT SAFE}}(t) = 1 - \chi_{\text{SAFE}}(t)$

The answer is then a conjunction

REL	NAME	χ
	D1	0.5
	D2	0.2
	D3	0.2

REL (n) is obtained by taking each x (i.e. each name, e.g. D1) in turn and finding the membership level of the corresponding y (i.e. S.F. = = 1.3) in NOT SAFE (i.e. 0.7) and finding the minimum of that value and the membership of (x,y) (i.e. D1, 1.3) in QUALITY which is (0.7 $\wedge$ 0.5) or 0.5.

b) Again four designs D1, D2, D3, D4 are given and the following virtual relations:

GOOD (x) ← SAFE (x,y) AND ECONOMIC (x,z)

SAFE (x,y) ← LS (x,y) AND HE (x,y)

ECONOMIC (x,y) ← FC (x,y) AND MAINT (x,y)

where LS refers to limit state safety, HE to human error.

FC refers to first cost and MAINT to maintenance costs and
← is "if".

The base relations are

LS	Name	notional prob.	χ		HE	Name	not. prob.	χ
	D1	NOT LOW	1			D1	LOW	1
	D2	LOW	1			D2	VERY LOW	1
	D3	LOW	1			D3	LOW	1
	D4	VERY LOW	1			D4	NOT LOW	1

FC	Name	cost	χ		MAINT	Name	cost	χ
	D1	EC	1			D1	EC	1
	D2	NOT EC	1			D2	EC	1
	D3	VERY EC	1			D3	EC	1
	D4	EC	1			D4	NOT EC	1

where LOW refers to a fuzzy low notional probability
EC refers to a fuzzy economic cost and NOT and VERY
are modifiers

LOW	$-\log_{10}pf$	χ		EC	$£/m^2$	χ		VERY	t	χ
	3	1			5	1			0.6	0.1
	4	0.9			6	0.9			0.7	0.3
	5	0.8			7	0.8			0.8	0.5
	6	0.5			8	0.5			0.9	0.8
	7	0.1			9	0.1			1	1

Question: Which designs are good?
            WHICH {x, GOOD (x) }

            The problem generator returns   LS (x,y)
                                              HE (x,y)
                                              FC (x,z)
                                           MAINT (x,z)

The solution procedure will be illustrated by writing out only that part
of each relation which refers to D1.

Thus   LS (Name = D1) $-\log_{10}$pf χ HE (D1) $-\log_{10}$pf χ

                                3   0   3   1

                                4   .1   4   .9

                                5   .2   5   .8

                                6   .5   6   .5

                                7   .9   7   .1

giving   similarly

            SAFE (Name = D1) $-\log_{10}$pf χ ECONOMIC (D1) $\pounds/m^2$ χ

                                3   0   5   1

                                4   .1   6   .9

                                5   .2   7   .8

                                6   .5   8   .5

                                7   .1   9   .1

Finally the maximum membership of SAFE x ECONOMIC is the membership of
D1 in GOOD i.e. 0.5

The total solution is   GOOD   Name χ

                                             D1   0.5

                                             D2   0.5

                                             D3   1

                                             D4   0.2

c)   Imagine we have evidence from three existing and reasonably
successful structures (E1, E2. E3), from three failures (F1, F2, F3)
and two alternative proposals (D1, D2). We wish to set up the evidence
in a computer using FRIL so that we may ask various important questions.
In this very simple example the evidence is in the form of notional
probabilities of failure and degrees of proneness to failure through
human error. As this is an illustrative example two further designs
will be included (D3, D4) which have extreme values of these measures.
All the information is stored in base relations, the names of which
are reasonably self explanatory if LS stands for Limit State, HE for
human error, EX for existing structures, F for failure, DS for proposed
designs. The base relations are:

EX-LS	Name	$-\log pf$	χ	EX-HE	Name	Proneness	χ
	E1	5	1		E1	0.1	1
	E2	6	1		E2	0.2	1
	E3	7	1		E3	0.4	1

DS-LS				DS-HE			
	D1	4	1		D1	0.1	1
	D2	7	1		D2	0.4	1
	D3	0	1		D3	1	1
	D4	10	1		D4	0	1

F-LS				F-HE			
	F1	0	1		F1	0.7	1
	F2	6	1		F2	0.9	1
	F3	0	1		F3	0.3	1

The degree of proneness to failure through human error is measured on a
scale 0,1 and 0 represents no proneness to failure (D4) and 1 represents
 certain failure.

    Three rules and four set theoretic base relations will be used in
the example.

    The first rule expresses the fact that for a design to be safe, it
should be similar to existing designs and it should not be the case that
it is similar to failures.

$$\text{DS-SAFE } (x) \leftarrow \text{EX-DS-SIM } (x) \text{ AND NOT } \{\text{F-DS-SIM } (x) \}$$

The relation expressing similarity between the designs and the existing structures needs also to be re-written.

$$\text{EX-DS-SIM } (x) \leftarrow \text{DS-LS } (x,a) \text{ AND EX-LS } (y,b) \text{ AND SIM PROB } (a,b)$$

$$\text{AND DS-HE } (x,c) \text{ AND EX-HE } (y,d) \text{ AND SIM-PRON } (c,d)$$

This states that a relation containing a list of names of designs (x) which are similar to existing structures is made up of the list of names from the base relations DS-LS and **EX**-LS which have similar notional probabilities SIM-PROB and names from DS-HE and EX-HE which have similar degrees of proneness to failure.
Also in like manner

$$\text{F-DS-SIM } (x) \leftarrow \text{DS-LS } (x,e) \text{ AND F-LS } (y,f) \text{ AND GT-PROB } (e,f)$$

$$\text{AND DS-HE } (x,g) \text{ AND F-HE } (h,i) \text{ AND GT-PRON } (g,i)$$

The last base relations to be defined are

$$\text{SIM-PROB } (a,b) = \{(a,b) \mid \chi = 1 - \frac{|a-b|}{10}\}$$

which states that SIM-PROB is a relation consisting of a set of points (a,b) with a fuzzy membership level of

$$1 - \frac{|a-b|}{10}$$

$$\text{SIM-PRON } (a,b) = \{(a,b) \mid \chi = 1 - |a-b|\}$$

$$\text{GT-PROB } (a,b) = \{(a,b) \mid \chi = (1 - \frac{(a-b)}{10}) \wedge 1\}$$

$$\text{GT-PRON } (a,b) = \{(a,b) \mid \chi = (1 - (b-a)) \wedge 1\}$$

The last two base relations express a one sided similarity so that designs are only considered to be disimilar to failures if they are somewhat safer. Once the knowledge base is set up there are different questions which may be asked. Only the answers are given here.

(i) Which designs are safe?

$$\text{WHICH } \{x, \text{DS-SAFE}(x)\} \quad \text{produces} \quad \text{DS-SAFE}$$

Name	χ
D1	0.4
D2	0.5
D3	0.6

ii) How prone are safe designs to human error?

    WHICH {(x,y) DS-SAFE(x) AND DS-HE (x,y)}

produces	Name	Proneness	χ
	D1	0.1	.4
	D2	0.4	.5
	D4	0	.6

iii) How prone to human error are designs which are very safe in limit states?

    WHICH {(x,y), DS-LS (x,z) AND VERY SAFE (z) AND DS-HE (x,y)}

To answer this question we need to add other base relations which define what is meant by very safe in a limit state.

SAFE	$-\log_{10}\text{pf}$	χ	VERY	τ	χ
	5	0.2		0.5	0
	6	0.6		0.75	0.5
	7	0.8		0.75	1
	8	1			
	9	1			
	10	1			

where τ is a truth or dependability value and VERY SAFE is obtained by the process of Truth Functional Modification.
    The answer to the above query is:

Name	Proneness	χ
D2	0.4	0.3
D4	0	1

iv) How safe in the limit state are designs which are not prone to human error and are similar to existing structures?

    WHICH {(x,y), DS-LS (x,y) AND DS-HE (x,w) AND NOT PRONE (w) AND EX-DS-SIM (x)}

where	NOT PRONE	w	χ	produces	Name	$-\log_{10}$pf	χ
		0	1		D1	4	.9
		0.5	0.5		D2	7	.6
		1	0		D4	10	.6

3. Examples

3.1 Limestone Mines in England

The legacy of old limestone in the Black Country, many of which were abandoned long ago, has resulted in sporadic incidents of subsidence over the past hundred years or more. Ove Arup and Partners have reported on the extent and physical characteristics of these workings and have recommended some remedial actions (7). The report was commissioned jointly by the Department of the Environment and the few local authorities principally concerned. As well as gathering data Ove Arup and Partners presented a simple assessment of risk.

Actions that could be taken by the local authorities and government include infilling the mines, monitoring, further investigation, strengthening pillars etc. Within a limited budget it is necessary to assess cost and benefit for each action on each mine. In an attempt to assist in this decision problem an initial tentative formulation of the problem in FRIL has been made.

Data Tables in the Arup report have been stored on the computer for three mines only: 12c Castlefields; 21 Littleton Street; 30 Wrens Nest.

Thus Table 301 of the report is

No	Name	Date Open	Date Closed	Area $\times 10^3 m^2$	Seam	No. Collapses	χ
12c	Castlefields	1750	1909	200	$\cup$	15	1
21	Littleton St.	1843	1903	80	ℓ	1	1
30	Wrens Nest	1700	1924	120	$u+\ell$	25	1

and has been called the base relation mine-sumdat (a, b, c, d, e, f, g).

The other tables are stored in a similar manner. I have defined other base relations.

e.g. old date	χ	large area $\times 10^3 m^2$	χ	large-No.-collapses	No	χ
1800	1	200	1		1	0.05
1850	0.9	150	0.8		5	0.1
1900	0.8	100	0.2		10	0.4
1984	0	50	0.05		15	0.8
					20	0.9
					25	1.0

In like manner the following were defined deep (x), large-area-coll (x) **large-pillars(x)**.

The computer system FRIL allows one to sit at a terminal and ask
questions of the database as one thinks of them. The answer is always
in the form of a relation. Fuzzy logic was used in all inferences.
It must be stressed that all χ values used here are. fictional.
        For example:
(i) Which mines have had large number of collapses?

        WHICH $\{$(x,y) mine-sumdat (x y _ _ _ _ a) and large-No.-colls(a)$\}$

This query looks for pairs of elements (x,y) in the relations mine-sumdat
and large-No.-colls which have an element a in common; interpolation is
used if necessary.
    The result is :

No.	Name	χ
12c	Castlefields	0.8
21	Littleton St.	0.05
30	Wrens Nest	1

The calculation is straightforward in this case. For example for
Castlefields there have been 15 collapses (i.e. a = 15) with χ = 1.
The χ value of 15 in large-No.-colls is 0.8. The minimum of 0.8 and 1
is 0.8 which is the entry against this mine in the answer.
    Other questions and answers are as follows:-

ii) Which large mines have had large number of collapses in shallow
     upper Wenlock limestone with Ludlow shale roof?

No.	Name	χ
12c	Castlefields	0.8
30	Wrens Nest	0.44

iii) What recent collapses have occurred in old mines in Lower Wenlock
      Limestone with Nodular Bed roof material at shallow depths?

No.	Name	Date	Area Coll.	χ
21	Littleton St.	1861	2900	0.12
30	Wrens Nest	1883	1600	0.17
30	"	1883	1800	0.17
30	"	1883	100	0.17
30	"	1978	30	0.94

The preceding questions referred only to a data base. It is possible to
model the decision problem hierarchically by a series of virtual relations

or rules or knowledge base. Twenty rules model the infilling decision
problem. There is insufficient space to include them all here. At the
top or the hierarchy are the vaguest most fuzzy rules.

1) (Infill mine (or part) x using method y)
    If (Benefit high) and (costs acceptable)
    which in FRIL becomes
    infill (x,y) if infill-benefit (x,y) and infill-cost (x,y).

These are then developed by writing down the necessary conditions on
these statements e.g.

2) (Benefit high)
    If {(method y seems appropriate but need testing for underground
    conditions at x) and (conditions are typical of other risky mines)}
    OR {(high chance of serious surface event) and (method y already
    tested for similar underground conditions at x) and (consequences
    of not infilling would be serious given a surface event)}and {method
    y is suitable for mine x for other criteria}.

Each of these propositions has further necessary conditions which can
be identified and this process stops when it is felt that the propositions
are detailed wnough for either direct measurement or subjective estimation of the X values. These propositions then become base relations.
For example the proposition (conditions are typical of other risky mines)
becomes Cond-typ(x).

Cond-typ	Mine-No.	χ
	12c	0.6
	21	0.5
	30	0.1

The χ values represent a subjective belief that a particular mine
belongs to a set of mines which are typical of all other risky mines.
     Clearly from this brief outline it is only possible to get a
flavour of the technique. Some questions and answers are:-

(i) Question: Which mines will benefit from infilling?

Answer:	Littleton St.	21	Colliery waste	0.4
	Littleton St.	21	Sand	0.4
	Castlefields	12c	Colliery waste	0.6
	Castlefields	12c	Sand	0.3
	Wrens Nest	30	Colliery waste	0.1
	Wrens Nest	30	Sand	0.1

Comment: If answer surprises the questioner, it is possible the
          check through the logic to find out the reason. Whether
          surprised or not one can identify areas of conflict or
          difficulty which may need further discussion of informa-
          tion.

(ii) Question: Which mines have a high chance of crown holing and are
shallow and and old and have had a large No. of collapses
and large areas of roof collapse?

    Answer: Littleton St.   0.05
                   Wrens Nest   0.52

(iii) Question: Which mines have high chance of crown holes and, long
term economic consequences if mine not infilled?

    Answer: Littleton St.   0.5
                   Wrens Nest   0.7
                   Castlefields   0.9

3.2 Safety of a Nuclear Reactor

The work on this problem is also in its very early stages. It is based
on the probabilistic reliability analysis of the PWR at Indian Point
Power Plant, New York (8). Thus both the system logic and quantitative
assessments of probabilities are available. The methods used in the
original analysis are well known and based on fault trees and event trees.
The first task has been to express these trees in logic terms. The
fourteen plant event trees can each be expressed in logic form by an
average of about 10 virtual relations for each tree. Again insufficient
detail can be included here and only a flavour of the technique can be
gained.

For plant event tree 1 the initiating event is a large LOCA which
itself is caused by random pipe breaks. There are eight protection
systems and operator functions to contain LOCA. They are TK, refueling
water storage tank; SA-1, safety injection actuation; LP-1, low pressure
injection, CS, containment spray; NA, addition of sodium hydroxide; CF-1,
containment fan coolers; R-1, recirculation cooling; RS, recirculation
spray. There are also eight degraded core states each of which derive
from various scenarios or sequences of success or failure of the
protection system. Associated with these success or failures are the
conditional probabilities that a system succeeds given a scenario up to
the point of demand. Also available are the probabilities that initial
events, such as pipe fracture, occur. The χ values are thus interpreted in this model as probabilities and a (plus, times) logic is used
for operations OR and AND.

Base Relations which have been specified for example are

System Name	Name	χ	Scenario	Event Tree No.	Scenario No.	Damage State	χ
	TK	1					
	SA-1	1					
	LP-1	1		1	1	AEFC	1

System Name	Name	χ	Scenario	Event Tree No.	Scenario No.	Damage State	χ
	CS	1		1	2	AEFC	1
	NA	1		1	3	AEFC	1
	CF-1	1		1	4	AEC	1
	R-1	1			etc		
	RS	1					

Init-event-freq.	Name	χ		Cond-probs.	Set	Name	χ
	pipe-fail	appop			1	pipe-fail	approp
	valve-fail	probab-			1	valve-fail	condit-
	vessel-fail	ility			1	vessel-fail	ional
	other				1	other	proba-
					2	pipe-fail	bility
					etc		

Scen-cond-probs-rel	Name	Event	Scenario	Cond-Prob Set No.	χ
	TK	1	–	1	1
	SA-1	1	1	2	1
	SA-1	1	1	3	1
			etc		

Virtual relations are:

  1 Scen-cond-probs(a, b, c, d, e) ← Scen-cond-probs-rel (a, b, c, x) and Cond-probs (x, d) and Scenarios (b, c, e)

  2 AEFC (a, b.c) ← Scen-cond-probs (TK, a, b, c, AEFC) and Scen-cond-probs (SA-1, a, b, c, AEFC) and

        NOT   ("   "   ")(LP-1, a, b, c, AEFC) and
            ("   "   ")(CS, a, b, c, AEFC) and
            ("   "   ")(CF-1, a, b, c, AEFC) and
        N(a, b, c) and equal (a, 1)

  3 N(a, b, c) ← Scen-cond-probs (R-1, a, b, c, AEFC) OR
          N1 (a, b, c)

  4 N1(a, b, c) ←NOT (Scen-cond-probs(R-1, a, b, c, AEFC)) and
         NOT (Scen-cond-probs(RS, a, b, c, AEFC))

Rules 2-4 then express plant event tree 1 for the outcome AEFC are of the eight degraded core states. Similar rules can then be written for the other states.

Queries will then be asked as follows

1 How likely are the various initiating events and subsequent
courses of action to result in failure state AEFC for a
large LOCA?

Which ((a, b, c) (plus times) AEFC (1,c,d) and
Init-events-freq (d))

2 What scenarios commence with pipe failure and large LOCAS
and result in failure state AEC?

Which ((a, b) (plus times) AEC (1, c, d) and Init-events-
freq (d) and where (d = pipe-fail))

This work is still in its problem formulation stage and will
be reported on more fully in a later paper.

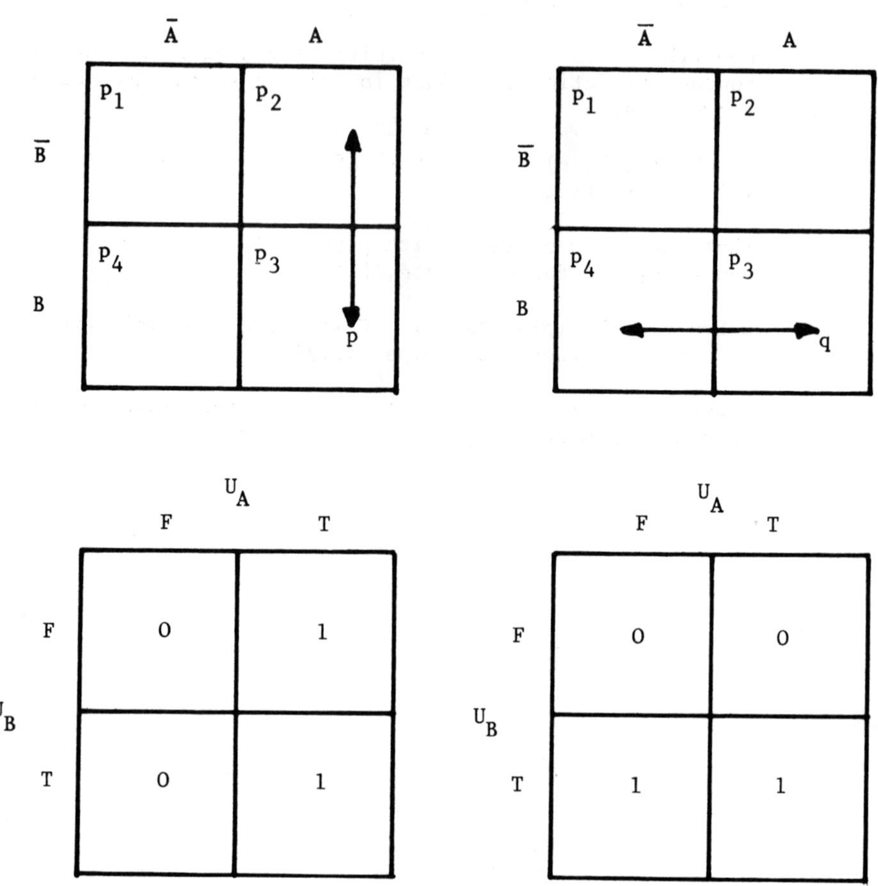

Fig. 1   Proportional Votes and Truth Values for the Voting Model

	$\overline{A}$	A
$\overline{B}$	$1 - p$	$p - q$
B	0	q

$$p > q$$

	$\overline{A}$	A
$\overline{B}$	$1 - p$	0
B	0	p

$$p = q$$

	$\overline{A}$	A
$\overline{B}$	$1 - q$	0
B	$q - p$	p

$$p < q$$

Fig. 2 Distribution of Votes under the Mutual Dependence Assumption

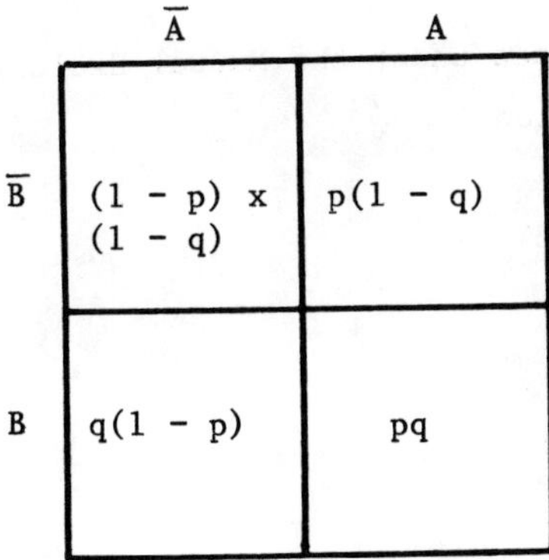

	$\overline{A}$	A
$\overline{B}$	$(1 - p) \times (1 - q)$	$p(1 - q)$
B	$q(1 - p)$	pq

Fig. 3 Distribution of Votes under the Statistical Independence Assumption

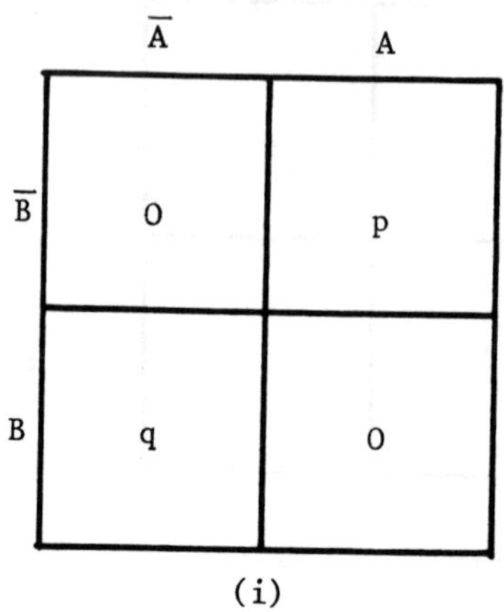

Fig. 4 Distribution of Votes under the Mutual Exclusion and Maximum Perversity Assumption

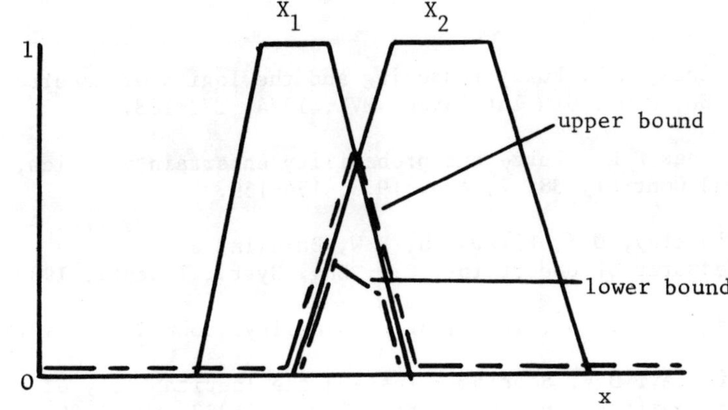

Fig. 5 Bounds on Conjunction

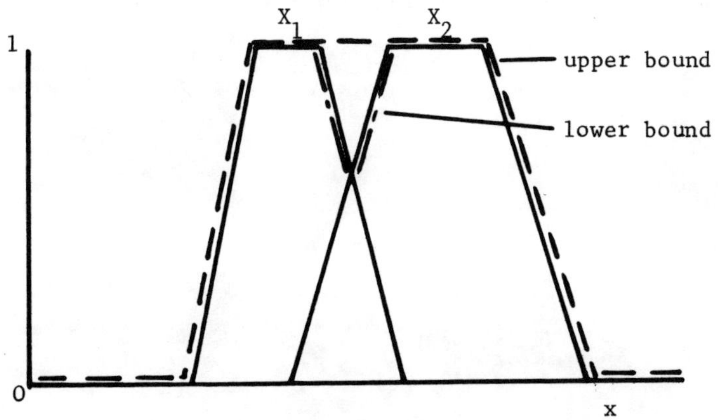

Fig. 6 Bounds on Disjunction

References

1 Gaines, B R, Fuzzy reasoning and the logics of uncertainty, IEEE, Proc. 6th Int. Symp, MVL, 1974, 179-188.

2 Gaines B R, Fuzzy and probability uncertainty logics, Inf. and Control, 38, 2, Aug. 1978, 154-169.

3 Blockley, D I, Pilsworth, B W, Baldwin, J F, Measures of uncertainty, Civ. Eng. Syst., 1 Sept., 1983, 3-9

4 De Finetti, B., Theory of Probability, John Wiley, 1974

5 Lindley, D V, Scoring Rules and the Inevitability of Probability, Int. Statistics Review, 1982, 50, 1-26.

6 Bellman, R., Giertz, H, On the analytic formalism of the the theory of fuzzy sets Inf. Sci., 1973, 5, 149-156.

7 Ove Arup & Partners, Limestone mines in the West Midlands, Dept. of the Environment, London, 1983.

8 Probabilistic Safety Study of Indian Point Units 2 and 3, Consolidated Edison Co. NY, Power Authority NY.

INTERACTION OF STRUCTURAL AND COMPONENT RELIABILITY

G. I. Schuëller
Professor of Engineering Mechanics
Institut für Mechanik
Universität Innsbruck
Technikerstrasse 13
A-6020 Innsbruck, Austria

ABSTRACT: The purpose of this lecture is to give an overview of the various methods available to treat the problem of the interaction between structural and systems, i.e. component reliability. Advantages and drawbacks are pointed out. A numerical example, treating the failure probability of the primary piping of a PWR is given.

1. INTRODUCTION

Risk analyses of large technological systems such as nuclear power plants [1,2], chemical plants, offshore platforms, large airplanes and shipping vessels, etc. receive increasing attention by the public. In this context one important aspect is the determination of the reliability of such systems. As they generally consist of a number of various sub systems, such as electronic, mechanical and structural systems, their respective interaction has also to be taken into account. To analyze such systems two methodologies of reliability analyses have been developed so far, i.e. the systems- or component reliability approach for determing the reliability approach. While the first method is based on discrete states, the latter approach utilizes continuous state considerations.

For the case of nuclear power plants, electronic systems for example, may consist of the protective system, which activates the ECCS, scram system, EFWS, EPS, etc. As mechanical systems, small pipes, valves, hangers may be considered. Pressure vessels, containments, heat exchangers, large pipes, etc. may be classified as structural systems.

In cases of external hazards, such as earthquakes or accidental aircraft impact it is quite obvious that structures and components are affected simultaneously. Consequently the various systems interact [3-7]. In this context it may also be referred to the example of the reliability of the containment structural system which is directly affected by the frequency of occurrence of the LOCA. This in turn depends directly on the reliability of the primary piping (structural of mechanical system). It will be shown in the numerical example that this reliability

A. C. Lucia (ed.), Advances in Structural Reliability, 73–85.
© *1987 by ECSC, EEC, EAEC, Brussels and Luxembourg.*

estimate is directly related to the reliability of various components of the protective system, i.e. the electronic system.

2. METHOD OF ANALYSIS

2.1 General

Although the definitions of the basic states, i.e. "failure" or "survival" are identical for both component and structural relaibility, the methods of analysis, i.e. the procedures to estimate or predict failure rates are of quite different nature. Methods of systems analysis are generally based on discrete independent state descriptions (binary approach) while structural reliability utilizes continuous state descriptions; correlation between various structural members or components may also be considered. Moreover, in the latter case multimode failure, i.e. bending, buckling, fracture, may be treated by the analysis.

2.2 Binary Approach

For analyzing systems, such as electronic or mechanical networks, the so-called event- and fault tree methodology has been developed. The event tree method enables the prediction of a sequence of undesired events or operational states of the system including all possible branches. Examples are given in Fig. 1 and Fig. 2.

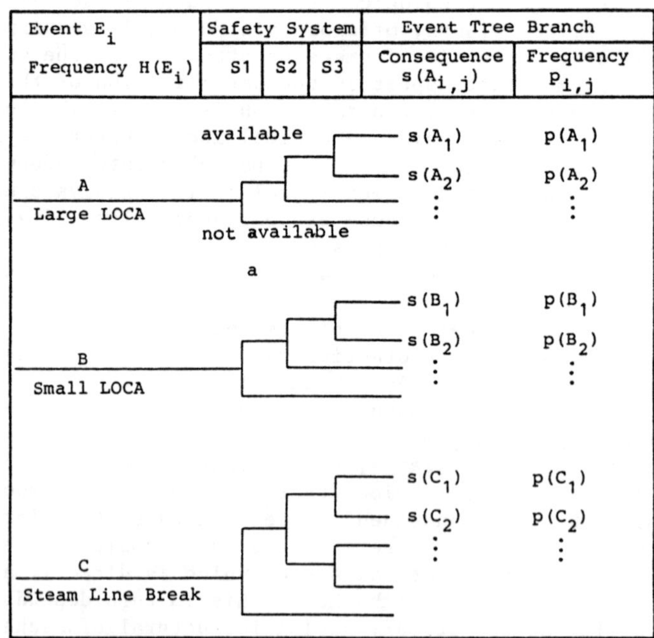

Fig. 1: Sample of event tree for three initiating events

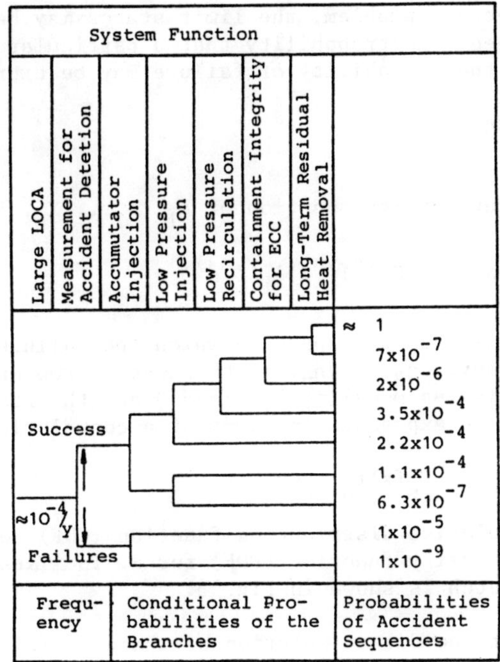

Fig. 2: Event tree for large LOCA

The transition probabilities between the states of the event trees are
determined by utilizing fault tree analysis. Fault trees, which are a
graphical representation of logical relations of components and subsystems, generally yield transition probabilities of a system moving
from one state to another in terms of average values. Considering the
large size of most of the reactor systems a more detailed analysis, i.e.
the description of state transitions in terms of distributions requires
– despite application of importance sampling simulation procedures for
practical application – too much computational efforts.

2.3 Continuous State Approach [8]

The fundamental concepts of structural reliability theory are based on
the derivation of mathematical models from principles of mechanics and
experimental data which relate structural resistance and load variables.
This relation might be defined by a limit state- or failure surface of
the following form:

$$g(\underline{X}) = g(X_1, X_2, \ldots, X_n) = 0 \tag{2.1}$$

where the X represent the random variables of structural resistance and
loading. The limit state is reached, i.e. failure occurs when the condition $g(\underline{X}) \leq 0$ is met. Limit states are a matter of definition. They
might be defined as ultimate load failure, uncerviceability, etc..

Depending on the respective problem, the limit states may be considered individually or combined. The probability that a particular limit state will be reached, i.e. the probability of failure may be expressed as

$$P_f = \int_D f_{\underline{X}}(\underline{n}) \, d\underline{n} \tag{2.2}$$

where $f_{\underline{X}}(\underline{n})$ is the joint density function of X_1, $X_2 \ldots, X_n$, i.e.

$$f_{\underline{X}}(\underline{n}) = f_{X_1, X_2, \ldots, X_n}(n_1, n_2, \ldots, n_n),$$

D the domain in the n-dimensional space in which the failure condition is satisfied and $d\underline{n} = dn_1,\ dn_2 \ldots, dn_n$. In the case of two stochastically independent random variables modeling the load S and the resistance R the above equation can be expressed in terms of a convolution integral:

$$P_f = P(R \leqq S) = \int_{-\infty}^{\infty} F_R(x) f_S(x) \, dx \tag{2.3}$$

where $F_R(x)$ is the cumulative distribution function (CDF) for R and $F_S(x)$ the probability density function (PDF) for S. To illustrate the problem a schematic sketch is shown in Fig. 3.

It is obvious that the direct solution of the equ. (2.2) is quite difficult if not to say in most cases intractable and this for mainly two reasons: first, in general the scarce data base is insufficient to estimate the parameters of the joint density function and second, there are great numerical difficulties to perform the multidimensional integration over the domain D, particularly for irregular shapes of D. In order to circumvent this problem, first order approximations which are based on the second-statistical moments have been developed to various degrees of sophistication. Although the procedure can not avoid the data problem, it can be simplified, however, by introducing correlations between the various variables involved. The method which has been introduced in its basic form, i.e. for two variables, by *Freudenthal* utilizes the standardized form of the random variables, i.e.

$$x_1 = \frac{X_i - E(X_i)}{\sigma_i} \tag{2.4}$$

The limit state equation is then

$$g_1(\underline{x}) = g_1(x_1, x_2, \ldots, x_n) \leqq 0 \tag{2.5}$$

The problem is now to find the so called design point x_* (see Fig. 4). It can be found by existing optimization procedures, such as the Lagrangian multiplier, etc. appropriate library routine for optimization are available in most computer centers. It has been shown by *Freudenthal* and *Shinozuka* that for correlated or non correlated Gaussian variables the design point is also the point of maximum likelihood. For non

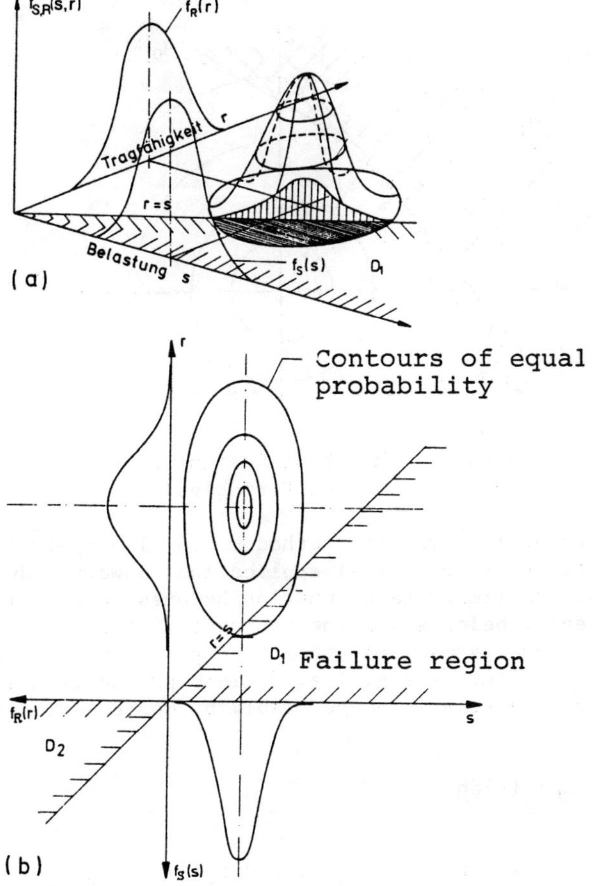

Fig. 3: Schematic Sketch for Calculation of Failure
Probability p_f for Case of Two Random Variables

Gaussian variables the following transformation may be introduced:

$$\underline{u}_i = \phi^{-1}[F_i(x_i)] \qquad (2.6)$$

in which $F_i(\cdot)$ is the CDF of x_i. The transformation of the vector $\underline{x}$ into $\underline{u}$ requires also the transformation of the limit state function $g_i(\underline{x})$

$$x_i = F_i^{-1}[\phi(u_i)] \qquad (2.7)$$

and

$$g_i(\underline{x}) = g_i\{F^{-1} \phi(\underline{u})\} = h_i(\underline{u}) \leq 0 \qquad (2.8)$$

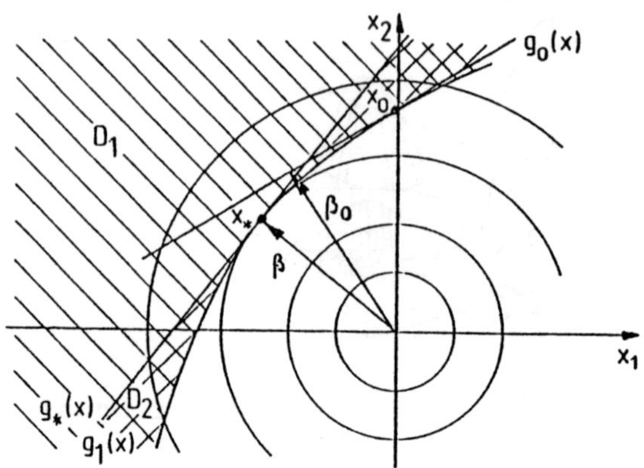

Fig. 4: Schematic Sketch of Design Point
(for Uncorrelated Variables)

Therefore, in general terms, the method would also apply for extreme
value distributions. In practical application, however, due to the
transformation, the limit state function becomes quite complicated. In
this case, however, being still the shortest distance to the origin, the
design point looses its maximum likelihood property. Finally, for the
calculation of the limit state, i.e. failure probabilities the following
integral has to be evaluated – preferably by Monte Carlo (MC) simulation
procedures –

$$P_f = {}_D\!\int f_{\underline{x}}(\underline{\eta})\,d\underline{\eta} \qquad (2.9)$$

where D is the domain, where $g_1(\underline{x}) \leqq 0$ and $f_{\underline{x}}(\eta) = f_{x_1,x_2\ldots,x_n}(\eta_1,\eta_2\ldots,\eta_1)$
is the joint density function of $\underline{x}$. The above equation is similar to
equ. (2.2). For MC evaluation for example the following expression is
suggested by *Shinozuka*:

$$P_f = \frac{A}{N} \sum_{i=1}^{N} \frac{\delta_i}{(2\pi)^{n/2}} \exp\left(-\frac{1}{2}\underline{n}_i^T\underline{n}_i\right) \qquad (2.10)$$

with $\delta_i = \begin{cases} 0 & \text{if } g(\underline{n}_i) > 0 \\ 1 & \text{if } g(\underline{n}_i) \leqq 0, \end{cases}$

A is the area of the domain of integration and n the number of simulated samples. By doing this, the advantage of importance sampling becomes quite obvious. Needless to say that like any approximate method
the first order second moment method has a limited possibility of application as well as accuracy. Aside some comparisons of selected problems

with exact methods only recently a first attempt has been made [9] to
illuminate these aspects.

In practical application, however, structures consist of a number
of members, i.e. one has to analyze a multimember system. Due to the
random properties of loading and resistance, deterministic failure mode
analysis is not applicable since each of the possible failure modes has
a certain probability of occurrence and therefore contributes a certain
amount to the total failure probability - which is the sum of the contributions of all failure modes. It is quite natural that attempts have
been made to find the stochastically most relevant failure modes, i.e.
the modes with the major contributions to the total failure probability
without the necessity of analyzing all possible failure modes [10]. For
the development of these procedures the utilization of the properties
of event tree analysis is quite convenient. Finally it should be
stressed again, that structural reliability analysis has to include the
analysis of the possible occurrence of various types of failure - depending on the load history - i.e. yield and buckling [11] or yield and
fracture [12], fatigue, etc..

3. NUMERICAL EXAMPLE

3.1 General

It is shown in [13] that a structure may be considered from two points
of view with respect to its function: it may be either looked at as
part of the operational system or as part of the standby safety system.
The primary piping may serve as an example. Its loading is to be determined from the operational parameters and from transients resulting
from function or malfunction of the relevant safety system. Moreover,
as in the case of the LOCA, the loads depend on the size of the pipe
break in different event trees. The loads which follow transients result mostly from the abnormal operational states. Reactor trip or loss
of electric power may serve as examples for this event.

If a structure is considered as part of the safety system it must
perform a certain function only in case of an accident. For example a
large leak in the primary piping, certain pressure and temperature
distributions will be quite different for the small leak. Therefore the
resulting failure probabilities for both cases will be quite different.
The analytical assessment of structural reliability is obviously based
on various paths of the different event trees [14].

3.2 Structural Model

In Fig. 5 the primary piping of a PWR to be analyzed is sketched
schematically. The structural model of LOOP2 - which is analyzed in
this context - is shown in Fig. 6. For static and dynamic analyses
linear elastic considerations are sufficient.

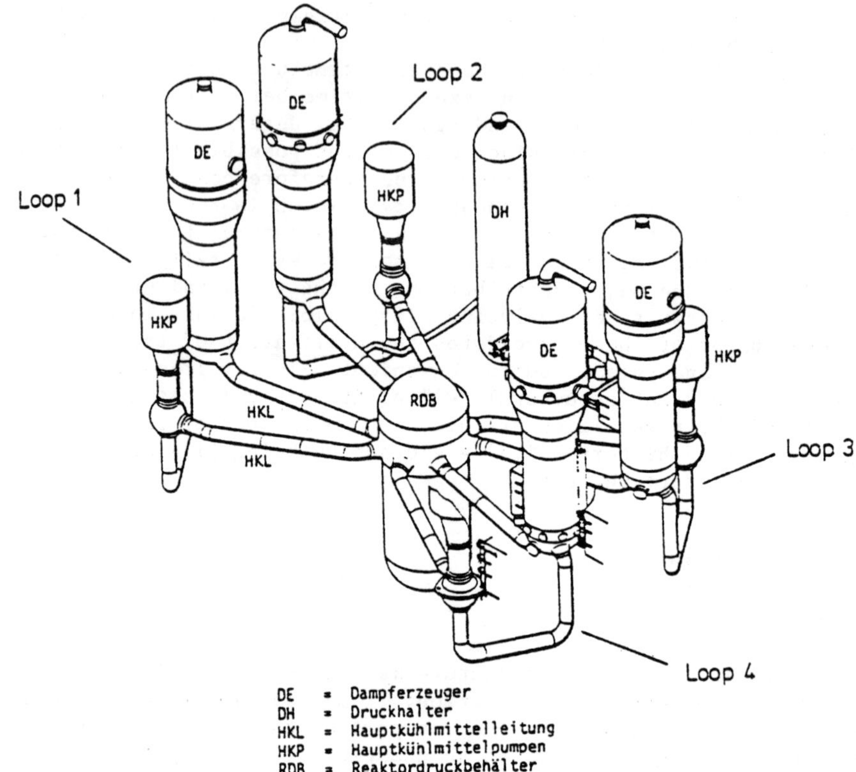

DE = Dampferzeuger
DH = Druckhalter
HKL = Hauptkühlmittelleitung
HKP = Hauptkühlmittelpumpen
RDB = Reaktordruckbehälter

Fig. 5: Schematic Sketch of the Primary
Piping System of a PWR

3.3 Load- and Structural Analysis

The frequency of occurrence of the various load cases to be taken into
account, i.e. due to transients, operation of the system, external
hazards, such as earthquakes are given in Table 1.

The results of the time history analysis in terms of initial
stresses and maximum stress variations of the various load types are
shown in Table 2. The seismic analysis includes a particular reliability model which is beyond the scope of this lecture [14].

A simulation procedure then yields the "most likely" load sequence
during the design life of the structure.

3.4 Failure Mode Analysis

As the primary piping contains a number shop and field welds, possible
initial cracks, i.e. flaws have to be taken into account in the analysis in terms of an initial crack probability distribution. Its parameters are to be statistically estimated from experimental data, i.e.

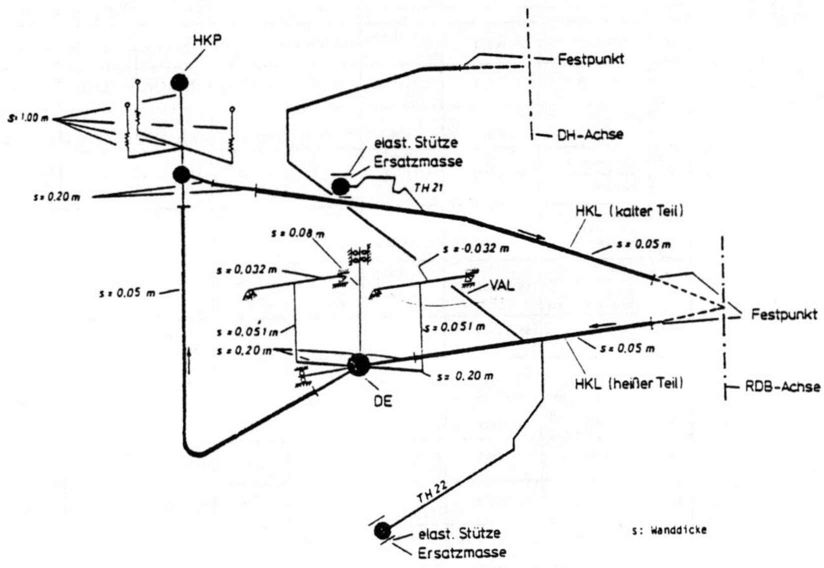

Fig. 6: Structural Model of LOOP 2

observations. The failure criterion has to consider both ductile (yield) and brittle (fracture) failure. Growth rates may be calculated for example by the well known Forman-relation,

$$\frac{da}{dN} = \frac{C(\Delta K^n - \Delta K_o^n)}{(1-R)K_c - \Delta K} \tag{3.1}$$

where a is the crack length and N the number of load cycles. The parameters along with their statistical properties, are given in Table 3.

4. RELIABILITY ANALYSIS

In the reliability analysis all aspects discussed in the previous section are combined to predict the failure rate. As failure mechanism the crack growth – initiating from an initial crack – around the circumference of the primary pipe is considered. If the crack has grown sufficiently long and deep, brittle fracture of the entire pipe has to be considered. Failure may also occur – in the presence of a long initial crack – due to a single (peak) earthquake overload (ductile failure). Again, it is pointed out, that most load and resistance para-

load case number	Plant condition	Stress level $|kN/m^2|$	Δσ in axial direction $|kN/m^2|$	frequency					
				Task force report	German Risk Study	Specification Biblis B RE-L 628	Specification Grafenrheinfeld RE-L 10 00b	Empirical Data	Others
1	2	3	4	5	6	7	8	9	10
1a	Start up	29.215,8	$\Delta\sigma_1=+11.906,3$ $\Delta\sigma_2=+126.521,6$	120/LT≙3/a	100/LT≙2,5a	120/LT≙3/a	100/LT≙2,5a	μ=4,8,σ=4,6	
1b	Shut down	152.891,7	-123.675,9	120/LT≙3/a	100/LT≙2,5/a	120/LT≙3/a	100/LT≙2,5a	μ=4,8,σ=4,6	
2a	Scram	159.895,6	- 7.878.3	400/LT	280/LT≙7/a	400/LT≙10/a	400/LT≙10/a	μ=3,5,σ=2,5	
2b	turbine trip without steam by-pass	159.895,6	15.429,2			80/LT≙2/a	40		
2c	load rejection with steam by-pass	159.895,6	4.214,7 $\Delta\sigma_1=2390,2$			400/LT≙10/a	400		
2d	load rejection to internal consumption (pre-operational test)	159.895,6	2x 4.105,1 $\Delta\sigma_3=-3.353,4$						KWU E 100.404
3/4	Load step= 10 %	159.895,6	4.547,2	10^5/LT≙2500/a		10^5/LT≙2500a	10^5/LT≙2500a		
5a	Load ramp 60 % to 100 %	158.620,1	Δσ=+1275,5	3×10^4/LT≙750/a		$1,4\cdot10^4$			
5b	Load ramp 80 % to 100 %	159.165,4	Δσ=+ 630,2			$1,4\cdot10^4$	$1,3\cdot10^5$/LT≙3250/a		
6a	Load ramp 100 % to 60 %	159.895,6	$\Delta\sigma_1=+ 227,5$ $\Delta\sigma_2=-1275,5$			$1,4\cdot10^4$			
6b	Load ramp 100 % to 80 %	159.895,6	$\Delta\sigma_1=+327,5$ $\Delta\sigma_2=-630,2$			$1,4\cdot10^4$			
7	Loss of feed-water	159.895,6	17.410,9 $\Delta\sigma_3=-1.302,2$		0,8/a				
8	Loss of offsite power	159.895,6	9.617 $\Delta\sigma_3=+517,2$	80/LT≙2/a	4/LT≙0,1/a		20/LT≙0,5/a	0,1/a	
9	failure of one reactor coolant pump (defect loop)	159.895,6	14.359,8 $\Delta\sigma_3=-9.858,7$	80/LT≙2/a		80/LT≙2/a	80/LT≙2/a		
10	Vibrations during normal operation	159.895,6	1.695,6					$8,1\cdot10^9$/LT $2,5\cdot10^8$/a	KWU R52/19/76
11	Leakage Test	20.720,4 (dead load)	$\Delta\sigma_1=98.585,5$			120/LT≙3/a	100/LT≙2,5/a		
12	Pressure Test	20.720,4 (dead load)	$\Delta\sigma_2=124.173,4$	20/LT≙0,5/a		30/LT≙0,8/a	30/LT≙0,8/a		

LT = life time   a = year

Table 1: Δσ, Stress Level and Frequencies of the
Different Plant Conditions

meters are random variables. The reliability analysis is based on the
"Two Criteria" or "Failure Assessment Diagramm" (FAD). In this context
the limit state equation as defined by Burdekin/Stone is utilized
(see Fig. 7). This relation is described by

$$K_r = \frac{S_r}{\sqrt{\dfrac{8}{\pi^2}}\, \ln \sec\left(\dfrac{\pi}{2}\cdot S_r\right)} \tag{4.1}$$

where $K_r = \dfrac{K}{K_{Ic}}$ and $S_r = \dfrac{\sigma}{\sigma_F}$ (σ_F = stress at failure).

Due to the presence of the large number of random input parameters, equ.
(4.1) is solved by Monte Carlo (importance sampling) simulation procedures.

Load-Case-Nr.	Load Case	Initial Stress σ_A N/mm^2	Max. Stress Difference N/mm^2	
			increasing	decreasing
1	Start up	29	127	− 12
2	Shut down	153	−	−124
3	Scram	160	1	− 8
4	Turbine Trip without steam by-pass	160	7	− 16
5/6	Load Step ± 10 %	160	2	− 4,5
7 a	Load Ramp 60/100 %	159	1	−
7 b	Load Ramp 80/100 %	159	0,5	−
8 a	Load Ramp 100/80 %	160	−	− 1
8 b	Load Ramp 100/60 %	160	−	− 1,5
9	Loss of Feed Water	160	18	− 12
10	Loss of Offsite Power	160	10	− 7
11	Failure of one Reactor Coolant Pump (defect loop)	160	3	− 14
12	Leakage – Test	21	99	−
13	Pressure – Test	21	124	−
14	Earthquake	160	see [14]	
15	Vibrations during Normal Operation	160	1,7	− 1,7

Table 2: Initial Stresses and Maximum Stress
Variations of the Various Load Cases

PARAMETER	MEAN VALUE	COEFFICIENT OF VARIATION	TYPE OF DISTRIBUTION
YIELD STRESS	$438\ Nmm^{-2}$	0.04	WEIBULL
TENSILE STRESS	$602\ Nmm^{-2}$	0.055	WEIBULL
TOUGHNESS	$3500\ Nmm^{-3/2}$	0.1	WEIBULL
C...CRACK PROPAGATION FACTOR	$10^{-10}N^{-2.3}mm^{4.45}$	0.5	LOGNORMAL
ΔK_0 THRESHOLD VALUE	$100\ Nmm^{-3/2}$	−	−
K_c CRITICAL VALUE	$8870\ Nmm^{-3/2}$	−	−
FORMAN EXPONENT	3.3	−	−
WALL THICKNESS	54 mm	−	−
PIPE DIAMETER	800 mm	−	−
a...INITIAL CRACK DEPTH	3.65 mm	0.93	SPECIAL TYPE
a/c GEOMETRY FACTOR	0.55	0.33	NORMAL

Table 3: Parameters Utilized in equ. (3.1)

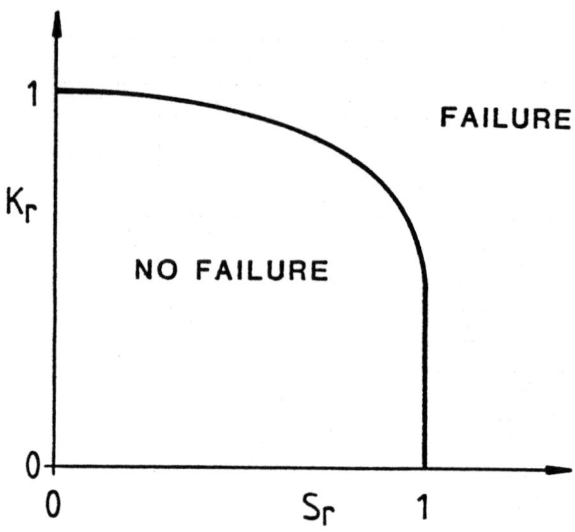

Fig. 7: Failure Assessment Diagram (FAD)

5. RESULTS

Utilizing realistic parameters, i.e. a threshold value ΔK_o of $80 \div 100$ N/mm$^{3/2}$ a failure rate of $5 \cdot 10^{-6}$ – for a design life of 40 years – is to be expected.

The analysis shows clearly how the systems failure rates – indicated in Table 1 – directly relate to the crack growth rate and consequently affect the structural failure rate. A sensitivity analysis may be easily performed.

6. REFERENCES

1 Reactor Safety Study, Washington 1400, U.S. NRC, Washington D.C., October 1975.

2 Deutsche Risikostudie – Kernkraftwerke, Verlag TÜV Rheinland, Köln, 1979.

3 BECHER, P.E., SCHMITT, W., SCHUELLER, G.I., "On the Interactions of Systems and Structural Reliability with Respect to Rare Events", OECD-CSNI SINDOC (77) 137, Paris, August 1977.

4 SCHUELLER, G.I., KAFKA, P., SCHMITT, W., "Some Aspects of the Interaction between Systems- and Structural Reliability", Proc. 5th Int. Conference on Structural Mechanics in Reactor Technology, Berlin, August 12-19, 1979, Vol M8/1, pp. 1-10.

5 BENJAMIN, J.R., SCHUELLER, G.I., WITT, F.J. (Editors), Proceedings 2nd International Seminar Structural Reliability,Mech. Comp. Subassembl. Nucl. Power Plants, Special Volume, Nucl.Eng.Design 59, 1980, 168p.

6 BENJAMIN, J.R., SCHUËLLER, G.I., WITT, F.J., "Critical Review of Second Intern. Seminar on Structural Reliability of Mechanical Components and Subassemblies of Nuclear Power Plants", Reliability Eng. No. 2, 1981, pp. 125-134.

7 SCHUËLLER, G.i., BENJAMIN, J.R., COSTES, D., WITT, J.F. (Editors), "Proceedings of the 3rd Intern. Seminar on the Reliability of Nuclear Power Plants", Intern. Journal NED - Nucl. Eng. Design, Vol. 71, No. 3, 1982, pp. 263-444.

8 SCHUËLLER, G.I., "Application of Extreme Values in Structural Engineering", Proc., NATO ASI Conf. on Statistical Extremes and Applications, Vimeiro, Portugal, NATO ASI Series, D.Reidel Publ. Comp., 1984.

9 STIX, R., SCHUËLLER, G.I., "Problemstellung bei der Berechnung der Versagenswahrscheinlichkeit", Internal Working Report No. 3, Institut für Mechanik, University of Innsbruck, to appear 1985.

10 GRIMMELT, M.J., SCHUËLLER, G.I., MUROTSU, Y., "On the Evaluation of Collapse Probabilities", Proc. ASCE-Specialty Conf., Purdue University, Lafayette, 1983, pp. 859-862.

11 GRIMMELT, M.J., SCHUËLLER, G.I., "A Method to Dertermine Reliability of Structures under Combined Loading", Proc. (ICASP-4) 4th Int. Conf. on Applications of Statistics and Probability in Soil and Structural Eng., Pitagora Editrice, Bologna, 1983, pp. 261-271.

12 OSWALD, G.F., SCHUËLLER, G.I., "Reliability of Deteriorating Structures", Proc. ICASP-4, 4th Int. Conf. on Applications of Statistics and Probability in Soil and Structural Eng., Pitagora Editrice, Bologna, 1983, pp. 597-608.

13 KAFKA, P., "Interface between Structural and System Analysis", JournalNucl. Eng. Design, 71, (1982), pp. 355-357.

14 SCHUËLLER, G.I., PRADLWARTER, H.J., HAMPL, N.C., MICHEL, T.H., "Zuverlässigkeitsbeurteilung für den Primärkreislauf am Beispiel des Druckwasserreaktors", Report No. 3-84, February, 1984, Institut für Mechanik, Universität Innsbruck.

RELIABILITY OF STRUCTURAL SYSTEMS

P. Thoft-Christensen
Aalborg University Centre
Sohngaardsholmsvej 57
DK-9000 Aalborg, Denmark

1. INTRODUCTION

During the last 10 years significant progress has been made in connection with estimation of the reliability of structural systems. Such a reliability analysis of a structural system consists of at least two parts, namely identification of critical (significant) failure modes and estimation of the failure probabilities of the failure modes.

In this paper critical failure modes are identified by the β-unzipping method (see Thoft -Christensen [1], [2] and Thoft-Christensen & Sørensen [3], [4]). The β-unzipping method is quite general in the sense that it can be used for two-dimensional and three-dimensional, framed and trussed structures, for structures with ductile and brittle elements and also in connection with a number of different failure mode definitions.

In this paper the structural system is modelled by a series system, where the elements are parallel systems (failure modes). The failure probability of such a series system is then estimated by

1) estimating the reliability indices for each element in the parallel systems (failure modes),

2) estimating the reliability indices for the parallel systems (failure modes),

3) approximation of the safety margin for each parallel system by an equivalent linear safety margin,

4) estimating the correlation between the parallel systems (failure modes),

5) estimating the reliability index for the series system.

In this paper a brief presentation of the application of the β-unzipping method is given. The method is illustrated by simple examples.

2. CLASSIFICATION OF FAILURE MODES

Clearly, the definition of failure modes for a structural system is of great importance in esti-

A. C. Lucia (ed.), Advances in Structural Reliability, 87–112.
© *1987 by ECSC, EEC, EAEC, Brussels and Luxembourg.*

mating the reliability of the structural system. In this section failure modes are classified in a systematic way convenient for the subsequent reliability estimate. A very simple estimate of the reliability of a structural system is based on failure of a single failure element, namely the failure element with the lowest reliability index (highest failure probability) of all failure elements. Failure elements are structural elements or cross-sections where failure can take place. The number of failure elements will usually be considerably higher than the number of structural elements. Such a reliability analysis is in fact not a system reliability analysis, but from a classification point of view it is convenient to call it *system reliability analysis at level 0*. Let a structure consist of n failure elements and let the reliability index (see e.g. Thoft-Christensen & Baker [5]) for failure element i be β_i, then the system reliability index β_S^0 at level 0 is

$$\beta_S^0 = \min_{i=1,n} \beta_i \tag{1}$$

Clearly, such an estimate of the system reliability is too optimistic. A more satisfactory estimate is obtained by taking into account the possibility of failure of any failure element by modelling the structural system as a series system with the failure elements as elements (see figure 1). The probability of failure for this series system is then estimated on the basis of the reliability indices β_i, i = 1, 2, . . . , n, and the correlation between the safety margins for the failure elements. This reliability analysis is called *system reliability analysis at level 1*. In general it is only necessary to include some of the failure elements in the series system (namely those with the smallest β-indices) to get a good estimate of the system failure probability P_f^1 and the corresponding generalized reliability index β_S^1, where

$$\beta_S^1 = - \Phi^{-1}(P_f^1) \tag{2}$$

and where Φ is the standardized normal distribution function. The failure elements included in the reliability analysis are called *critical failure elements*.

The modelling of the system at level 1 is natural for a statically determinate structure, but failure in a single failure element in a structural system will not always result in failure of the total system, because the remaining elements may be able to sustain the external loads due to redistribution of the load effects. This situation is characteristic of statically indeterminate structures. For such structures *system reliability analysis at level 2* or higher levels may be reasonable. At level 2 the systems reliability is estimated on the basis of a series system where the elements are parallel systems each with two failure elements - socalled *critical pairs of failure elements* (see figure 2). These critical pairs of failure elements are obtained by modifying the structure by assuming in turn failure in the critical failure elements and adding fictitious loads corresponding to the load-carrying capacity of the elements in failure.

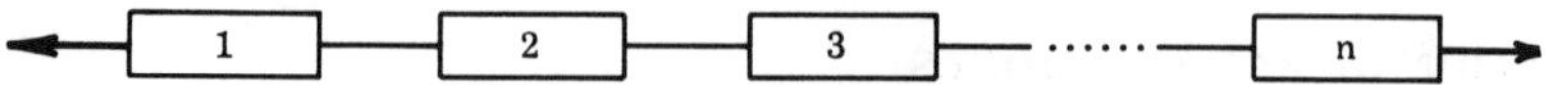

Figure 1. System modelling at level 1.

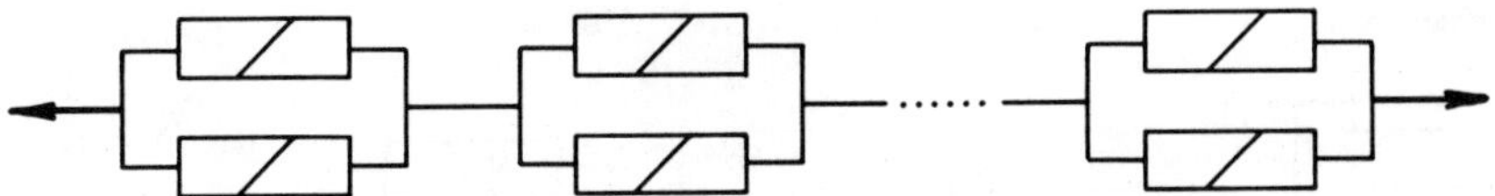

Figure 2. System modelling at level 2.

Let e.g. element i be a critical failure element, then the structure is modified by assuming failure in element i and the load-carrying capacity of the failure element is added as fictitious loads if the element is ductile. If the failure element is brittle, no fictitious loads are added. The modified structure is then analysed elastically and new β-values are calculated for all the remaining failure elements. Failure elements with low β-values are then combined with failure element i so that a number of critical pairs of failure elements are defined.

The procedure sketched above is now continued by analysing modified structures where failure is assumed in critical pairs of failure elements. In this way *critical triples of failure elements* are identified and a *reliability analysis at level 3* can be made on the basis of a series system, where the elements are parallel systems each with three failure elements (see figure 3). By continuing in the same way reliability estimates at level 4, 5, etc. can be performed, but in general analysis beyond level 3 is of minor interest.

Many recent investigations in structural systems theory concern structures which can be modelled as elastic-plastic structures. In such cases failure of the structure is usually defined as formation of a mechanism. When this failure definition is used it is of great importance to be able to identify the most significant failure modes because the total number of mechanisms is usually much too high to be included in the reliability analysis. The β-unzipping method can be used for this purpose simply by continuing the procedure described above until formation of a mechanism has taken place. However, this will be very expensive due to the great number of reanalyses needed. It turns out to be much better to base the unzipping on reliability indices for fundamental mechanisms and linear combination of fundamental mechanisms. When system failure is defined as formation of a mechanism the probability of failure of the structural system is estimated by modelling the structural system as a series system with the significant mechanisms as elements (see figure 4). Reliability analysis based on the mechanism failure definition is called *systems reliability analysis at mechanism level*.

For real structures a mechanism will often involve a relatively large number of yield hinges and the deflections at the moment of formation of a mechanism can usually not be neglected. Therefore, the failure definition must be combined with some kind of deflection failure definition.

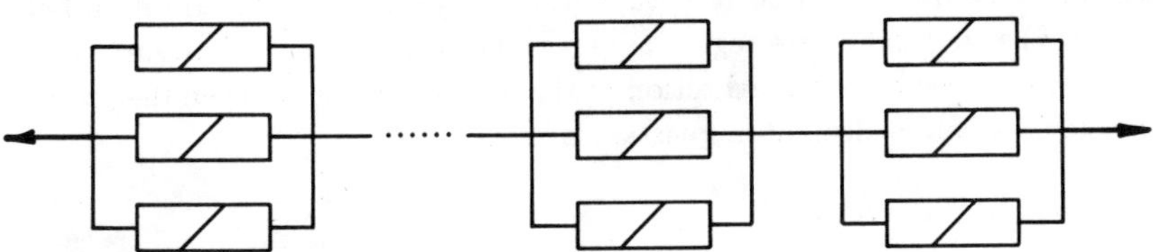

Figure 3. System modelling at level 3.

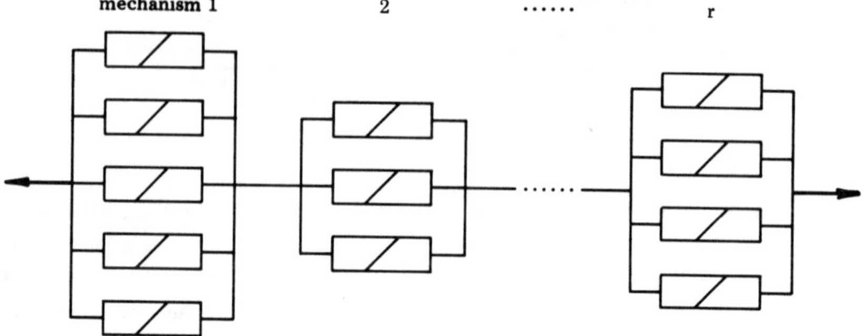

Figure 4. System modelling at mechanism level.

3. RELIABILITY OF FUNDAMENTAL SYSTEMS

In this paper it is assumed that all basic variables (load variables and strength variables) are normally distributed. All geometrical quantities and elasticity coefficients are assumed deterministic. This assumption significantly facilitates the estimation of the failure probability, but in general basic variables cannot with a satisfactory degree of accuracy be modelled by normally distributed variables. To overcome this problem a number of different transformation methods have been suggested. The most well-known method was suggested by Rackwitz & Fiessler [6]. One drawback to these methods is that they increase the computational work considerably due to the fact that they are iterative methods. A more simple (but also less accurate) method called the multiplication factor has been proposed by P. Thoft-Christensen [2]. The multiplication factor method does not increase the computational work. Without loss of generality it is assumed that all basic variables are standardized, i.e. the mean value is 0 and the variance 1.

The β-unzipping method is in this paper only used for trussed and framed structures but it can easily be modified to other classes of structures. The structure is considered at a fixed point in time, so that only static behaviour has been treated. It is assumed that failure in a structural element (section) is either pure tension/compression or failure in bending. Combined failure criteria have also been used in connection with the β-unzipping method, but only little experience has been obtained for the time being.

Let the vector $\overline{X} = (X_1, \ldots, X_n)$ be the vector of the standardized normally distributed basic variables with the joint probability density function φ_n and let failure of a failure element be determined by a failure function $f: \omega \to R$, where ω is the n-dimensional basic variable space. Let f be defined in such a way that the space ω is divided into a failure region $\omega_f = \{\overline{x}: f(\overline{x}) \leqslant 0\}$ and a safe region $\omega_s = \{\overline{x}: f(x) > 0\}$ by the failure surface $\partial\omega = \{\overline{x}: f(\overline{x}) = 0\}$, where the vector $\overline{x}$ is a realization of the random vector $\overline{X}$. Then the probability of failure P_f for the failure element in question is given by

$$P_f = P(f(\overline{X}) \leqslant 0) = \int_{\omega_f} \varphi_n(\overline{x})d\overline{x} \qquad (2)$$

If the function f is linearized in the socalled design point with distance β to the origin of the coordinate system, then an approximate value for P_f is given by

$$P_f \approx P(\alpha_1 X_1 + \ldots + \alpha_n X_n + \beta \leqslant 0)$$

$$= P(\alpha_1 X_1 + \ldots + \alpha_n X_n \leqslant -\beta) = \Phi(-\beta) \tag{3}$$

where $\bar{\alpha} = (\alpha_1, \ldots, \alpha_n)$ is the directional cosinus of the linearized failure surface. β is the Hasofer-Lind reliability index. Φ is the standardized normal distribution function. The random variable

$$M = \alpha_1 X_1 + \alpha_2 X_2 + \ldots + \alpha_n X_n + \beta \tag{4}$$

is the *linearized safety margin* for the failure element.

Next consider a *series system* with k elements. An estimate of the failure probability P_f^s of this series system can be obtained on the basis of the linearized safety margin of the form (4) for the k elements.

$$P_f^s = P(\bigcup_{i=1}^{k} (\bar{\alpha}_i \bar{X} + \beta_i \leqslant 0)) = P(\bigcup_{i=1}^{k} (\bar{\alpha}_i \bar{X} \leqslant -\beta_i))$$

$$= 1 - P(\bigcap_{i=1}^{k} (\bar{\alpha}_i \bar{X} > -\beta_i) = 1 - P(\bigcap_{i=1}^{k} (-\bar{\alpha}_i \bar{X} < \beta_i) = 1 - \Phi_k(\bar{\beta} ; \bar{\bar{\rho}}) \tag{5}$$

where $\bar{\alpha}_i$ and β_i are the directional cosinus and the reliability index for failure element i, i = 1, ... k and where $\bar{\beta} = (\beta_1, \ldots, \beta_k)$. $\bar{\bar{\rho}} = \{\rho_{ij}\}$ is the correlation coefficient matrix given by $\rho_{ij} = \bar{\alpha}_i^T \bar{\alpha}_j$ for all i $\neq$ j. Φ_k is the standardized k-dimensional normal distribution function.

For a *parallel system* with k elements an estimate of the failure probability P_f^p can be obtained in the following way

$$P_f^p = P(\bigcap_{i=1}^{k} (\bar{\alpha}_i \bar{X} + \beta_i \leqslant 0)) = P(\bigcap_{i=1}^{k} (\bar{\alpha}_i \bar{X} \leqslant -\beta_i)) = \Phi_k(\bar{\beta} ; \bar{\bar{\rho}}) \tag{6}$$

where the same notation as above is used.

It is important to note the approximation behind (5) and (6) namely the linearization of the general non-linear failure surfaces in the distinct design points for the failure elements. The main problem in connection with application of (5) and (6) is numerical calculation of the n-dimensional normal distribution function Φ_n for n $\geqslant$ 3. This problem will be treated later in this paper where a number of methods to get approximate values for Φ_n is mentioned.

4. IDENTIFICATION OF CRITICAL FAILURE MODES

A number of different methods to identify critical failure modes has been suggested (see e.g. Ferregut-Avila [7], Moses [8], Gorman [9], Ma & Ang [10], Klingmüller [11], Murotsu et al. [12] and Kappler [13]). In this paper the β-unzipping method [1] - [4] is used.

At *level 1* the system reliability is defined as the reliability of a series system with n elements the n failure elements. Therefore, the first step is to calculate β-values for all failure elements and then use equation (5). As mentioned earlier, equation (5) cannot be used directly. However, excellent upper and lower bounds - called Ditlevsen bounds - exist (see Ditlevsen [14] and Kounias [15]). The well-known simple bounds

$$\max_{i=1}^{n} \Phi(-\beta_i) \leqslant P_f \leqslant 1 - \prod_{i=1}^{n} \Phi(\beta_i) \tag{7}$$

are useful when the safety margins M_i, $i = 1, 2, \ldots, n$ are almost perfectly correlated (the lower bound) or when the correlation between any pair of safety margins is very small (the upper bound).

Dunnet & Sobel [16] have shown that

$$P_f = 1 - \int_{-\infty}^{\infty} \varphi(t) \prod_{i=1}^{n} \Phi\left(\frac{\beta_i - \sqrt{\rho}\, t}{\sqrt{1 - \rho}}\right) dt \tag{8}$$

if all correlation coefficients ρ_{ij} are equal ($= \rho$) and positive.

When the correlation coefficients ρ_{ij} are unequal a simple approximation for P_f can be obtained from (8) by putting $\rho = \bar{\rho}$, where $\bar{\rho}$ is the average correlation coefficient (see Thoft-Christensen & Sørensen [17]) defined by

$$\bar{\rho} = \frac{1}{n(n-1)} \sum_{i, j = 1, i \neq j}^{n} \rho_{ij} \tag{9}$$

Usually for a structure with n failure elements, the estimate of the failure probability of the series system with n elements can be calculated with sufficient accuracy by only including some of the failure elements, namely those with the smallest reliability indices. One way of selecting is to include only failure elements with β-values in an interval $[\beta_{min}, \beta_{min} + \Delta\beta_1]$, where β_{min} is the smallest reliability index of all failure element indices and where $\Delta\beta_1$ is a prescribed positive number. The failure elements chosen to be included in the system reliability analysis at level 1 are called *critical failure elements*. If two or more critical failure elements are perfectly correlated, then only one of them is included in the series system of critical failure elements.

At *level 2* the system reliability is estimated as the reliability of a series system where the elements are parallel systems each with 2 failure elements (see figure 2) - socalled *critical pairs of failure elements*. Let the structure be modelled by n failure elements and let the number of critical failure elements at level 1 be n_1. Let the critical failure element ℓ have the lowest reliability index β of all critical failure elements. Failure is then assumed in failure element ℓ and the structure is modified by removing the corresponding failure element and adding a pair of socalled fictitious loads F_ℓ (normal forces or moments). If the removed failure element is brittle, then no fictitious loads are added. However, if the removed failure element ℓ is ductile then the fictitious load F_ℓ is a stochastic load given by $F_\ell = \gamma_\ell R_\ell$, where R_ℓ is the load-carrying capacity of failure element ℓ and where $0 < \gamma_\ell \leqslant 1$.

The modified structure with the loads $P_1, \ldots, P_k$ and the fictitious load F_ℓ (normal force or moment) is then reanalysed and influence coefficients a_{ij} with respect to $P_1, \ldots, P_k$ and $a'_{i\ell}$ with respect to F_ℓ are calculated. The load effect (force or moment) in the remaining failure elements is then described by a stochastic variable. The load effect in failure element i is called $S_{i|\ell}$ (load effect in failure element i given failure in failure element ℓ) and

$$S_{i|\ell} = \sum_{j=1}^{k} a_{ij}P_j + a'_{i\ell}F_\ell \tag{10}$$

The corresponding safety margin $M_{i|\ell}$ then is

$$M_{i|\ell} = \min(R_i^+ - S_{i|\ell}, \ R_i^- + S_{i|\ell}) \tag{11}$$

where R_i^+ and R_i^- are the stochastic variables describing the (yield) strength capacity in »tension» and »compression» for failure element i. In the following $M_{i|\ell}$ will be approximated by either $R_i^+ - S_{i|\ell}$ or $R_i^- + S_{i|\ell}$ depending on the corresponding reliability indices. The reliability index for failure element i, given failure in failure element ℓ, is

$$\beta_{i|\ell} = \mu_{M_{i|\ell}}/\sigma_{M_{i|\ell}} \tag{12}$$

In this way new reliability indices are calculated for all failure elements (except the one where failure is assumed) and the smallest β-value is called $\beta_{\min}$. The failure elements with β-values in the interval $[\beta_{\min}, \beta_{\min} + \Delta\beta_2]$, where $\Delta\beta_2$ is a prescribed positive number, are then in turn combined with failure element ℓ to form a number of parallel systems.

The next step is then to evaluate the failure probability for each critical pair of failure elements. Consider a parallel system with failure elements ℓ and r. During the reliability analysis at level 1 the safety margin M_ℓ for failure element ℓ is determined and the safety margin $M_{r|\ell}$ for failure element r has the form (11). From these safety margins the reliability indices $\beta_1 = \beta_\ell$ and $\beta_2 = \beta_{r|\ell}$ and the correlation coefficient $\rho = \rho_{\ell,r|\ell}$ can easily be calculated. The probability of failure for the parallel system then is

$$P_f = \Phi_2(-\beta_1, -\beta_2 ; \rho) \tag{13}$$

The same procedure is then in turn used for all critical failure elements and further critical pairs of failure elements are identified. In this way the total series system used in the reliability analysis at level 2 is determined (see figure 2). The next step is then to estimate the probability of failure for each critical pair of failure elements (see (13)) and also to determine a safety margin for each critical pair of failure elements. When this is done generalized reliability indices for all parallel systems in figure 2 and correlation coefficients between any pair of parallel systems are calculated. Finally, the probability of failure P_f for the series system (figure 2) is estimated. The socalled equivalent linear safety margin introduced by Gollwitzer & Rackwitz [18] is used as approximations for safety margins for the parallel systems.

An important property by the β-unzipping method is the possibility of using the method when brittle failure elements occur in the structure. When failure occurs in a brittle failure element then the β-unzipping method is used in exactly the same way as presented above, the only difference being that no fictitious loads are introduced. If e.g. brittle failure occurs in a tensile bar in a trussed structure then the bar is simply removed without adding fictitious tensile loads. Likewise, if brittle failure occurs in bending, then a yield hinge is introduced, but no (yielding) fictitious bending moments are added.

The method presented above can easily be generalized to higher levels $N > 2$. *At level 3* the estimate of the system reliability is based on so-called *critical triples of failure elements*, i.e. a set of three failure elements. The critical triples of failure elements are identified by the β-unzipping method and each triple forms a parallel system with three failure elements. These parallel systems are then elements in a series system (see figure 3). Finally, the estimate of the reliability of the structural system at level 3 is defined as the reliability of this series system.

Assume that the critical pair of failure elements (ℓ, m) has the lowest reliability index $\beta_{\ell,m}$ of all critical pairs of failure elements. Failure is then assumed in the failure elements ℓ and m adding for each of them a pair of fictitious loads F_ℓ and F_m (normal forces or moments).

The modified structure with the loads $P_1, \ldots, P_k$ and the fictitious loads F_ℓ and F_m are then reanalysed and influence coefficients with respect to $P_1, \ldots, P_k$ and F_ℓ and F_m are calculated. The load effect in each of the remaining failure elements is then described by a stochastic variable $S_{i|\ell,m}$ (load effect in failure element i given failure in failure elements ℓ and m) and

$$S_{i|\ell,m} = \sum_{j=1}^{k} a_{ij} P_j + a'_{i\ell} F_\ell + a'_{im} F_m \tag{14}$$

The corresponding safety margin $M_{i|\ell,m}$ then is

$$M_{i|\ell,m} = \min(R_i^+ - S_{i|\ell,m}, R_i^- + S_{i|\ell,m}) \tag{15}$$

where R_i^+ and R_i^- are the stochastic variables describing the load-carrying capacity in »tension» and »compression» for failure element i. In the following $M_{i|\ell,m}$ will be approximated by either $R_i^+ - S_{i|\ell,m}$ or $R_i^- + S_{i|\ell,m}$ depending on the corresponding reliability indices. The reliability index for failure element i, given failure in failure elements ℓ and m, is then given by

$$\beta_{i|\ell,m} = \mu_{M_{i|\ell,m}} / \sigma_{M_{i|\ell,m}} \tag{16}$$

In this way new reliability indices are calculated for all failure elements (except ℓ and m) and the smallest β-value is called β_{min}. These failure elements with β-values in the interval $[\beta_{min}, \beta_{min} + \Delta\beta_3]$, where $\Delta\beta_3$ is a prescribed positive number, are then in turn combined with failure elements ℓ and m to form a number of parallel systems.

The next step is then to evaluate the failure probability for each of the critical triple of failure elements. Consider the parallel system with failure elements ℓ, m, and r. During the reliability analysis at level 1 the safety margin M_ℓ for failure element ℓ is determined and during the reliability analysis at level 2 the safety margin $M_{m|\ell}$ for the failure element m is determined. The safety margin $M_{r|\ell,m}$ for safety element r has the form (15). From these safety margins the reliability indices $\beta_1 = \beta_\ell$, $\beta_2 = \beta_{m|\ell}$ and $\beta_3 = \beta_{r|\ell,m}$ and the correlation matrix $\bar{\bar{\rho}}$ can easily be calculated. The probability of failure for the parallel system then is

$$P_f = \Phi_3(-\beta_1, -\beta_2, -\beta_3; \bar{\bar{\rho}}) \tag{17}$$

An equivalent safety margin $M_{i,j,k}$ can be determined by the procedure mentioned above. When the equivalent safety margins are determined for all critical triples of failure elements the correlation between them two and two can easily be calculated. The final step is then to arrange all the critical triples as elements in a series system (see figure 3) and estimate the probability of failure P_f and the generalized reliability index β_S for the series system.

The β-unzipping method can be used in exactly the same way as described in the preceding text to estimate the system reliability at levels $N > 3$. However, a definition of failure modes based on a fixed number of failure elements greater than 3 will hardly be of practical interest.

The application of the β-unzipping method presented above can also be used when failure is defined as formation of a mechanism. However, it is much more efficient to use the β-unzipping method in connection with fundamental mechanisms. Experience has shown that such a procedure is less computer time consuming than unzipping based on failure elements.

If unzipping is based on failure elements, then formation of a mechanism can be unveiled by the fact that the corresponding stiffness matrix is singular. Therefore, the unzipping is simply continued until the determinant of the stiffness matrix is zero. By this procedure a number of mechanisms with different numbers of failure elements will be identified. The number of failure elements in a mechanism will often be quite high so that several re-analyses of the structure are necessary.

As emphasized above it is more efficient to use the β-unzipping method in connection with fundamental mechanisms. Consider an elasto-plastic structure and let the number of potential failure elements (e.g. yield hinges) be n. It is then known from the theory of plasticity that the number of fundamental mechanisms is $m = n - r$, where r is the degree of redundancy. All other mechanisms can then be formed by linear combinations of the fundamental mechanisms. Some of the fundamental mechanisms are so-called joint mechanisms. They are important in the formation of new mechanisms by linear combinations of fundamental mechanisms, but they are not real failure mechanisms. Real failure mechanisms are by definition mechanisms which are not joint mechanisms.

Let the number of loads be k. The safety margin for fundamental mechanism i can then be written

$$M_i = \sum_{j=1}^{n} |a_{ij}| R_j - \sum_{j=1}^{k} b_{ij} P_j \tag{18}$$

where a_{ij} and b_{ij} are the influence coefficients. R_j is the yield strength of failure element j and P_j is load number j. a_{ij} is the rotation of yield hinge j corresponding to the yield mechanism i and b_{ij} is the corresponding displacement of load j. The numerical value of a_{ij} is used in the first summation at the right-hand side of (18) to make sure that all terms in this summation are non-negative.

The total number of mechanisms for a structure is usually too high to include all possible mechanisms in the estimate of the system reliability. It is also unnecessary to include all mechanisms because the majority of them will in general have a relatively small probability of occurrence. Only the most critical or most significant failure modes should be included. The problem is then how the most significant mechanisms (failure modes) can be identified. In this section it is shown how the β-unzipping method can be used for this purpose. It is not possible to prove that the β-unzipping method identifies all significant mechanisms, but experience with structures where all mechanisms can be taken into account seems to confirm that the β-unzipping method gives reasonably good results. Note that since some mechanisms are excluded the estimate of the probability of failure by the β-unzipping method is a lower bound for the correct probability of failure. The corresponding generalized reliability index determined by the β-unzipping method is therefore an upper bound of the correct generalized reliability index. However, the difference between these two indices is usually negligible.

The first step is to identify all fundamental mechanisms and calculate the corresponding reliability indices. Fundamental mechanisms can be automatically generated by a method suggested by Watwood [19], but when the structure is not too complicated the fundamental mechanisms can be identified manually.

The next step is then to select a number of fundamental mechanisms as starting points for the unzipping. By the β-unzipping method this is done on the basis of the reliability index β_{min} for the real fundamental mechanism that has the smallest reliability index and on the basis of a preselected constant ϵ_1 (e.g. ϵ_1 = 0.50). Only real fundamental mechanisms with β-indices in the interval $[\beta_{min}; \beta_{min} + \epsilon_1]$ are used as starting mechanisms in the β-unzipping method. Let $\beta_1 \leqslant \beta_2 \leqslant \ldots \leqslant \beta_f$ be an ordered set of reliability indices for f real fundamental mechanisms 1, 2, ... , f, selected by this simple procedure.

The f fundamental mechanisms selected as described above are now in turn combined linearly with all m (real and joint) mechanisms to form new mechanisms. First the fundamental mechanism 1 is combined with the fundamental mechanisms 2, 3, . . . , m and reliability indices $\beta_{1,2}, \ldots, \beta_{1,m}$ for the new mechanisms are calculated. The smallest reliability index is determined, and the new mechanisms with reliability indices within a distance ϵ_2 from the smallest reliability index are selected for further investigation. The same procedure is then used on the basis of the fundamental mechanisms 2, . . . , f and a failure tree as the one shown in figure 5 is constructed. Let the safety margins M_i and M_j of two fundamental mechanisms i and j combined as described above (see (18)) be

$$M_i = \sum_{r=1}^{n} |a_{ir}| R_r - \sum_{s=1}^{k} b_{is} P_s \tag{19}$$

$$M_j = \sum_{r=1}^{n} |a_{jr}| R_r - \sum_{s=1}^{k} b_{js} P_s \tag{20}$$

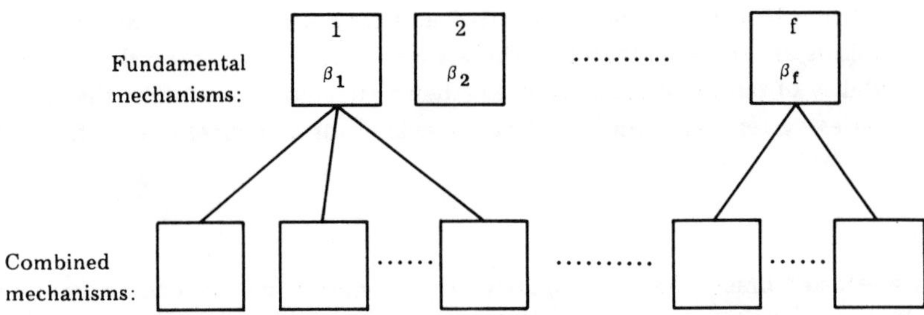

Figure 5. Construction of new mechanisms.

The combined mechanism i ± j then has the safety margin

$$M_{i \pm j} = \sum_{r=1}^{n} |a_{ir} \pm a_{jr}| R_r - \sum_{s=1}^{k} (b_{is} \pm b_{js}) P_s \qquad (21)$$

where + or − is chosen dependent on which sign will result in the smallest reliability index. From the linear safety margin (21) the reliability index $\beta_{i \pm j}$ for the combined mechanism can easily be claculated.

More mechanisms can be identified on the basis of the combined mechanisms in the second row of the failure tree in figure 5 by adding or subtracting fundamental mechanisms. Note that in some cases it is necessary to improve the technique by modifying (21), namely when a new mechanism requires not only a combination with 1 X but a combination with k X a new fundamental mechanism. The modified version is

$$M_{i + kj} = \sum_{r=1}^{n} |a_{ir} + ka_{jr}| R_r - \sum_{s=1}^{k} (b_{is} + kb_{js}) P_s \qquad (22)$$

where k is chosen equal to e.g. −1, 1, −2, +2, −3 or +3 dependent on which value of k will result in the smallest reliability index. By (22) it is easy to calculate the reliability index $\beta_{i + kj}$ for the combined mechanism i + kj.

By repeating this simple procedure the failure tree for the structure in question can be constructed. The maximum number of rows in the failure tree must be chosen and can typically be m + 2, where m is the number of fundamental mechanisms. A satisfactory estimate of the system reliability index can usually be obtained by using the same ϵ_2-value for all rows in the failure tree.

During the indentification of new mechanisms it will often occur that a mechanism already identified will turn up again. If this is the case, then the corresponding branch of the failure tree is terminated just one step earlier so that the same mechanism does not occur more than once in the failure tree.

The final step is the application of the β-unzipping method in evaluating the reliability of an elasto-plastic structure at mechanism level is to select the significant mechanisms from the mechanisms identified in the failure tree. This selection can, in accordance with the selection-criteria used in making the failure tree, e.g. be made by first identifying the smallest

β-value, β_{min} of all mechanisms in the failure tree and then selecting a constant ϵ_3. The significant mechanisms are then by definition those with β-values in the interval $[\beta_{min}; \beta_{min} + \epsilon_3]$. The probability of failure of the structure is then estimated by modelling the structural system as a series system with the significant mechanisms as elements (see figure 4).

5. EXAMPLE 1

Consider the two-storied braced frame in figure 6. The geometry and the loading are shown in the figure. This example is taken from [2] where all detailed calculations are shown. The area A and the moment of interia I for each structural member are shown in table 1. In the same table the expected values of the yield moment M and the tensile strength capacity R for all structural members are also stated. The compression strength capacity of one structural member is assumed to be one half of the tensile strength capacity.

The expected values of the loading are

$$E[P_1] = 100 \text{ kN}$$
$$E[P_2] = 350 \text{ kN}$$

For the sake of simplicity the coefficient of variation for any load or strength is assumed to be $V[\ \cdot\ ] = 0.1$. All elements are assumed to be perfectly ductile and made of a material having the same modulus of elasticity $E = 0.21 \cdot 10^9 \text{ kN/m}^2$.

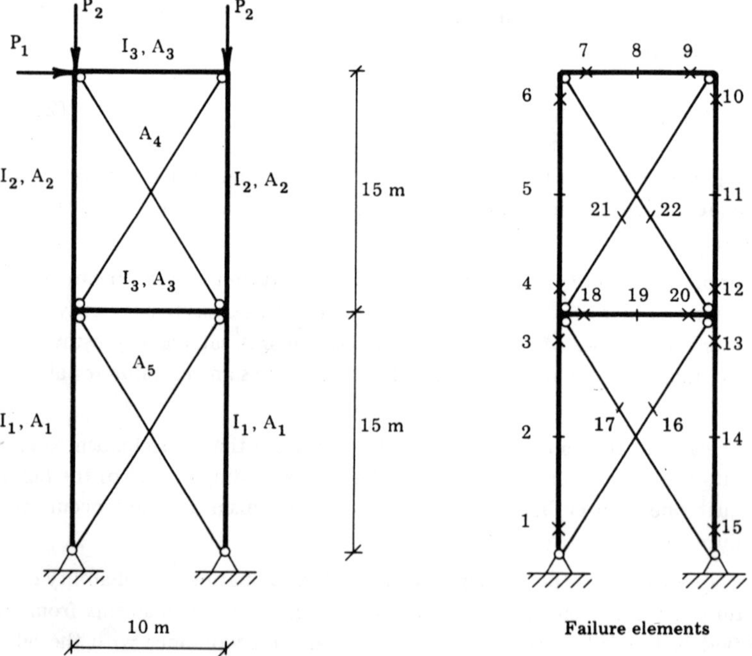

Figure 6. Two-storied braced frame.

	$A \cdot 10^3 \, m^2$	$I \cdot 10^6 \, m^4$	$E[M]$, kNm	$E[R]$, kN
1	4.59	57.9	135	1239
2	2.97	29.4	76	802
3	4.59	4.25	9.8	1239
4	2.6	0	0	702
5	2.8	0	0	756

Table 1. Section data and expected strength values.

Failure element	2	3	4	5	6	7	8	9	10	11
β-value	8.13	9.96	9.92	4.97	9.98	9.84	9.79	9.86	9.98	1.81
Failure element	12	13	14	16	17	18	19	20	21	22
β-value	9.94	9.97	1.80	9.16	4.67	9.91	8.91	9.91	8.04	3.34

Table 2. Reliability indices at level 0 (and level 1).

The failure elements are shown in figure 1. $\times$ indicates a potential yield hinge and | failure in tension/compression. The total number of failure elements is 22, namely $2 \times 6 = 12$ yield hinges in 6 beams and 10 tension/compression failure possibilities in the 10 structural elements. The following pairs of failure elements $(1, 3)$, $(4, 6)$, $(7, 9)$, $(10, 12)$, $(13, 15)$ and $(18, 20)$ are assumed fully correlated. All other pairs of failure elements are uncorrelated. Further, the loads P_1 and P_2 are uncorrelated.

The β-values for all failure elements are shown in table 2. Failure element 14 has the lowest reliability index $\beta_{14} = 1.80$ of all failure elements. Therefore, at level 0 the system reliability index is $\beta_S^0 = 1.80$.

Let $\Delta\beta_1 = 0$. It then follows from table 2 that the critical failure elements are 14, 11, 22, and 17. The corresponding correlation matrix (between the safety margins in the same order) is

$$\bar{\bar{\rho}} = \begin{bmatrix} 1.00 & 0.24 & 0.20 & 0.17 \\ 0.24 & 1.00 & 0.21 & 0.16 \\ 0.20 & 0.21 & 1.00 & 0.14 \\ 0.17 & 0.16 & 0.14 & 1.00 \end{bmatrix} \qquad (23)$$

The Ditlevsen bounds for the system probability of failure P_f^1 (see figure 7) give

$$0.06843 \leqslant P_f \leqslant 0.06849$$

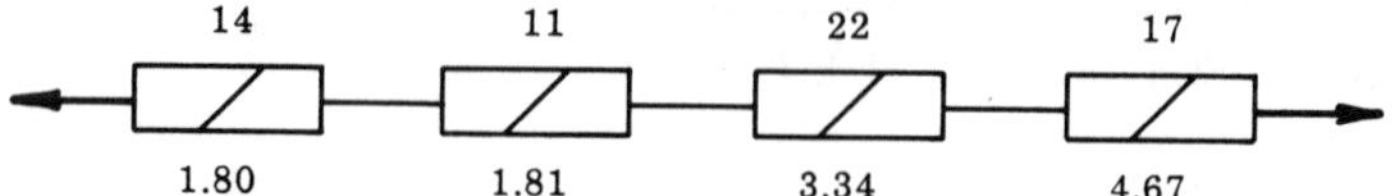

Figure 7. Series system used in estimating the system reliability at level 1.

Therefore, a (good) estimate for the system reliability index at level 1 is $\beta_S^1 = 1.49$. It follows from (23) that the coefficients of correlation are rather small. Therefore, the simple upper bound in (7) can be expected to give a good approximation. One gets $\beta_S^1 = 1.48$. A third estimate can be obtained by (8) and (9). The result is $\beta_S^1 = 1.49$.

At level 2 it is initially assumed that the ductile failure element 14 fails (in compression) and fictitious loads equal to 0.5 R_{14} are added (see figure 8). This modified structure is then analysed elastically and new reliability indices are calculated for all the remaining failure elements (see table 3). Failure element 11 has the lowest β-value 1.87. With $\Delta\beta_2 = 1.00$ failure element 11 is the only failure element with a β-value in the interval [1.87, 1.87 + $\Delta\beta_2$]. Therefore, in this case only one critical pair of failure elements is obtained by initiating the unzipping with failure element 14.

Based on the safety margin M_{14} for failure element 14 and the safety margin $M_{11|14}$ for failure element 11, given failure in failure element 14, the correlation coefficient can be calculated as $\rho = 0.28$. Therefore, the probability of failure for this parallel system is

$$P_f = \Phi_2(-1.80, -1.87 \; ; 0.28) = 0.00347 \tag{24}$$

and the corresponding generalized index

$$\beta_{14,11} = 2.70 \tag{25}$$

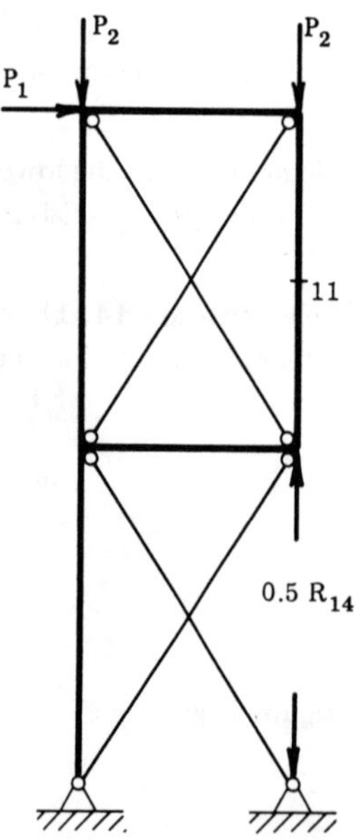

Figure 8. Modified structure when failure takes place in failure element 14.

Failure element	2	3	4	5	6	7	8	9	10	11
β-value	9.67	10.02	9.99	5.01	9.99	9.92	9.97	9.93	9.99	1.87

Failure element	12	13	16	17	18	19	20	21	22
β-value	10.00	10.03	5.48	3.41	10.24	8.87	10.25	9.08	4.46

Table 3

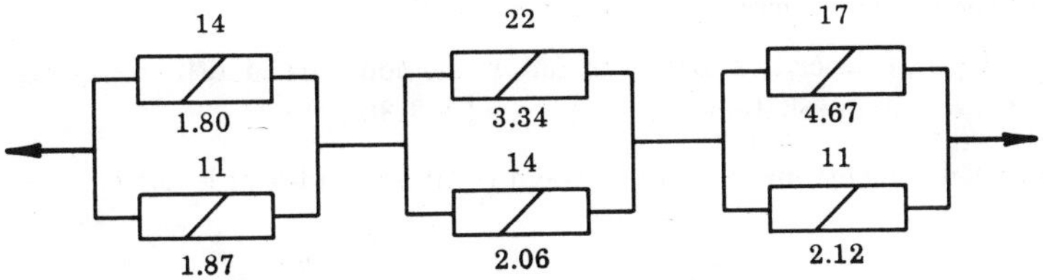

Figure 9. Series system used to estimate the reliability at level 2.

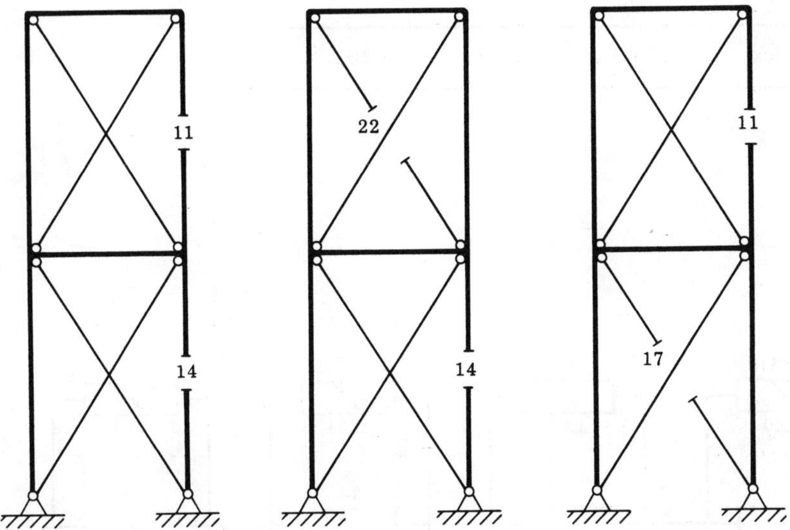

Figure 10. Failure modes at level 2.

The same procedure can be performed with the three other critical failure elements 11, 22 and 17. The final result is shown in figure 9 and the corresponding failure modes in figure 10. Generalized reliability indices and approximate equivalent safety margins for each parallel system in figure 9 are calculated. Then the correlation matrix $\bar{\bar{\rho}}$ can be calculated

$$\bar{\bar{\rho}} = \begin{bmatrix} 1.00 & 0.56 & 0.45 \\ 0.56 & 1.00 & 0.26 \\ 0.45 & 0.26 & 1.00 \end{bmatrix} \tag{26}$$

The Ditlevsen bounds for the failure probability of the series system, see figure 9, are

$$0.3488 \cdot 10^{-2} \leqslant P_f^2 \leqslant 0.3488 \cdot 10^{-2}$$

Therefore, estimate of the system reliability at level 2 is $\beta_S^2 = 2.70$.

At level 3 (with $\Delta\beta_3 = 1.00$), four critical triples of failure elements are identified (see figure 11) and an estimate of the system reliability at level 3 is $\beta_S^3 = 3.30$.

It is of interest to note that the estimates of the system reliability index at levels 1, 2, and 3 are very different.

Level	1	2	3
Reliability index	1.49	2.70	3.30

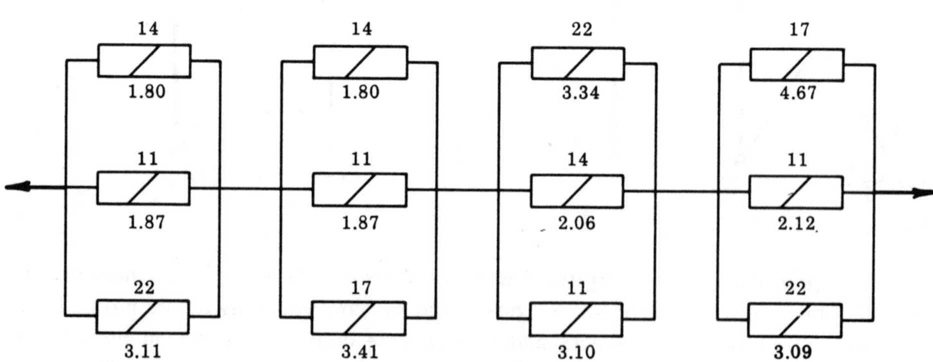

Figure 11. Reliability modelling at level 3 of the two-storied braced frame.

6. EXAMPLE 2

Consider the same structure as in example 1, but now the failure elements 2, 5, 11, and 14 are assumed brittle (see figure 6). All other data are unchanged. By a linear elastic analysis the same reliability indices for all (brittle and ductile) failure elements as in table 2 are calculated. Therefore, the critical failure elements are 14 and 11, and the estimate of the system reliability at level 1 is unchanged, i.e. $\beta_S^1 = 1.49$.

The next step is to assume brittle failure in failure element 14 and remove the corresponding part of the structure without adding fictitious loads (see figure 12, left). The modified structure is then linear elastically analysed and reliability indices are calculated for all remaining failure elements. Failure element 17 now has the lowest reliability index, namely the negative value

$$\beta_{17|14} = -6.01 \tag{27}$$

This very low negative value indicates that failure takes place in failure element 17 instantly after failure in failure element 14. The failure mode identified in this way is a mechanism and it is the only one when $\Delta\beta_2 = 1.00$. It can be mentioned that $\beta_{16|14} = -3.74$ so that failure element 16 also fails instantly after failure element 14.

Again, by assuming brittle failure in failure element 11 (see figure 12, right) only one critical pair of failure elements is identified, namely the pair of failure elements 11 and 22, where

$$\beta_{22|11} = -4.33 \tag{28}$$

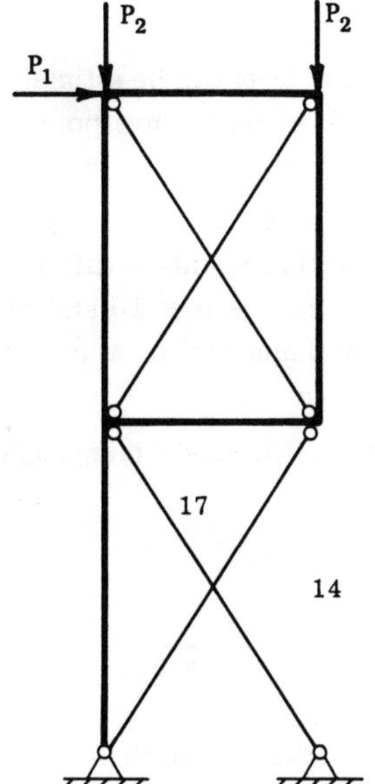

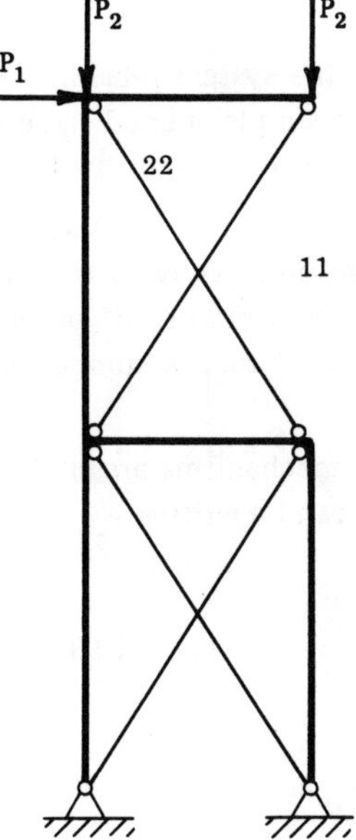

Figure 12. Modified structures.

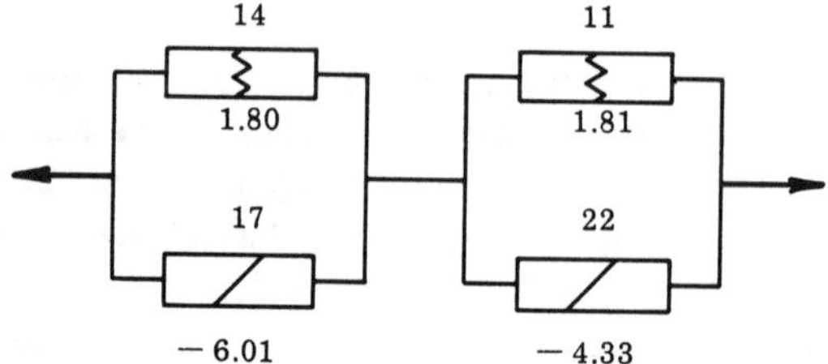

Figure 13. Modelling at level 2.

This failure mode is not a mechanism. The series system used in calculating an estimate of the system reliability at level 2 is shown in figure 13. Due to the small reliability indices (27) and (28) the strength variables 17 and 22 do not affect the safety margins for the two parallel systems in figure 13 significantly. Therefore, the reliability index at level 2 is unchanged from level 1, namely $\beta_S^2 = 1.49$.

As expected this value is much lower than the value 2.70 (see example 1) calculated for the structure with only ductile failure elements. This fact stresses the importance of the reliability modelling of the structure.

It is of interest to note that the β-unzipping method was capable of disclosing that the structure cannot survive failure in failure element 14. Therefore, when brittle failure occurs it is often reasonable to define failure of the structure as failure of just one failure element. This is equivalent to estimating the reliability of the structure at level 1.

7. EXAMPLE 3

In this example it is shown how the system reliability at mechanism level can be estimated in an efficient way. Consider the simple framed structure in figure 14 with corresponding expected values and coefficients of variation for the basic variables in table 4.

The load variables are P_i, i = 1, . . . , 4 and the yield moments are R_i, i = 1, . . . , 19. Yield moments in the same line are considered fully correlated and the yield moments in different lines are mutually independent. The number of potential yield hinges is n = 19 and the degree of redundancy is r = 9. Therefore, the number of fundamental mechanisms is n − r = 10.

One possible set of fundamental mechanisms are shown in figure 15. The safety margins M_i for the fundamental mechanisms can be written

$$M_i = \sum_{j=1}^{19} |a_{ij}| R_j - \sum_{i=1}^{4} b_{ij} P_j \quad , \quad i = 1, \ldots, 10 \tag{29}$$

where the influence coefficients a_{ij} and b_{ij} are determined by considering the mechanisms in the deformed state.

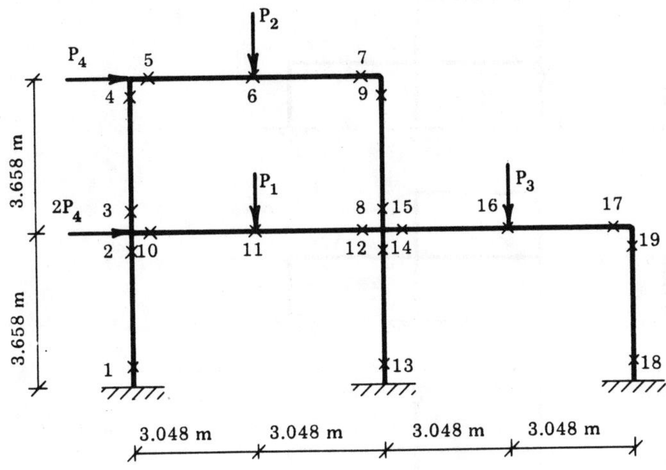

Figure 14. Geometry, loading and potential yield hinges (×).

Variables	Expected values	Coefficients of variation
P_1	169 kN	0.15
P_2	89 kN	0.25
P_3	116 kN	0.25
P_4	31 kN	0.25
$R_1, R_2, R_{13}, R_{14}, R_{18}, R_{19}$	95 kNm	0.15
R_3, R_4, R_8, R_9	95 kNm	0.15
R_5, R_6, R_7	122 kNm	0.15
R_{10}, R_{11}, R_{12}	204 kNm	0.15
R_{15}, R_{16}, R_{17}	163 kNm	0.15

Table 4

i	1	2	3	4	5	6	7	8	9	10
β_i	1.91	2.08	2.17	2.26	4.19	10.75	9.36	9.36	12.65	9.12

Table 5. Reliability indices for the fundamental mechanisms.

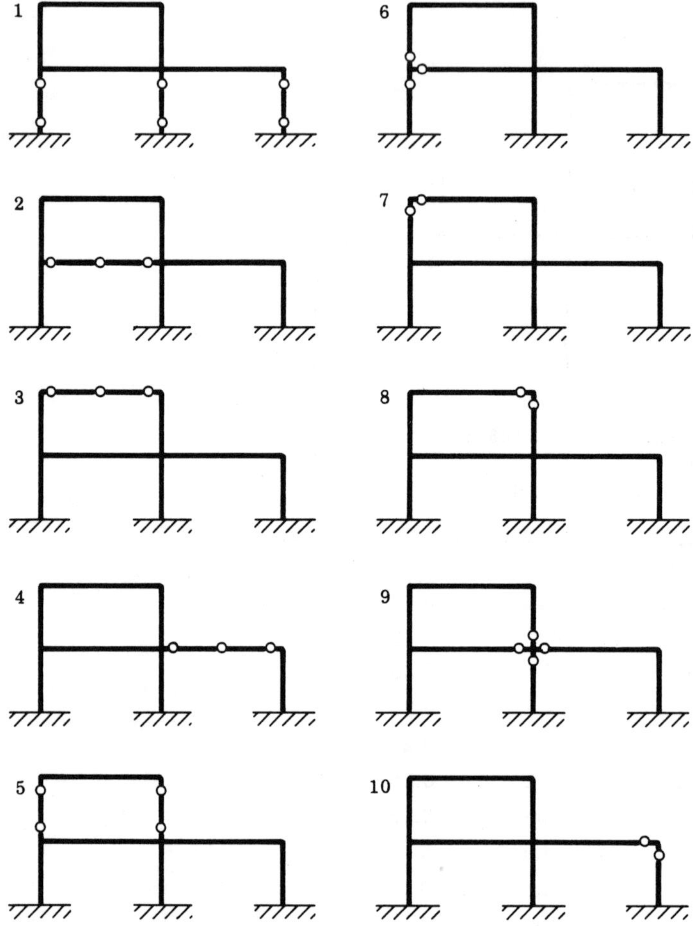

Figure 15. Set of fundamental mechanisms.

The reliability indices β_1, i = 1, . . . , 10 for the 10 fundamental mechanisms can be calculated from the safety margins taking into account the correlation between the yield moments. The result is shown in table 5.

With ϵ_1 = 0.50 the fundamental mechanisms 1, 2, 3, and 4 are selected as starting mechanisms in the β-unzipping and combined in turn with the remaining fundamental mechanisms. As an example consider the combination 1 + 6 of mechanisms 1 and 6. The linear safety margin M_{1+6} is obtained from the linear safety margins M_1 and M_6 by addition taking into account the signs of the coefficients. The corresponding reliability index is β_{1+6} = 3.74.

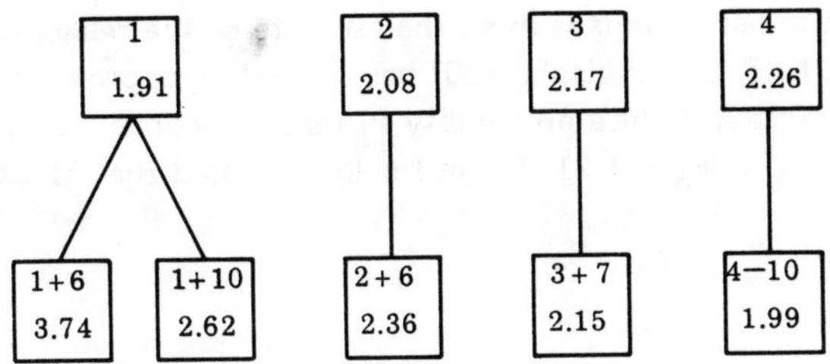

Figure 16. The first two rows in the failure tree.

With ϵ_2 = 1.20 the following new mechanisms $1 + 6, 1 + 10, 2 + 6, 3 + 7$, and $4 - 10$ are identified by this procedure. The failure tree at this stage is shown in figure 16. It contains 4 + 5 = 9 mechanisms. The reliability indices and the fundamental mechanisms involved are shown in the same figure.

This procedure is now continued as explained earlier by adding or subtracting fundamental mechanisms. If the procedure is continued 8 times (up to 10 fundamental mechanisms in one mechanism) and if the significant mechanisms are selected by ϵ_3 = 0.31, then the system modelling at mechanism level will be a series system where the elements are 12 parallel systems. These 12 parallel systems (significant mechanisms) and corresponding reliability indices are shown in table 6. The correlation matrix is

$$\bar{\bar{\rho}} = \begin{bmatrix} 1.00 & 0.65 & 0.89 & 0.44 & 0.04 & 0.91 & 0.59 & 0.97 & 0.87 & 0.81 & 0.36 & 0.92 \\ 0.65 & 1.00 & 0.55 & 0.00 & 0.09 & 0.67 & 0.00 & 0.63 & 0.54 & 0.58 & 0.00 & 0.71 \\ 0.89 & 0.55 & 1.00 & 0.35 & 0.45 & 0.83 & 0.61 & 0.88 & 0.98 & 0.95 & 0.31 & 0.83 \\ 0.44 & 0.00 & 0.35 & 1.00 & 0.00 & 0.03 & 0.00 & 0.45 & 0.37 & 0.04 & 0.89 & 0.08 \\ 0.04 & 0.09 & 0.45 & 0.00 & 1.00 & 0.05 & 0.00 & 0.04 & 0.44 & 0.48 & 0.00 & 0.05 \\ 0.91 & 0.67 & 0.83 & 0.03 & 0.05 & 1.00 & 0.73 & 0.87 & 0.80 & 0.88 & 0.00 & 0.98 \\ 0.59 & 0.00 & 0.61 & 0.00 & 0.00 & 0.73 & 1.00 & 0.58 & 0.60 & 0.65 & 0.00 & 0.64 \\ 0.97 & 0.63 & 0.88 & 0.45 & 0.04 & 0.87 & 0.58 & 1.00 & 0.90 & 0.77 & 0.48 & 0.87 \\ 0.87 & 0.54 & 0.98 & 0.37 & 0.44 & 0.80 & 0.60 & 0.90 & 1.00 & 0.90 & 0.41 & 0.78 \\ 0.81 & 0.58 & 0.95 & 0.04 & 0.48 & 0.88 & 0.65 & 0.77 & 0.90 & 1.00 & 0.00 & 0.88 \\ 0.36 & 0.00 & 0.31 & 0.89 & 0.00 & 0.00 & 0.00 & 0.48 & 0.41 & 0.00 & 1.00 & 0.00 \\ 0.92 & 0.71 & 0.83 & 0.08 & 0.05 & 0.98 & 0.64 & 0.87 & 0.78 & 0.88 & 0.00 & 1.00 \end{bmatrix}$$

The probability of failure P_f for the series system with the 12 significant mechanisms as elements can then be estimated by the usual techniques. The Ditlevsen bounds are

$$0.08646 \leqslant P_f \leqslant 0.1277$$

If the average value of the lower and upper bounds is used, the estimate of the reliability index β_S at mechanism level is $\beta_S = 1.25$. Hohenbichler [20] has derived an approximate method to calculate estimates for the system failure probability P_f (and the corresponding reliability index β_S). The estimate of β_S is $\beta_S = 1.21$. It can finally be noted that Monte Carlo simulation gives $\beta_S = 1.20$.

No.	Significant mechanisms	β
1	$1 + 6 + 2 + 5 + 7 + 3 - 8$	1.88
2	1	1.91
3	$1 + 6 + 2 + 5 + 7 + 3 + 4 - 8 - 10$	1.94
4	$3 + 7 - 8$	1.98
5	$4 - 10$	1.99
6	$1 + 6 + 2$	1.99
7	2	2.08
8	$1 + 6 + 2 + 5 + 7 + 3 + 8$	2.09
9	$1 + 6 + 2 + 5 + 7 + 3 + 8 + 9 + 4 - 10$	2.11
10	$1 + 6 + 2 + 5 + 9 + 4 - 10$	2.17
11	3	2.17
12	$1 + 6 + 2 + 5$	2.18

Table 6

8. STRUCTURAL OPTIMIZATION AND SYSTEM RELIABILITY

In *classical structural optimization* for trussed and framed structures the *design variables* are usually the cross-sectional areas A_i, i = 1, . . . , k, where k is the number of structural elements, so that each structural member is characterized by only one number. This is fully satisfactory for trussed structures. However, when bending occurs in the structural members, the plastic resistance moments W_i, i = 1, . . . , k, and the second moments of area I_i, i = 1, . . . , k are important. To maintain the great advantage of having only one design variable for each structural member it is assumed that

$$W_i = k_1 A_i^{3/2} \quad \text{and} \quad I_i = k_2 A_i^2 \tag{30}$$

where k_1 and k_2 are constants.

A natural choice of *objective function* for a structure is the cost of the structure, not only the cost of the structural material used, but also fabrication, transportation, etc. It is often assumed that the total cost of a structure with a satisfactory degree of accuracy is proportional to the weight of the structure. For structures where only one material (steel, concrete, etc.) is used the weight is proportional to

$$W = \sum_{i=1}^{k} \ell_i A_i \tag{31}$$

where ℓ_i is the length of structural member i.

The *constraints* in classical structural optimization signify that the stresses should everywhere be smaller than some prescribed values and likewise that the displacements should be smaller than some prescribed values. Further that all cross-sectional areas are greater than or equal to zero. The stresses and the displacements will for statically indeterminate structures depend on the areas. Therefore, the constraints can be given by

$$g_j(\overline{A}) \geqslant 0 \qquad j = 1, \ldots, m$$

$$A_i \geqslant 0 \qquad i = 1, \ldots, k \tag{33}$$

where $\overline{A} = (A_1, \ldots, A_k)$ and $g_j, j = 1, \ldots, m$, are functions expressing stresses and displacements.;

The classical optimization problem can be formulated in the following way. Determine the cross-sectional areas A_i, $i = 1, \ldots, k$ so that the weight function $W(\overline{A})$ is minimum under the constraints (32) and (33).

In classical structural optimization structural reliability is not taken into account. If a structure has to be designed from a reliability point of view at least two different structural optimization problems can be formulated. The first one is concerned with *element reliability*. In this case the constraints (32) are replaced by

$$\beta_i(\overline{A}) \geqslant \beta_i^0 \qquad i = 1, \ldots, n \tag{34}$$

where β_i is the reliability index for failure element i and β_i^0 is the corresponding target value. n is the number of failure elements.

The second optimization problem is related to *system reliability*. In this case the constraints (32) are replaced by only one constraint, namely

$$\beta_S(\overline{A}) \geqslant \beta_S^0 \tag{35}$$

where β_S is the system reliability index and β_S^0 the corresponding target value. In some situations it might be useful to combine the constraints (34) and (35). Note that use of (35) instead of (32) reduces the number of constraints drastically. So in this sense the optimization problem is simpler. However, calculation of the system reliability index β_S is now the main problem.

The structural system reliability optimization problem can be formulated in the following way. Determine the cross-sectional areas A_i, $i = 1, \ldots, n$ for all failure elements so that the weight function $W(\overline{A})$ is minimum under the constraints (35) and (33). Clearly, the well-known optimization methods can also be used for this optimization problem and the β-unzipping method can be used in calculating the system reliability index β_S. From a computational point of view, calculation of β_S is expensive. When solving the optimization problem it is therefore important to reduce the number of recalculations of β_S.

The following optimization procedure has been used with some success although a number of problems have still not been completely solved.

1)   Choose some starting values for the cross-sectional areas.

2)   Identify significant failure modes by the β-unzipping method and calculate equivalent safety margins for the failure modes.

3)   Find a suboptimal design by an optimization method, e.g. SUMT (Sequential Unconstrained Minimization Technique) without updating the system of failure modes and the directional cosinus of the equivalent safety margins. The system reliability index is calculated (evaluated) during this suboptimal design by a simple approximate method.

4)   Check the convergence by a suitable stopping rule. If the stopping rule is satisfied, go to 5), if not, go to 2).

5)   Repeat 2), 3), and 4), but use a more accurate method to evaluate β_S. When the stopping rule is satisfied, the optimization procedure is stopped.

This procedure has been used in a number of examples with different system levels. Due to the updating of significant failure modes some convergence problems occur, namely when new failure modes are introduced. This can be avoided if updating takes place at each iterative step. However, such a procedure is much too expensive. Research in this field is still going on, so only some tentative conclusions can be made by now. One conclusion is that the penalty function methods seem to be suitable for this kind of problem. Further, that the updating problem mentioned above seems to be negligible at level 1 and at mechanism level. By choosing the different β-unzipping parameters in an appropriate way the convergence can be improved. It seems also to be an approvement to use A_i^{-1} as design variables. A more detailed description of this procedure is given with examples in Thoft-Christensen & Sørensen [21].

REFERENCES

[1]   Thoft-Christensen, P.: *The β-unzipping method.* Institute of Building Technology and Structural Engineering, Aalborg University Centre, Aalborg, Report 8207, 1982 (not published).

[2]   Thoft-Christensen, P.: *Reliability Analysis of Structural Systems by the β-unzipping Method.* Institute of Building Technology and Structural Engineering, Aalborg University Centre, Aalborg, Report 8401, March 1984.

[3]   Thoft-Christensen, P. & Sørensen, J. D.: *Calculation of Failure Probabilities of Ductile Structures by the β-unzipping Method.* Institute of Building Technology and Structural Engineering, Aalborg University Centre, Aalborg, Report 8208, 1982.

[4]   Thoft-Christensen, P. & Sørensen, J. D.: *Reliability Analysis of Elasto-Plastic Structures*. Proc. 11. IFIP Conf. »System Modelling and Optimization», Copenhagen, July 1983. Springer-Verlag, 1984, pp. 556-566.

[5]   Thoft-Christensen, P. & Baker, M. J.: *Structural Reliability and Its Applications*. Springer-Verlag, Berlin-Heidelberg-New York, 1982.

[6]   Rackwitz, R. & Fiessler, B.: *An Algorithm for Calculation of Structural Reliability under Combined Loading*. Berichte zur Sicherheitstheorie der Bauwerke, Lab. f. Konstr. Ingb., Tech. Univ. München, München, 1977.

[7]   Ferregut-Avila, C. M.: *Reliability Analysis of Elasto-Plastic Structures*. NATO Advanced Study Institute, P. Thoft-Christensen (ed.), Martinus Nijhoff, The Netherlands, 1983, pp. 445-451.

[8]   Moses, F.: *Structural System Reliability and Optimization*. Computers & Structures, Vol. 7, 1977, pp. 283-290.

[9]   Gorman, M. R.: *Reliability of Structural Systems*. Report No. 79-2, Case Western Reserve University, Ohio, Ph.D. Report, 1979.

[10]   Ma, H.-F. & Ang, A. H.-S.: *Reliability Analysis of Ductile Structural Systems*. Civil Eng. Studies, Structural Research Series No. 494, University of Illinois, Urbana, August 1981.

[11]   Klingmüller, O.: *Anwendung der Traglastberechnung für die Beurteilung der Sicherheit von Konstruktionen*. Forschungsberichte aus dem Fachbereich Bauwesen. Gesamthochschule Essen, Heft 9, Sept. 1979.

[12]   Murotsu, Y., Okada, H., Yonezawa, M., Kishi, M.: *Identification of Stochastically Dominant Failure Modes in Frame Structures*. Proc. Int. Conf. on Appl. Stat. and Prob. in Soil and Struct. Eng., Universita di Firenze, Italia. Pitagora Editrice 1983, pp. 1325-1338.

[13]   Kappler, H.: *Beitrag zur Zuverlässigkeitstheorie von Tragwerken unter Berücksichtigung nichtlinearen Verhaltens*. Dissertation, Technische Universität München, 1980.

[14]   Ditlevsen, O.: *Narrow Reliability Bounds for Structural Systems*. Journal of Structural Mechanics, Vol. 7, No. 4, 1979, pp. 453-472.

[15]   Kounias, E. G.: *Bounds for the Probability of a Union, with Applications*. The Annals of Mathematical Statistics, Vol. 39, 1968, pp. 2154-2158.

[16]   Dunnett, C. W. L. & Sobel, M.: *Approximations to the Probability of Integral and Certain Percentage Points of a Multivariate Analogue of Student's t-distribution*. Biometrika, Vol. 42, pp. 258-260.

[17]   Thoft-Christensen, P. & Sørensen, J. D.: *Reliability of Structural Systems with Correlated Elements*. Applied Mathematical Modelling. Vol. 6, 1982, pp. 171-178.

[18]   Gollwitzer, S. & Rackwitz, R.: *Equivalent Components in First-Order System Reliability*. Reliability Engineering, Vol. 5, 1983, pp. 99-115.

[19]   Watwood, V. B.: *Mechanism Generation for Limit Analysis of Frames.* ASCE, Journal of the Structural Division, Vol. 105, No. ST1, Jan. 1979, pp. 1-15.

[20]   Hohenbichler, M.: *An Approximation to the Multivariate Normal Distribution Function.* Proc. 155th Euromech on Reliability of Structural Eng. Systems, Lyngby, Denmark, June 1982, pp. 79-110.

[21]   Thoft-Christensen, P. & Sørensen, J. D.: *Optimization and Reliability of Structural Systems.* NATO ASI on Computational Mathematical Programming, July 23 to August 2, 1984, Bad Windsheim, Germany F. R.

STOCHASTIC MODELING OF FATIGUE CRACK GROWTH

Henrik O. Madsen
A.S.Veritas Research
P.O.Box 300
N-1322 Høvik
Norway

ABSTRACT. A stochastic model for stable fatigue crack growth is proposed for constant amplitude loading and extended to variable amplitude loading. The model is based on a linear elastic fracture mechanical description of crack growth and is developed with consideration to experimentally observed results. Statistical estimation of model parameters is discussed. The model is combined with a first or second order reliability method to give the reliability against crack growth beyond a critical crack size or against brittle fracture.

1. INTRODUCTION

In many metallic structures flaws are inherent due to e.g. notches, welding defects and voids. Macro cracks can originate from these flaws and under time varying loading grow to a critical size causing catastrophic failure. The conditions governing the fatigue crack growth are the geometry of the structure and crack initiation site, the material characteristics, the environmental conditions and the loading. In general, these conditions are of random nature. The appropriate analysis and design methodologies should therefore be based on probabilistic methods.

In recent years considerable research efforts have been reported on probabilistic modeling of fatigue crack growth based on a fracture mechanics approach. In particular, stable crack growth has been studied and it is this aspect that will be covered here. These notes present a stochastic model for the crack growth phase for which linear elastic fracture mechanics (LEFM) is applicable. The model is first formulated for constant amplitude loading and then extended to variable amplitude loading. The characteristics of the model are compared with results from experimental tests. Estimation of parameters in the crack growth model is discussed and the distribution of statistical estimators is analyzed. The crack growth model can be used to predict future crack growth from observations. This is particularly useful for formulation of inspection and repair programs and for formulation of criteria for damage tolerant design. The probability that the crack length exceeds a critical size during some time interval is of interest. It is demonstrated how this probability can be calculated by a first or second order reliability method (FORM or SORM) both for an individual crack and for a system of independent cracks.

113

A. C. Lucia (ed.), Advances in Structural Reliability, 113–137.
© *1987 by ECSC, EEC, EAEC, Brussels and Luxembourg.*

2. CRACK GROWTH MODELS

2.1. Crack Growth under Constant Amplitude Loading

In a fracture mechanical analysis the fatigue damage under time varying loading is measured by the size of the dominant crack. Mathematical functions for the crack propagation rate have been proposed, which are of the general form, see e.g. [1]

$$\frac{da}{dt} = \eta_1(a,K,\Delta K,S,R) \tag{1}$$

where a is the crack size at time t, η_1 is a non-negative function, K is the stress intensity factor and ΔK is the stress intensity factor range, S is the far-field stress range and R is the stress ratio for the far-field stress. Thus $R = \sigma_{min}/\sigma_{max}$, where σ_{min} and σ_{max} are the minimum and maximum values of the far-field stress in a load cycle.

The crack size is described as a scalar, e.g. the crack depth of a surface crack or the crack length of a through crack. Surface cracks are generally assumed to be semi-elliptical and thus described by the crack depth a and the surface length $2c$. In Eq.(1) a should therefore ideally be a vector leading to a set of coupled first order differential equations. This has been done in [2] and compared to test results. The test results indicate that the ratio a/c independent of its initial value tends to a fixed function of a. In these notes the crack size is described as a scalar with the ratio a/c being constant. This facilitates the presentation and is sufficient in most cases.

Eq.(1) is here only applied in a simplified form

$$\frac{da}{dN} = \eta_2(a,\Delta K,\sigma_m) \tag{2}$$

where time is now measured in terms of stress cycles N, and where σ_m is the average far-field stress in a stress cycle.

$$\sigma_m = \frac{\sigma_{min}+\sigma_{max}}{2} = \frac{1}{2}\frac{1+R}{1-R}S \tag{3}$$

Compared to Eq.(1) the dependence on S is only included in the dependence on ΔK. Similarly, the dependence on K is only included in the dependence on σ_m. σ_m can include residual stresses from manufacturing which can vary over the thickness. Only the case of a constant σ_m over the thickness is included here.

A simple relation which is sufficient for most purposes is the Paris' law

$$\frac{da}{dN} = C(\Delta K)^m \quad , \qquad \Delta K > 0 \tag{4}$$

This relation was proposed in [3] based on test results. The relation is also the result of various mechanical and energy based models, see e.g. [1,3]. The equation is used for all $\Delta K > 0$, i.e. no threshold value is considered. C and m are material constants and C can also depend on the average stress value in a stress cycle. For some materials the constant C is independent of the average stress. For other materials experimental results indicate that a change in the average stress results in a parallel shift of the $da/dN - \Delta K$ curve on a double logarithmic diagram, see Figure 1. The effect of the average far-field stress is therefore introduced in Eq.(4) in the following simple form

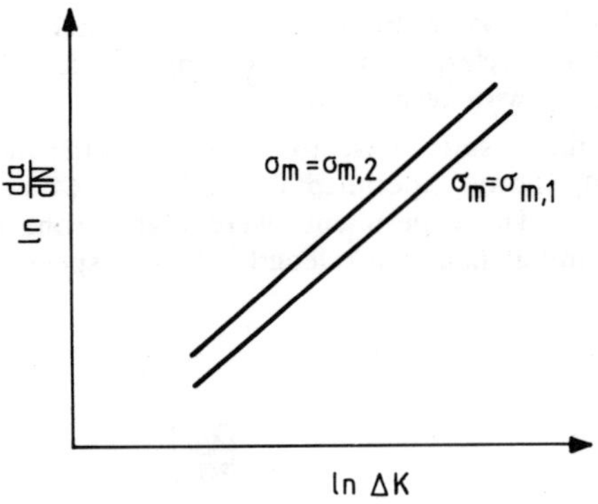

Figure 1. $da/dN - \Delta K$ relations.

$$\frac{da}{dN} = C\, h(\sigma_m)(\Delta K)^m \quad , \qquad \Delta K > 0 \tag{5}$$

where $h(\ )$ is a monotonic function and C is a true material constant independent of the loading.

The stress intensity factor is calculated by linear elastic fracture mechanics theory. ΔK is expressed in the form

$$\Delta K = Y(a)\, S\, \sqrt{\pi a} \tag{6}$$

where $Y(a)$ is a factor that depends on geometry, including the crack geometry and size. Inserting Eq.(6) in Eq.(5) yields

$$\frac{da}{dN} = C\, h(\sigma_m)\, Y(a)^m\, S^m\, (\sqrt{\pi a})^m \tag{7}$$

The solution to the differential equation is obtained by separating the variables and integrating.

$$\psi(a) = C\, h(\sigma_m)\, S^m\, N \tag{8}$$

where the damage function $\psi(a)$ is

$$\psi = \psi(a) = \int_{a_0}^{a} \frac{dz}{Y(z)^m\, (\sqrt{\pi z})^m} \tag{9}$$

and a_0 is the initial crack size. It follows from Eq.(8) that if failure is defined by the crack size exceeding a critical value a_c, then the following equation is valid at failure

$$N\, S^m = \frac{\psi(a_c)}{C\, h(\sigma_m)} = constant \tag{10}$$

This relation is in agreement with the $S - N$ relation generally applied in fatigue calculations. One way to define a damage index, D, in terms of the crack size is

$$D = \frac{\psi(a)}{\psi(a_c)} \tag{11}$$

Using this definition it follows from Eq.(8) that damage increases linearly from 0 to 1 with the number of stress cycles. The crack length after N stress cycles, a_N, is obtained by solving Eqs.(8,9) with respect to a.

Numerous experimental results exist for crack growth under constant amplitude loading. Figure 2 from [4] shows experimental results for 64 center cracked specimens made of 2024-T3 aluminum. The experiments were highly controlled and performed by the same laboratory. The initial half crack length of each specimen was $a_0 = 9\,mm$.

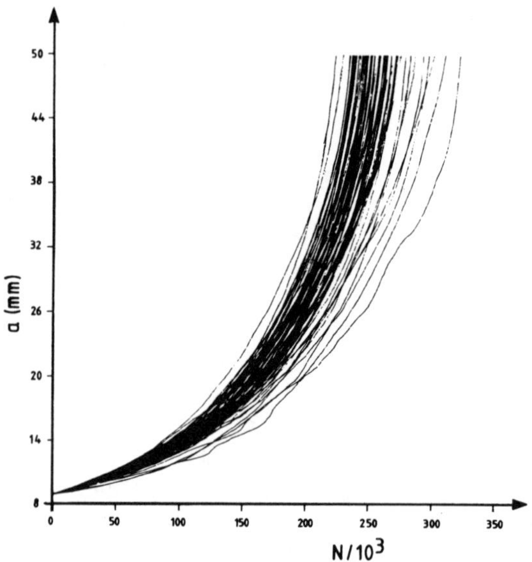

Figure 2. Experimental results.

Let the geometry factor be taken corresponding to an infinite plate, i.e. $Y(a) = 1$. The crack length after N stress cycles is obtained from Eq.(8) as

$$a(N) \;=\; \begin{cases} [a_0^{\frac{2-m}{2}} + \dfrac{2-m}{2} C\,h(\sigma_m)\,\pi^{m/2}\,S^m\,N]^{\frac{2}{2-m}} & m \neq 2 \\[2ex] a_0 \exp(C\,h(\sigma_m)\,\pi\,S^2\,N) & m = 2 \end{cases} \qquad (12)$$

Figure 3 shows the crack length as a function of N for fixed values of C, σ_m and a_0. Although the experimental curves resemble this curve, they are all different, they are slightly irregular and they show a good deal of intermingling. In the experiments the only non-deterministic factor in Eq.(12) is the material constant C. An attempt is therefore made to 'randomize' C in such a way that sample functions become similar to those experimentally obtained, [4-9].

First, let C be a random variable varying independently from specimen to specimen. Sample functions then appear as in Figure 4. The general behavior is still correct and different sample functions are obtained. The sample curves are, however, quite smooth and no intermingling takes place.

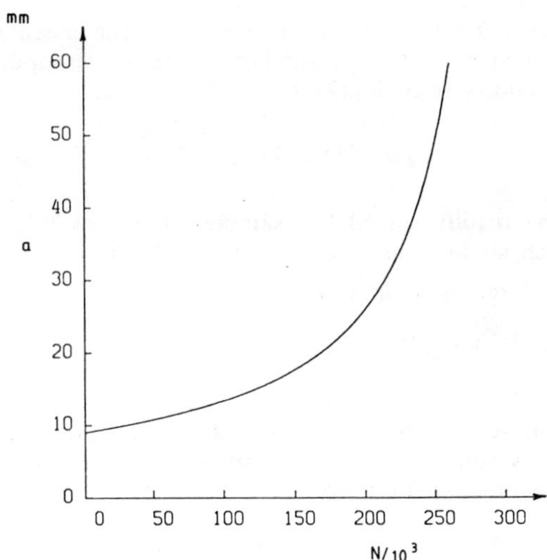

Figure 3. *C* deterministic.

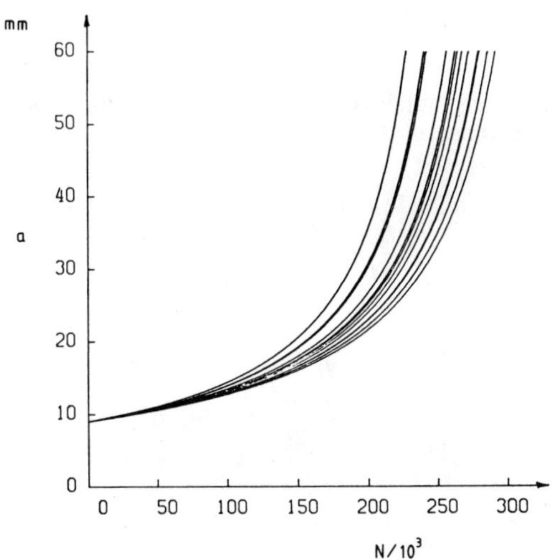

Figure 4. *C* random variable.

Second, C is written as

$$C = C(a) = C_1 + C_2(a) \tag{13}$$

Here C_1 is a random variable describing random variations between mean values in different specimens, while $C_2(a)$ is a random process describing variations from the mean value along the crack path within each specimen. The mean value of $C_2(a)$ is zero and the process is assumed to be stationary. The corresponding stochastic differential equation for a follows from Eq.(7) as

$$\frac{da}{dN} = (C_1 + C_2(a))\, h(\sigma_m)\, Y(a)^m\, S^m\, (\sqrt{\pi a}\,)^m \tag{14}$$

This equation may not be readily solved but sample curves for a can be obtained by simulation. Results of such simulations resemble the results of Figure 2 well.

An alternative way of representing C in Eq.(13) is

$$C = C_1 + C_2(N) \tag{15}$$

Since C varies with a it also varies with $N = N(a)$ and the representation in Eq.(15) is relevant. $C_2(N)$ is a mean zero stochastic process. $C_2(N)$ is further taken as a stationary process although this is somewhat contradictory with the assumption of stationarity of $C_2(a)$. The covariance function of $C_2(N)$ is

$$Cov[C_2(N), C_2(N+n)] = E[C_2(N)\, C_2(N+n)] = R_{22}(n) \tag{16}$$

and a scale of fluctuation can be defined as

$$N_{fl} = \frac{\displaystyle\int_0^\infty n\mid R_{22}(n)\mid dn}{\displaystyle\int_0^\infty \mid R_{22}(n)\mid dn} \tag{17}$$

Figure 5 shows sample functions obtained by simulation for the model in Eq.(15). The sample curves now resemble those of Figure 2 very well.

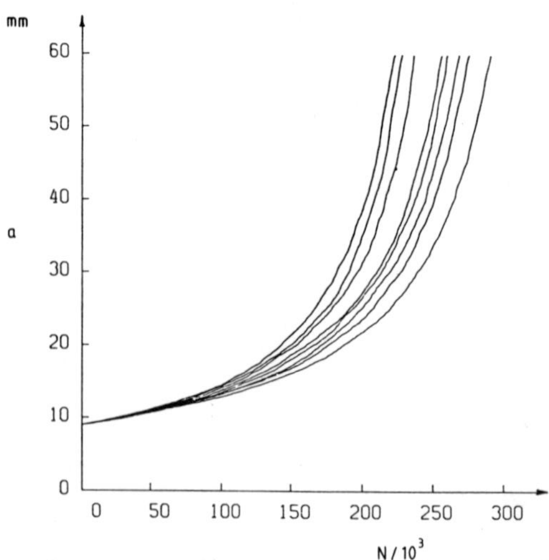

Figure 5. C_i random sequence.

In [8,9] C_1 is a fixed quantity. Sample curves obtained with a scale of fluctuation which gives realistic spread in a for large N, appears to be too irregular. Although the modeling of C_1 as a random variable makes parameter estimation more difficult it is recommended here.

2.2. Approximation of $a(t)$ by a Markov Process

With C from Eq.(15), the differential form of Eq.(8) is

$$d\psi = [C_1 + C_2(N)] h(\sigma_m) S^m \, dN \tag{18}$$

This equation has been analyzed in [8,9] for fixed C_1. This section is based on the analysis in [8,9]. The results are thus based on a given value of C_1. If the scale of fluctuation of $C_2(N)$ is short compared to the characteristic scale of fluctuation of ψ, then ψ is close to a diffusive Markov process which is governed by an Ito's stochastic differential equation

$$d\psi(N) = m(\psi,N)dN + \sigma(\psi,N)dB(N) \tag{19}$$

where m is called the drift coefficient, σ the diffusion coefficient and $B(N)$ is the Wiener's process.

When the Markov approximation is justified, Stratonovich's stochastic averaging method provides the required formulas to compute the drift and diffusion coefficients. In the case of Eq.(18)

$$m = C_1 h(\sigma_m) S^m \tag{20}$$

$$\sigma^2 = (h(\sigma_m)S^m)^2 \, 2 \int_{-\infty}^{0} R_{22}(n)dn = (h(\sigma_m)S^m)^2 v^2 \tag{21}$$

where v^2 is the area under the covariance function

$$v^2 = \int_{-\infty}^{\infty} R_{22}(n)dn \tag{22}$$

The replacement of Eq.(18) by Eq.(19) amounts to substituting $C_2(N)$ by a white noise process. This introduces an error associated with the probability for $d\psi$, and thereby da, to become negative. This error is negligible so long as the tendency for drift (mean crack growth rate) dominates the tendency for diffusion (variation around the mean crack growth rate). The probability of an increment in one stress cycle being negative can thus be large but the probability that the accumulated increment for a large number of stress cycles is negative is very small.

The transition probability density $f(\psi,N \mid \psi_0,N_0)$ of the Markov process $\psi(N)$ is a conditional probability density which describes the distribution of $\psi(N)$ under the condition that $\psi(N_0)=\psi_0$ at an earlier value $N=N_0$. It is governed by the following Fokker-Planck equation

$$\frac{\partial f}{\partial N} + m\frac{\partial f}{\partial \psi} - \frac{1}{2}\sigma^2\frac{\partial^2 f}{\partial \psi^2} = 0 \tag{23}$$

subject to the initial condition

$$f(\psi,0 \mid 0,0) = \delta(\psi) \tag{24}$$

where $\delta(\ )$ denotes Dirac's delta function. The backward equation corresponding to Eq.(23) does not have a physical meaning as described in [9] but the backward equation is not needed in this approach.

The solution of Eq.(23) that satisfies the initial condition in Eq.(24) is

$$f(\psi,N \mid 0,0) = \frac{1}{\sqrt{2\pi}v \sqrt{N}h(\sigma_m)S^m} \exp(-\frac{(\psi - C_1 h(\sigma_m)S^m N)^2}{2v^2 N(h(\sigma_m)S^m)^2}) \tag{25}$$

i.e. a normal distribution with mean value $C_1 h(\sigma_m)S^m N$ and standard deviation $v\sqrt{N}h(\sigma_m)S^m$. The same result is obtained in [5,6] where $C_2(N)$ is modeled directly as a white noise process.

The corresponding conditional density for a is

$$f(a,N \mid a_0,0) \tag{26}$$

$$= \frac{1}{\sqrt{2\pi}v \sqrt{N}h(\sigma_m)S^m Y(a)^m(\sqrt{\pi a})^m} \exp(-\frac{(\int_{a_0}^{a} \frac{dz}{Y(z)^m(\sqrt{\pi z})^m} - C_1 h(\sigma_m)S^m N)^2}{2v^2 N(h(\sigma_m)S^m)^2})$$

The probability that the crack size a is less than a_0 is

$$\int_{-\infty}^{a_0} f(a,N \mid a_0,0)da = \int_{-\infty}^{0} f(\psi,N \mid 0,0)d\psi = \Phi(-\frac{C_1\sqrt{N}}{v}) \tag{27}$$

which is generally negligible for large values of N of interest. $\Phi(\ )$ denotes the standardized normal distribution function.

If a crack size a_1 is observed after N_1 stress cycles then due to the Markovian property

$$f(a,N \mid a_1,N_1,a_0,0) = f(a,N \mid a_1,N_1) \tag{28}$$

$$= \frac{1}{\sqrt{2\pi}v \sqrt{N-N_1}h(\sigma_m)S^m Y(a)^m(\sqrt{\pi a})^m} \exp(-\frac{(\int_{a_1}^{a} \frac{dz}{Y(z)^m(\sqrt{\pi z})^m} - C_1 h(\sigma_m)S^m(N-N_1))^2}{2v^2(N-N_1)(h(\sigma_m)S^m)^2})$$

In an incremental form Eq.(18) can be written as

$$\Delta\psi_i = (C_1 + C_{2i})h(\sigma_m)S^m \tag{29}$$

where $\Delta\psi_i$ is the increment of ψ in the ith stress cycle and where C_{2i} is a white noise sequence, i.e a sequence of mutually independent and identically distributed random variables. The distribution of C_{2i} is normal with zero mean value and with variance v^2. In terms of a this equation is

$$\Delta a_i = (C_1 + C_{2i})Y(a_i)^m h(\sigma_m)S^m(\sqrt{\pi a_i})^m \tag{30}$$

Application of these equations leads to the same distribution for a as in Eq.(26), [6].

2.3. Crack Growth under Variable Amplitude Loading

The model in Eq.(30) can be directly extrapolated to variable amplitude loading where the appropriate values of σ_m and S are inserted for each stress cycle. It must be emphasized that this is an extrapolation beyond experimental experience and that possible sequence effects are neglected.

The crack growth law is written as

$$\Delta a_i = (C_1 + C_{2i}) \, Y(a_i)^m \, h(\sigma_{m,i}) \, S_i^m (\sqrt{\pi a_i})^m \tag{31}$$

where S_i and $\sigma_{m,i}$ are the stress range and the average far-field stress in the ith stress cycle, respectively. The result for $\Delta \psi_i$ is analogous and the crack length a_N after N stress cycles is then given through

$$\psi(a_N) = \sum_{i=1}^{N} (C_1 + C_{2i}) \, h(\sigma_{m,i}) \, S_i^m \tag{32}$$

$$= C_1 W_1 + W_2$$

where the loading history is summarized in two random variables W_1 and W_2.

$$W_1 = \sum_{i=1}^{N} h(\sigma_{m,i}) \, S_i^m \tag{33}$$

$$W_2 = \sum_{i=1}^{N} C_{2i} \, h(\sigma_{m,i}) \, S_i^m \tag{34}$$

For a large fixed N and for a general class of load processes the distribution of (W_1, W_2) will be approximately joint normal. The expected values, variances and the covariance are

$$E[W_1] = \sum_{i=1}^{N} E[h(\sigma_{m,i}) \, S_i^m] \tag{35}$$

$$E[W_2] = 0 \tag{36}$$

$$Var[W_1] = \sum_{i=1}^{N} \sum_{j=1}^{N} Cov[h(\sigma_{m,i}) \, S_i^m, h(\sigma_{m,j}) \, S_j^m] \tag{37}$$

$$Var[W_2] = \sum_{i=1}^{N} \sum_{j=1}^{N} Cov[C_{2i} \, h(\sigma_{m,i}) \, S_i^m, C_{2j} \, h(\sigma_{m,j}) \, S_j^m] \tag{38}$$

$$Cov[W_1, W_2] = 0 \tag{39}$$

In the next section, these statistics are derived for stationary narrow band Gaussian processes and for stationary Poisson pulse processes. For stationary narrow band processes σ_m is constant and for Poisson pulse processes only the case $h(\sigma_m) \equiv 1$ is considered. Eqs.(35,37,38) then reduce to

$$E[W_1] = N \, h(\sigma_m) \, E[S_i^m] \tag{40}$$

$$Var[W_1] = h(\sigma_m)^2 \sum_{i=1}^{N} \sum_{j=1}^{N} Cov[S_i^m, S_j^m] \tag{41}$$

$$= h(\sigma_m)^2 \left\{ N\, Var[S_i^m] + 2 \sum_{k=1}^{N-1} (N-k)\, Cov[S_i^m, S_{i+k}^m] \right\}$$

$$Var[W_2] = h(\sigma_m)^2 \sum_{i=1}^{N} \sum_{j=1}^{N} Cov[C_{2i} S_i^m, C_{2j} S_j^m] \tag{42}$$

$$= h(\sigma_m)^2\, N\, v^2 \left\{ E[S_i^m]^2 + Var[S_i^m] \right\}$$

2.4. Random Stress Processes

A *stationary Gaussian process* is completely characterized by the mean value and the covariance function

$$E[X(t)] = \mu_X \tag{43}$$

$$Cov[X(t+\tau), X(t)] = C_X(\tau) \tag{44}$$

The standard deviation is denoted by σ_X.

$$\sigma_X^2 = C_X(0) \tag{45}$$

The covariance function can be represented in terms of the one-sided non-negative spectral density function $S_X(\omega)$ as

$$C_X(\tau) = \int_0^{\infty} S_X(\omega) \cos(\omega\tau)\, d\omega \tag{46}$$

The spectral moments λ_j are defined as

$$\lambda_j = \int_0^{\infty} \omega^j S_X(\omega)\, d\omega \qquad j = 0,1,2, \cdots \tag{47}$$

The mean rate of upcrossings of the mean level is

$$v_0 = \frac{1}{2\pi} \sqrt{\lambda_2/\lambda_0} \tag{48}$$

and the mean rate of local maxima is similarly

$$v_M = \frac{1}{2\pi} \sqrt{\lambda_4/\lambda_2} \tag{49}$$

The regularity factor α is defined as the ratio

$$\alpha = \frac{v_0}{v_M} = \frac{\lambda_2}{\sqrt{\lambda_0 \lambda_4}} \quad ; \quad 0 < \alpha \leqslant 1 \tag{50}$$

The parameter α characterizes the bandwidth of the spectral density function with $\alpha \approx 1$ for a narrow band process. A bandwidth measure analogous to α is

$$\delta = \frac{\lambda_1}{\sqrt{\lambda_0 \lambda_2}} \quad ; \quad 0 < \delta \leqslant 1 \tag{51}$$

The crack growth is described in terms of stress cycles. When the process is narrow banded, there is only one local maximum between successive upcrossings of the mean level. A stress cycle is thus defined as the load history between two consequetive local maxima. The pulse shape in a stress cycle varies from cycle to cycle but this is not assumed to have any effect on the crack growth. The amplitude of a stress cycle is defined in terms of an envelope process. In short time intervals the sample curves are approximately sinusoidal around an equilibrium value with an amplitude called the envelope. An envelope process $R_1(t)$ is defined in [10]

$$R_1(t)^2 = (X(t) - \mu_X)^2 + \left(\frac{\dot{X}(t)}{2\pi v_0}\right)^2 \tag{52}$$

where $\dot{X}(t)$ is the time derivative of $X(t)$. $R_1(t)$ has a Rayleigh distribution and the moments are

$$E[R_1^m] = (\sqrt{2})^m \sigma_X^m \Gamma(1 + \frac{m}{2}) \tag{53}$$

where $\Gamma(\ )$ denotes the gamma function.

A definition similar to Eq.(52) which is less strict on the requirements of differentiability of the process is, [11]

$$R_2(t)^2 = (X(t) - \mu_X)^2 + (X(t)\hat{} - \mu_X)^2 \tag{54}$$

where a 'hat' denotes a Hilbert transform

$$X(t)\hat{} - \mu_X = \int_{-\infty}^{\infty} \frac{X(s) - \mu_X}{t - s} ds \tag{55}$$

The distribution of $R_2(t)$ is the same as the distribution of $R_1(t)$. In the definitions the equilibrium position is taken as the mean value. The concept of a double envelope process which defines at each time both a random amplitude and a random equilibrium position is defined in [12] and has been used in damage accumulation analyses in [6].

The covariance function of the mth power of the envelope process is needed in the evaluation of Eqs.(41,42). The joint probability density of $R_1(t_1)$ and $R_1(t_2)$ is approximately

$$f_{R_1(t_1),R_2(t_2)}(r_1, r_2) = \frac{r_1 r_2}{1 - k(t_2 - t_1)^2} I_0\left(\frac{k(t_2 - t_1) r_1 r_2}{1 - k(t_2 - t_1)^2}\right) \exp\left(-\frac{1}{2} \frac{r_1^2 + r_2^2}{1 - k(t_2 - t_1)^2}\right) \tag{56}$$

where $I_0(\ )$ is the modified Bessel function of order zero, and

$$k(t_2 - t_1)^2 = \frac{1}{\sigma_X^2}\left[C_X(t_2 - t_1)^2 + \left(\frac{C_X'(t_2 - t_1)}{2\pi v_0}\right)^2\right] \tag{57}$$

Eq.(56) also applies for $R_2(t)$ but the definition of $k(t_2 - t_1)$ is

$$k(t_2 - t_1)^2 = \frac{1}{\sigma_X^2}\left[C_X(t_2 - t_1)^2 + \hat{C}_X(t_2 - t_1)^2\right] \tag{58}$$

The covariance function $C_{R_i^m}(\tau) = Cov[R_i(t+\tau)^m, R_i(t)^m]$, $i=1,2$ can then be determined directly from Eq.(56), [13]

$$C_{R_i^m}(\tau) = 2^m \sigma_X^{2m} \Gamma(1+\frac{m}{2}) (_2F_1(-\frac{m}{2},-\frac{m}{2};1;k(\tau)^2)-1) \tag{59}$$

where $_2F_1(\ )$ is the hypergeometric function. This is only an approximation due to the approximations leading to Eq.(56). Exact results are available for integer values of m. For $m=2$ and $R_1(t)$ the result is

$$C_{R_1^2}(\tau) = 2[\,C_X(\tau)^2 + 2\frac{C_X'(\tau)^2}{(2\pi v_0)^2} + \frac{C_X''(\tau)^2}{(2\pi v_0)^4}] \tag{60}$$

The stress ranges in Eqs.(40-42) are taken as twice the amplitudes defined through the envelopes. The mean values and covariances are then

$$E[S_i^m] = (2\sqrt{2})^m \sigma_X^m \Gamma(1+\frac{m}{2}) \tag{61}$$

$$Cov[S_i^m, S_j^m] = 4^m C_{R_1^m}(\frac{|\,i-j\,|}{v_0}) \tag{62}$$

where the mean distance between two stress cycles has been used as argument in the covariance function. The number of stress cycles N in a time period $[0,T]$ is taken as its mean value $v_0 T$.

A stochastic process

$$X(t) = \sum_{i=1}^{N(t)} w(t,\tau_i,S_i) \tag{63}$$

is called a *filtered Poisson process*, when $N(t)$ is a Poisson process, $\{S_i\}$ is a sequence of identically distributed and independent random variables and $w(t,\tau,s)$ is the response function which is defined to be zero for $t<\tau$, [14]. An intuitive interpretation of Eq.(63) is : if τ_i represents the time at which an event took place, then S_i represents the magnitude of a signal associated with the event - here the far-field stress range. $w(t,\tau_i,s)$ represents the value at time t of a signal of magnitude s originating at time τ_i and $X(t)$ represents the value at time t of the sum of signals arising from the events occurring in the interval $[0,t]$. Figure 6 shows some sample functions from filtered Poisson processes. Filtered Poisson processes are analyzed extensively in [13-15]. The analysis here is solely for 'sparse' filtered Poisson processes for which the possibility of pulse overlap can be ignored. The intensity of the Poisson process is denoted by μ. The expected value and covariances in Eqs.(40-42) for a time period T are, with $h(\sigma_m)\equiv 1$

$$E[W_1] = \mu T E[S_i^m] \tag{64}$$

$$Var[W_1] = \mu T (E[S_i^m]^2 + Var[S_i^m]) \tag{65}$$

$$Var[W_2] = \mu T v^2 (E[S_i^m]^2 + Var[S_i^m]) \tag{66}$$

2.5. Parameter Estimation in Paris' Law

The value of m is predicted from theoretical models as $m=2$ or $m=4$. Statistical

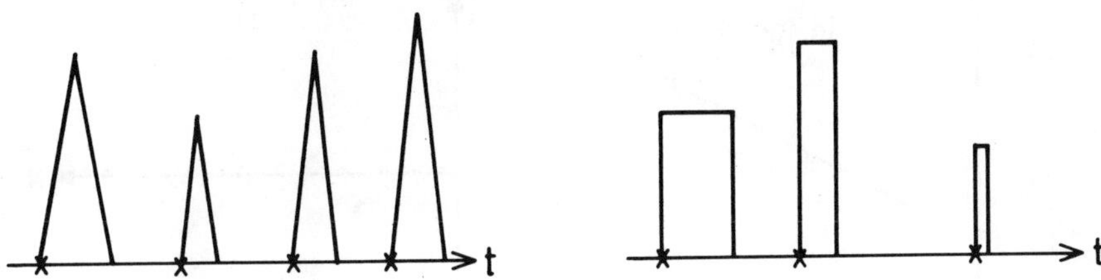

Figure 6. Sample functions from 'sparse' filtered Poisson processes.

estimation of m from experimental results generally results in other values and m should be treated as a random variable in addition to C.

Consider first the crack growth law in the original form of Eq.(14)

$$\frac{da}{dN} = C(\Delta K)^m \tag{67}$$

This equation is somewhat inconvenient with respect to dimensions and is therefore rewritten in the form

$$\frac{da}{dN} = C_0\left(\frac{\Delta K}{K_0}\right)^m \tag{68}$$

where K_0 is a fixed reference value of the same dimension as K. C_0 then has the same dimension as $\frac{da}{dN}$. The constant C in Eq.(67) is

$$C = C_0 K_0^{-m} \quad \Leftrightarrow \quad \ln C = \ln C_0 - m \ln K_0 \tag{69}$$

It follows that if the scatter in C_0 is negligible then $\ln C$ and m are linearly related. Otherwise $\ln C$ and m are expected to be negatively correlated. Several studies, e.g. [16], reports a high negative correlation between m and $\ln C$. This is also demonstrated in the study [17] where crack propagation data for 25 identical specimens under identical loading conditions have been collected. The least square estimates $\hat{m}$ and $\hat{C}$ for m and C have been computed for each specimen and a joint distribution for m and C has been estimated. The value C is representative of C_1 in Eq.(15) since an averaging has been carried out, see Figure 7. The 25 tests are summarized in Table 1 and the following sample statistics are obtained

$$\overline{m} = 2.85 \quad s_m = 0.284 \quad V_m = 0.100 \tag{70}$$

$$\overline{\ln C} = -20.164 \quad s_{\ln C} = 1.067 \quad r_{m,\ln C} = -0.971 \tag{71}$$

An 'over bar' denotes a sample mean, s denotes a sample standard deviation, V denotes a sample coefficient of variation and r denotes a sample correlation coefficient. A standard test does not reject an assumption of binormality for $(m,\ln C)$. Based on these results m and $\ln C$ are expressed in terms of two independent and standardized normal random variables U_1 and U_2 as

$$m = \hat{m} + 0.100\,\hat{m}\,U_1 \tag{72}$$

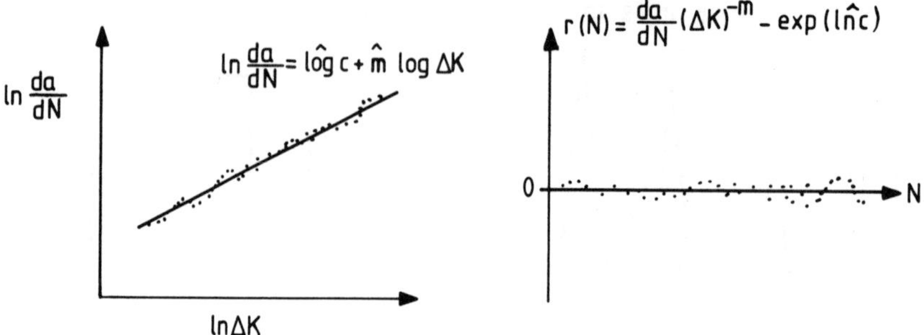

Figure 7. Illustration of parameter estimation.

	$\hat{m}$	$\ln \hat{C}$ $(C$ in $\dfrac{mm}{cycle}$ $(Mpa \sqrt{m})^{-m})$
1	2.728	-19.765
2	3.001	-20.702
3	2.887	-20.188
4	2.977	-20.868
5	2.380	-18.614
6	2.561	-19.090
7	2.853	-20.159
8	2.616	-19.250
9	2.928	-20.275
10	3.418	-22.360
11	2.934	-20.507
12	2.383	-18.260
13	3.163	-21.362
14	3.194	-21.592
15	3.053	-20.893
16	2.840	-20.046
17	2.608	-19.147
18	3.137	-21.206
19	2.420	-18.719
20	2.628	-19.360
21	2.572	-19.074
22	3.138	-21.244
23	2.671	-19.466
24	3.191	-21.421
25	3.000	-20.535

Table 1. Variability in m and C for 0.04% Carbon Steel (Plane Bending, Stress Ratio = 0)

$$\ln C = \ln \hat{C} + 1.067(-0.971\, U_1 + 0.239\, U_2) \tag{73}$$

thus giving a coefficient of variation of 10% for m, a standard deviation of 1.067 for $\ln C$ and a correlation coefficient of -0.971 between m and $\ln C$. The statistical uncertainty in the estimates in Eqs.(70,71) is thus ignored. In [18] a fatigue reliability analysis which includes this statistical uncertainty is reported.

When the crack growth law is in the form

$$\frac{da}{dN} = (C_1 + C_2(N))(\Delta K)^m \tag{74}$$

then least square estimates $\hat{C}$ and $\hat{m}$ for C and m are computed as before. Residuals

$$r(N) = \frac{da}{dN}(\Delta K)^{-\hat{m}} - \hat{C} \tag{75}$$

are then computed and v^2 is estimated by the residual variance. In practice, residuals are only given as average residuals over say n stress cycles. The estimate of v^2 is then the residual variance multiplied by n.

3. PROBABILISTIC DESIGN FOR FATIGUE CRACK GROWTH

3.1. Failure Criteria

Reliability problems for fatigue crack growth can be formulated as limit state problems and both serviceability and ultimate limit states can be defined. Two separate types of failure criteria can be envisioned, [20].

$$a_c - a \leqslant 0 \tag{76}$$

$$K_{IC} - K \leqslant 0 \tag{77}$$

In the first case a critical crack size a_c is selected perhaps based on serviceability considerations. In the second case failure occurs when the stress intensity factor K exceeds the fracture toughness K_{IC}. Then the crack growth becomes unstable and rapid failure occurs. Four cases should be considered corresponding to the two failure criteria and constant or variable amplitude loading.

Case 1: Crack through under constant amplitude loading

Case 2: Brittle fracture under constant amplitude loading

Case 3: Crack through under variable amplitude loading

Case 4: Brittle fracture under variable amplitude loading

3.2. Basic Variables

The uncertain parameters in the failure criteria are modeled in terms of the random *basic variables*. The selection of suitable models for the uncertainties must be done separately for each application, but some ideas shall be given here.

a_0 :   The initial crack size is modeled as a random variable. The distribution of a_0 is estimated from test results, see e.g. [19], or can possibly be directly related to the manufacturing, welding and control procedures. As mentioned in the introduction, the description of the initial defect size by a scalar is not always sufficient.

S : The far field stress range S is often uncertain due to uncertainties in the loading and in the global structural response. A representation of S as

$$S = I_S \, \hat{S} \tag{78}$$

is therefore relevant, with $\hat{S}$ being the best estimate for the stress range and I_S an uncertainty factor with mean value 1 and variance depending on the confidence in $\hat{S}$.

σ_m : The far field average stress is uncertain for the same reasons that S is uncertain and because of lack of knowledge of residual stresses. A suitable representation of σ_m can be

$$\sigma_m = I_R \, I_S \, \hat{\sigma}_m \tag{79}$$

where I_R is independent of I_S and represents uncertainties in the time independent stresses including the residual stresses.

C_1, m : The uncertainty representation of C_1 and m is dealt with in the previous section. The conditional distribution of C_1 is from Eqs.(72,73)

$$F_{C_1}(C_1 \mid m) = \Phi\left(\frac{1}{0.239}\left[\frac{\ln C_1 - \ln \hat{C}}{1.067} + 0.971\,\frac{m - \hat{m}}{0.100\hat{m}}\right]\right) \tag{80}$$

v : An estimate for v is given in the previous section. It appears relevant to model v as a random variable to account for uncertainties in the statistical estimate and uncertainties in the white noise process model. The mean value of v is taken as the estimate from the previous section and the standard deviation is selected according to the confidence in this estimate.

$h(\ )$: The dependence of the crack propagation rate on the average far-field stress is not known exactly but a possible model is

$$h(\sigma_m) = I_h \, \hat{h}(\sigma_m) \tag{81}$$

where $\hat{h}(\sigma_m)$ represents the best estimate and I_h is a model uncertainty factor.

$Y(a)$: The geometry factor is computed by linear fracture mechanics and is often expressed by simplified formulas. The uncertainty due to this simplification can be modeled as

$$Y(a) = \hat{Y}(a)(1 + I_Y \sigma_Y(a)) \tag{82}$$

Here $\hat{Y}(a)$ represents the value from the applied formula which is considered as the estimate. I_Y is a random variable with zero mean and unit variance. $\sigma_Y(a)$ expresses the confidence in $\hat{Y}(a)$ for various values of the crack size.

K_{IC} : The fracture toughness is modeled as a random variable taking into account uncertainties in the measurement of K_{IC} and in the failure criterion.

W_1, W_2 : The joint distribution of the loading derived random variables W_1 and W_2 is described in a previous section. The distribution of W_2 depends on v.

a : The distribution of the crack size a depends on several of the other random variables. For constant amplitude loading the conditional distribution function of a follows from Eq.(26)

$$F_{a_N}(a_N \mid a_0,S,\sigma_m,C_1,m,v,I_h,I_Y) = \Phi\left(\frac{\psi(a_N) - E[\psi(a_N)]}{D[\psi(a_N)]}\right) \quad (83)$$

with

$$\psi(a_N) = \int_{a_0}^{a_N} \frac{dz}{Y(z)^m (\sqrt{\pi z})^m} \quad (84)$$

$$E[\psi(a_N)] = C_1 h(\sigma_m) S^m N \quad (85)$$

$$D[\psi(a_N)] = v\, h(\sigma_m) S^m \sqrt{N} \quad (86)$$

For variable amplitude loading the crack length is directly expressed in terms of other basic variables through Eq.(32)

$$\psi(a) = C_1 W_1 + W_2 \quad (87)$$

3.3. Reliability Calculation

For each of the four cases a limit state function can be determined. The limit state function is a function of the basic variables, denoted $\mathbf{Z}$, and satisfies

$$g(\mathbf{z}) \begin{cases} < 0 & \text{for } \mathbf{z} \text{ in } \textit{failure set } F \\ = 0 & \text{for } \mathbf{z} \text{ on } \textit{limit state surface} \\ > 0 & \text{for } \mathbf{z} \text{ in } \textit{safe set } S \end{cases} \quad (88)$$

The space of possible realizations of the basic variables is thus divided into two sets called the *failure set F* and the *safe set S*. The separating surface is the *limit state surface,* see Figure 8 and [13]. The reliability is

$$P_R = P(g(\mathbf{Z}) > 0) \quad (89)$$

The limit state functions for the four cases are :

Case 1 : Crack through under constant amplitude loading.

$$g(a,a_0,S,\sigma_m,C_1,m,v,I_h,I_Y) = a_c - a(a_0,S,\sigma_m,C_1,m,v,I_h,I_Y) \quad (90)$$

Case 2 : Brittle fracture under constant amplitude loading.

$$g(a,a_0,S,\sigma_m,C_1,m,v,K_{IC},I_h,I_Y) = K_{IC} - Y(a)(\sigma_m + \tfrac{1}{2}S)\sqrt{\pi a} \quad (91)$$

with $a = a(a_0,S,\sigma_m,C_1,m,v,I_h,I_Y)$.

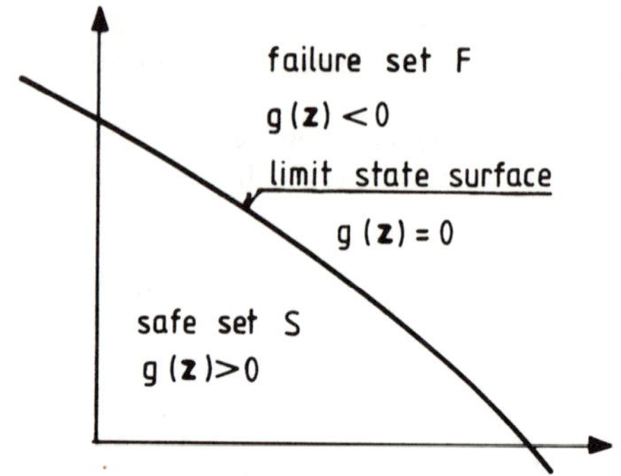

Figure 8. Limit state surface in z-space.

Case 3 : Crack through under variable amplitude loading.

$$g(a,a_0,S,\sigma_m,C_1,m,v,W_1,W_2,I_h,I_Y) = \psi(a) - C_1 W_1 - W_2 \tag{92}$$

Case 4 : Brittle fracture under variable amplitude loading.
Failure occurs if, see Eq.(91)

$$\sigma > \frac{K_{IC}}{Y(a)\sqrt{\pi a}} \tag{93}$$

where $\sigma = \sigma(t)$ is the far-field stress. This is illustrated in Figure 9. No failure occurs in the time period $[0,T]$ if the stress process $\sigma(t)$ does not cross over the time varying threshold $\xi(t) = K_{IC}/Y(a(t))\sqrt{\pi a(t)}$ in $[0,T]$.

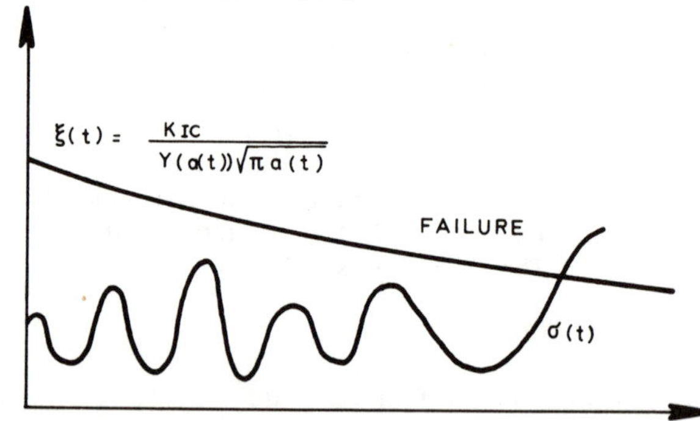

Figure 9. Illustration of failure event for brittle fracture under variable amplitude loading.

This probability is approximated by, see e.g. [13]

$$P_R = F_{T_f}(T) \approx F_{\sigma(0)}\left(\frac{K_{IC}}{Y(a_0)\sqrt{\pi a_0}}\right) \exp\left(-\frac{\int_0^T v_\sigma^+(\xi(t))dt}{F_{\sigma(0)}\left(\dfrac{K_{IC}}{Y(a_0)\sqrt{\pi a_0}}\right)}\right) \tag{94}$$

T_f is the random life time and $v_\sigma^+(\xi(t))$ is the mean upcrossing rate of the level $\xi(t)$ by the process $\sigma(t)$ at time t. This mean upcrossing rate is computed by Rice's formula, [21]

$$v_\sigma^+(\xi(t)) = \int_{\xi}^{\infty} (\dot\sigma - \dot\xi)\, f_{\sigma,\dot\sigma}(\xi(t),\dot\sigma)\, d\dot\sigma \tag{95}$$

where a 'dot' denotes a time derivative. In this application the time derivative $\dot\xi$ can be neglected. Eq.(95) therefore reduces to

$$v_\sigma^+(\xi(t)) = \int_{0}^{\infty} \dot\sigma\, f_{\sigma\dot\sigma}(\xi(t),\dot\sigma)\, d\dot\sigma \tag{96}$$

For a stationary Gaussian process Eq.(96) yields

$$v_\sigma^+(\xi(t)) = v_0 \exp\left(-\frac{1}{2}\left(\frac{\xi(t)-\mu_X}{\sigma_X}\right)^2\right) \tag{97}$$

and for a 'sparse' filtered Poisson process, the result is

$$v_\sigma^+(\xi(t)) = \mu(1 - F_S(\xi(t))) \tag{98}$$

The limit state function can be stated as

$$g(T_f,a_0,S,\sigma_m,C_1,m,v,W_1,W_2,K_{IC},I_h,I_Y) = T_f - T \tag{99}$$

The conditional distribution function for T_f is given in Eq.(94).

The reliability is computed by a first order or second order reliability method, FORM or SORM. The first step is to transform the set of basic variables into a set of independent and standardized normal variables U_i. The Rosenblatt transformation, [22], is used as suggested in [23]. Denoting the basic variables Z_i the transformation is written as

$$U_i = \Phi^{-1}(F_i(Z_i \mid Z_1,Z_2,..,Z_{i-1})) \qquad i=1,2,..,k \tag{100}$$

where k is the number of basic variables. $F_i(z_i \mid z_1,z_2,..,z_{i-1})$ is the distribution function of Z_i conditioned upon $(Z_1=z_1,Z_2=z_2,..,Z_{i-1}=z_{i-1})$.

$$F_i(z_i \mid z_1,z_2,..,z_n) = \frac{\int_{-\infty}^{z_i} f_{Z_1,Z_2,..,Z_{i-1},Z_i}(z_1,z_2,..,z_{i-1},t)\, dt}{f_{Z_1,Z_2,..,Z_{i-1}}(z_1,z_2,..,z_{i-1})} \tag{101}$$

For independent variables the transformation reduces to

$$U_i = \Phi^{-1}(F_{Z_i}(Z_i)) \tag{102}$$

An analysis shows, that for all four cases considered here, the necessary conditional distribution functions are directly available.

The limit state surface $g=0$ in u-space is illustrated in Figure 10. It divides the u-space into a safe set S and a failure set F. In a first order reliability analysis the limit state surface is approximated by its tangent hyperplane at the design point $\mathbf{u}^*$.

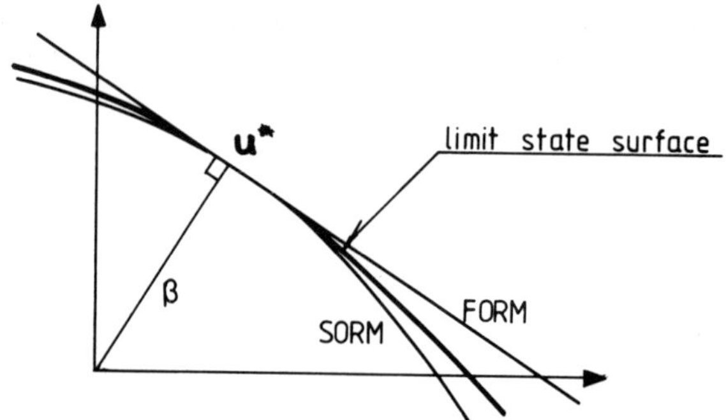

Figure 10. Limit state surface in u-space.

$\mathbf{u}^*$ is the point on the limit state surface closest to the origin, i.e. the point in the failure set with the largest probability density. In a second order reliability analysis the limit state surface is approximated by a hyperparaboloid surface with the same tangent hyperplane and curvatures at $\mathbf{u}^*$. The corresponding approximations to the reliability are

$$1 - P_R \approx \Phi(-\beta) \qquad\qquad FORM \qquad\qquad (103)$$

$$1 - P_R \approx \Phi(-\beta) \prod_{j=1}^{k-1}(1-\beta\kappa_j)^{-\frac{1}{2}} + \qquad SORM \qquad\qquad (104)$$

$$+ [\beta\Phi(-\beta)-\varphi(\beta)] \left\{ \prod_{j=1}^{k-1}(1-\beta\kappa_j)^{-\frac{1}{2}} - \prod_{j=1}^{k-1}(1-(\beta+1)\kappa_j)^{-\frac{1}{2}} \right\} +$$

$$+ (\beta+1)[\beta\Phi(-\beta)-\varphi(\beta)] \left\{ \prod_{j=1}^{k-1}(1-\beta\kappa_j)^{-\frac{1}{2}} - \mathrm{Re}\{ \prod_{j=1}^{k-1}(1-(\beta+i)\kappa_j)^{-\frac{1}{2}} \} \right\}$$

Here i is the imaginary unit, $\mathrm{Re}\{\ \}$ denotes the real part, β is the length of $\mathbf{u}^*$ and κ_j are the main curvatures at $\mathbf{u}^*$. The result for the second order reliability method is derived in [24]. The design point can be found by an iterative procedure, e.g. as described in [13].

3.4. An Example

As a simple illustration the formulas for crack growth derived in the preceding sections are combined with the failure criterion in Eq.(90). First order reliability methods (FORM) are applied and the probability of failure is computed. Only one very simple structure is considered as shown in Figure 11. It consists of a large panel with a center crack and loaded uniaxially perpendicular to the crack.

The geometrical parameter $Y(a)$ is equal to 1 independent of the crack length. For $h(\sigma_m) \equiv 1$ and $m = 2$ the combination of Eqs.(12,32) with Eq.(90) gives the failure criterion

$$a_c - a_0 \exp(C_1 \pi W_1 + \pi W_2) \leqslant 0 \qquad\qquad (105)$$

In the numerical calculations the critical crack size a_c is taken as 20 mm. The initial crack size a_0 is assumed normally distributed with expected value and variance as

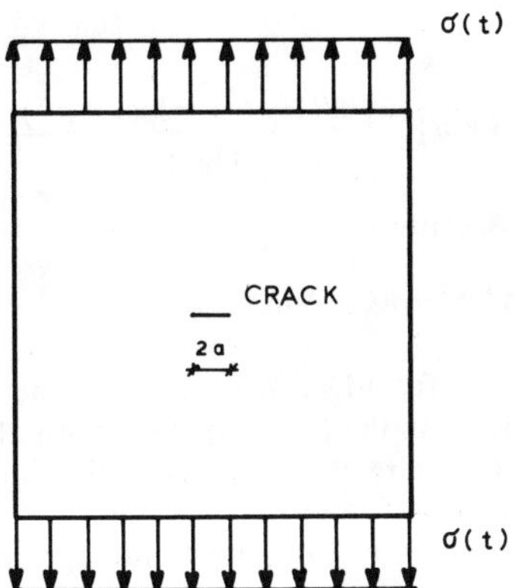

Figure 11. Panel with center crack.

$$E[a_0] = 2\,mm \quad , \quad Var[a_0] = 1\,mm^2 \tag{106}$$

The material parameters C_1 and C_{2i} are taken as normally distributed with expected values, variances and covariances as

$$E[C_1] = 1.0{\cdot}10^{-13} \; , \quad E[C_{2i}] = 0 \tag{107}$$

$$Var[C_1] = 9.0{\cdot}10^{-28} \; , \quad Var[C_{2i}] = 5\,Var[C_1] \tag{108}$$

$$Cov[C_1,C_{2i}] = 0 \; , \quad Cov[C_{2i},C_{2j}] = \rho^{\,|i-j|}\,Var[C_{2i}] \tag{109}$$

with $\rho = 0.9991$. The numerical values are based on the use of units N/mm^2 for stresses and mm for crack lengths. The model for the random sequence $\{C_{2i}\}$ is slightly more complicated than the model presented previously. The two models give essentially the same distribution of the crack length for large values of N, but in this model the probability of a negative crack increment in a stress cycle is negligible. The loading is assumed to be a constant amplitude loading with a period of 6 seconds and resulting in far-field stresses of stress range S. Due to uncertainties in the loading and the structural response, S is modeled as a random variable. The distribution is taken as a normal distribution with an expected value and variance of

$$E[S] = 60\,N/mm^2 \; , \qquad Var[S] = (10\,N/mm^2)^2 \tag{110}$$

The random variables W_1 and W_2 in Eq.(33) and Eq.(34) are for constant amplitude loading

$$W_1 = N\,S^2 \tag{111}$$

$$W_2 = \sum_{i=1}^{N} C_{2i}\,S^2 = C_2(N)\,S^2 \tag{112}$$

N is the number of stress cycles in the time period $[0,T]$.

$$N = T/6\,s \tag{113}$$

$C_2(N)$ is a normally distributed random variable with zero mean value and a variance of

$$Var[C_2(N)] \;=\; 5\,Var[C_1]\frac{N(1-\rho^2)-2\rho(1-\rho^N)}{(1-\rho)^2} \tag{114}$$

The failure criterion in Eq.(105) is rewritten as

$$a_c - a_0\exp(C_1\pi NS^2 + \pi C_2(N)S^2) \;\leqslant\; 0 \tag{115}$$

In u-space the design point $\mathbf{u}^*$ is found as the point on the failure surface closest to the origin. The failure surface is described by Eq.(115) with the equality sign. The design point has the coordinates $\beta\boldsymbol{\alpha}$, where β is the reliability index and where $\boldsymbol{\alpha}$ is a vector of direction cosines called sensitivity factors. The failure surface is approximated by its tangent hyperplane at the design point and the failure probability P_F is therefore approximately

$$P_F \;\approx\; \Phi(-\beta) \tag{116}$$

Table 2 shows results from the FORM calculation. The sensitivity factor for the material variability within each specimen in terms of $C_2(N)$ is very small. This implies that, at least for this example, the uncertainty due to C_{2i} can be neglected.

Table 2. FORM results						
T(years)	β	α_{a_0}	α_S	α_{C_1}	α_{C_2}	$\Phi(-\beta)$
32	6.97	0.558	0.662	0.501	<0.00001	$1.60\cdot10^{-12}$
64	4.61	0.513	0.680	0.523	<0.00001	$2.10\cdot10^{-6}$
128	2.58	0.481	0.689	0.542	<0.00001	$4.85\cdot10^{-3}$
256	0.88	0.459	0.690	0.561	<0.00001	0.189
512	-0.54	0.443	0.680	0.585	<0.00001	0.705
1024	-1.70	0.432	0.656	0.618	<0.00001	0.955
2048	-2.56	0.421	0.596	0.684	<0.00001	0.995

The last column gives the first order approximation to the failure probability, i.e. the probability $P(T_f \leqslant T)$ where T_f is the random life time. The expected life time is

$$E[T_f] \;=\; \int_0^\infty (1 - P(T_f \leqslant T))\,dT \;\approx\; \int_0^\infty \Phi(\beta(T))\,dT \;=\; 420\,years \tag{117}$$

So far, only a single crack has been considered, but extensions to a system of cracks are straight-forward as long as the cracks do not interact. In the latter case a new mechanical model must be introduced. If the structure fails when the first crack reaches the critical size, then the structure belongs to the class of weakest link structures for which efficient FORMs are available, [13]. Let there be n independent cracks of the same type and subjected to the same loading. In the standardized normal space (u-space) the linearized failure surface for a single crack has the equation

$$\beta - \sum_j \alpha_j u_j \;=\; 0 \tag{118}$$

A similar linearized failure surface can be determined for each crack. The normally distributed safety margin M_i for the ith crack is defined as

$$M_i = \beta_i - \sum_j \alpha_{ij} U_j \tag{119}$$

The reliability index β_i is the same for all cracks.

$$\beta_i = \beta = \frac{E[M_i]}{\sqrt{Var[M_i]}} \tag{120}$$

The correlation coefficient ρ_{ij} between the linearized safety margins is the same for each pair of cracks.

$$\rho_{ij} = \rho = \alpha_{C_1}^2 + \alpha_S^2 \tag{121}$$

The system reliability index β_S is, [13]

$$\beta_S = \Phi^{-1}\left(\int_{-\infty}^{\infty} \varphi(u)\, \Phi\left(\frac{\beta + \sqrt{\rho}\, u}{\sqrt{1-\rho}}\right)^n du \right) \tag{122}$$

with a corresponding first order approximation to the failure probability as in Eq.(116).

The system reliability index is also obtained in a simple way if the number of cracks is random and follows a Poisson distribution with intensity μ_1 (= mean number of cracks per unit length/ area/ volume of the structure). Let the length/ area/ volume be V. The system reliability index is

$$\beta_S = \Phi^{-1}\left(\sum_{n=0}^{\infty} \frac{(\mu_1 V)^n}{n!}\, e^{-\mu_1 V} \int_{-\infty}^{\infty} \varphi(u)\, \Phi\left(\frac{\beta + \sqrt{\rho}\, u}{\sqrt{1-\rho}}\right)^n du \right) \tag{123}$$

$$= \Phi^{-1}\left(\int_{-\infty}^{\infty} \varphi(u) \exp\left(-\mu_1 V \left(1 - \Phi\left(\frac{\beta + \sqrt{\rho}\, u}{\sqrt{1-\rho}}\right)\right)\right) du \right)$$

4. CONCLUSIONS

A probabilistic model for fatigue crack growth is formulated. The formulation is based on a linear elastic fracture mechanical approach and experimentally observed results for constant amplitude loading. The uncertainty in the material resistance is modeled both within each specimen and between specimens. The model is extrapolated to variable amplitude loading. Reliability problems for fatigue crack growth are formulated as limit state problems and first order reliability calculations are used to obtain the reliability both for an individual crack and for a system of independent cracks.

5. ACKNOWLEDGMENT

The author wishes to thank Det norske Veritas for its permission for these lecture notes to be prepared and presented. The opinions stated in the paper are those of the author and not necessarily those of Det norske Veritas.

6. REFERENCES

[1] P.E.Irving and L.N.McCartney: 'Prediction of Fatigue Crack Growth Rates: Theory, Mechanisms and Experimental Results', *Metal Science*, 1977, 351-361.

[2] J.C.Newman and I.S.Raju : 'An Empirical Stress-Intensity Factor Equation for the Surface Crack', *Engineering Fracture Mechanics*, **15**, 1981, 185-192.

[3] P.Paris and F.Erdogan: 'A Critical Analysis of Crack Propagation Laws', *Journal of Basic Engineering*, ASME, **85**, 1963, 528-534.

[4] D.A.Virkler, B.M.Hillberry and P.K.Goel : 'The Statistical Nature of Fatigue Crack Propagation', *Journal of Engineering Materials and Technology*, ASME, **101**, 1979, 148-153.

[5] R.Arone: 'On Reliability Assessment for a Structure with a System of Cracks', in *DIALOG 6-82*, Danish Engineering Academy, Lyngby, Denmark, 1983, 329-334.

[6] H.O.Madsen: 'Probabilistic and Deterministic Models for Predicting Damage Accumulation due to Time Varying Loading', *DIALOG 5-82*, Danish Engineering Academy, Lyngby, Denmark, 1983.

[7] H.O.Madsen : 'Stochastic Modeling and First Order Reliability Analysis of Fatigue Crack Growth', Paper presented at International Workshop on Stochastic Methods in Structural Mechanics, University of Pavia, Italy, June 9-12, 1983.

[8] Y.K.Lin and J.N.Yang: 'On Statistical Moments of Fatigue Crack Propagation', *Engineering Fracture Mechanics*, **18**, 1983, 243-256.

[9] Y.K.Lin and J.N.Yang : 'A Stochastic Theory of Fatigue Crack Propagation', Paper presented at the 24th Structures, Structural Dynamics and Materials Conference, Lake Tahoe, Nevada, May 2-4, 1983.

[10] S.H.Crandall and W.D.Mark : *Random Vibration in Mechanical Systems*, Academic Press, 1963.

[11] H.Cramer and M.R.Leadbetter : *Stationary and Related Stochastic Processes*, Wiley, New York, 1967.

[12] S.Krenk : 'A Double Envelope for Stochastic Processes', DCAMM Report No.134, The Technical University of Denmark, 1978.

[13] N.C.Lind, S.Krenk and H.O.Madsen: *Method of Structural Safety*, Prentice-Hall, 1985.

[14] E.Parzen : *Stochastic Processes*, Holden-Day, 1962.

[15] H.O.Madsen : 'Load Models and Load Combinations', Report No.113, Department of Structural Engineering, Technical University of Denmark, 1979.

[16] T.R.Gurney : 'An Analysis of Some Fatigue Crack Propagation Data for Steel Subjected to Pulsating Tension Loading', The Welding Institute Report 59/1978/E, 1978.

[17] S.Tanaka, M.Ichikawa and S.Akita : 'Variability of m and C in the Crack Propagation Law $da/dN = C(\Delta K)^m$', *International Journal of Fracture*, **17**, 1981, R121-R124.

[18] H.O.Madsen : 'Bayesian Fatigue Life Prediction', Paper presented at IUTAM Symposium on Probabilistic Methods in the Mechanics of Solids and Structures, Stockholm, Sweden, June 19-21, 1984.

[19] J.Dufresne : 'Probabilistic Application of Fracture Mechanics', in *Advances in Fracture Research (Fracture 81)*, **2**, Pergamon Press Oxford, 1981, 517-531.

[20] Series of Articles by the ASCE Committee on Fatigue and Fracture Reliability, *Journal of the Structural Division*, ASCE, **108**, 1982, 3-88.

[21] S.O.Rice : 'Mathematical Analysis of Random Noise', in N.Wax ed., *Selected Papers on Noise and Stochastic Processes*, Dover, 1954.

[22] M.Rosenblatt : 'Remarks on a Multivariate Transformation', *Annals of Mathematical Statistics*, **23**, 1954, 470-472.

[23] M.Hohenbichler and R.Rackwitz : 'Non-Normal Dependent Variables in Structural Reliability', *Journal of the Engineering Mechanics Division*, ASCE, **107**, 1981, 1227-1238.

[24] L.Tvedt : 'Two Second Order Approximations to the Failure Probability', VERITAS Report RDIV/20-004083, 1983.

CRACKED COMPONENTS IN RELIABILITY ASSESSMENT

D. Munz
University of Karlsruhe, Institute for reliability and
failure analysis
Postfach 6380
D-7500 Karlsruhe
FR Germany

ABSTRACT.   The fracture stress of cracked components is dependent on
the size and location of the cracks and of the material toughness. Due
to the variability of the toughness and a statistical distribution of
the size of the cracks a large amount of scatter in the failure stress
can be expected. After a brief review of deterministic fracture mechanics
the principles of probabilistic fracture mechanics are presented.
Especially emphasized is the scatter in the ductile-brittle transition
region of ferritic steels, where different failure modes are observed.

1.   DETERMINISTIC AND PROBABILISTIC RELIABILITY ASSESSMENT

In the design of mechanical components a loading parameter L is compared with a parameter R, which characterises the materials resistance
against failure. The loading parameter may be the stress, a stress
amplitude or the stress intensity factor. The resistance parameter may
be the yield strength, the ultimate tensile strength, the fatigue
strength or the fracture toughness. The reliability is assured in a
deterministic approach using lower bound values R_{LB} of materials properties and upper bound values L_{UB} of the loading parameter. Between
these two values a safety margin $M = R_{LB} - L_{UB}$ is determined as shown
in Fig. 1. This safety margin or the corresponding safety factor $S =
R_{LB}/L_{UB}$ is introduced for different reasons:

- Variability of material resistance. There can be a chance that the
  material resistance R is even lower than the assumed lower bound
  value.

- Degradation of the material resistance, for instance by radiation
  damage in nuclear reactor components or by microstructural changes
  in high temperature applications. Such a degradation has to be
  taken into account in selecting the lower bound value. However, the
  uncertainties in the extent of degradation can be covered by a safety
  factor.

A. C. Lucia (ed.), Advances in Structural Reliability, 139–164.
© *1987 by ECSC, EEC, EAEC, Brussels and Luxembourg.*

- Variability in the primary sources for the stresses (or other quantities characterising the loading) such as pressure, external forces or moments, temperature transients.

- Inaccuracies in the transformation of these primary sources into stresses, for instance in the calculation of thermal stresses or of stress intensity factors for stress gradients.

- Uncertainties in the physical model describing the failure event.

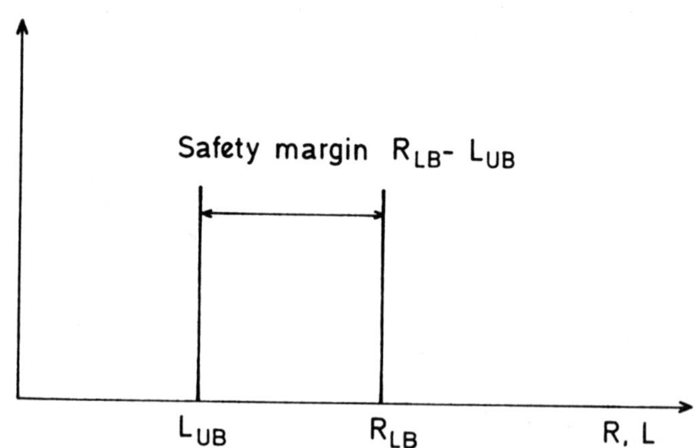

Fig. 1: Deterministic design with lower bound of material resistance R_{LB} and upper bound of load L_{UB}.

The safety factor chosen for a given component depends on the assessment of these uncertainties and variabilities and on the consequences of the failure of a component. For high reliability requirements, as in primary reactor components, large safety margins are used.

In a strictly deterministic design it is assumed that with sufficiently large safety factors failure should be excluded at all. After selection of the safety factors two contradictory questions may arise:

- Is there still a possibility of failure?

- Is the design too conservative?

No exact answer can be given within the deterministic approach. In a probabilistic approach the fixed lower and upper bound values are replaced by a statistical distribution of the load $f(L)$ and the material resistance $g(R)$, as shown in Fig. 2.

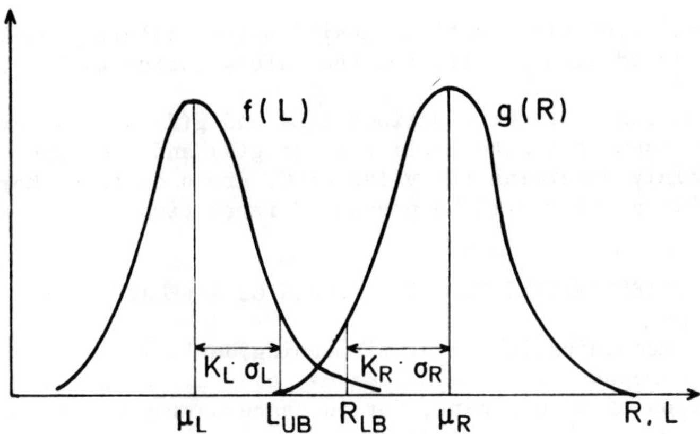

Fig. 2: Density distribution of load f(L) and of
material resistance g(R).

Then the probability of failure is the probability that in a component the load variable L exceeds the material resistance R. This probability is given by

$$Q = \int_{0}^{\infty} \int_{R}^{\infty} f(L) \; dL \cdot g(R) \; dR \tag{1}$$

It is possible to relate the safety factor to this failure probability. For known distribution functions f(L) and g(R) the bound
values R_{LB} and L_{UB} can be related to the mean (μ_L, μ_R) and standard
deviations (σ_L, σ_R) by (s. Fig. 2)

$$R_{LB} = \mu_R - K_R \sigma_R \tag{2}$$

$$L_{UB} = \mu_L + K_L \sigma_L \tag{3}$$

where K_R and K_L are chosen according to the safety assessment.

Then the safety factor is given by

$$S = \frac{R_{LB}}{L_{UB}} = \frac{\mu_R - K_R \sigma_R}{\mu_L + K_L \sigma_L} \tag{4}$$

Thus, for given distribution densities the failure probability can
be calculated with eq. (1) and the safety factor with eq. (4).

Unfortunately, the functions f(L) and g(R) usually are not known
exactly. Especially the lower tail of g(R) and the upper tail of f(L),
which mainly determine the value of Q, are uncertain. More details
about this problem will be presented in section 3.

2. THE DETERMINISTIC FRACTURE MECHANICS APPROACH

Fracture mechanics,in its usual meaning,deals with the behaviour of
cracked components.. These cracks may be flaws produced during the
fabrication of a component, for instance during welding or casting.
Cracks may also be formed during the operation of a component as
fatigue or stress corrosion cracks.

2.1 Qualitative description of failure of cracked components

The size of a crack in a component is characterised by one or two
size parameters (Fig. 3).

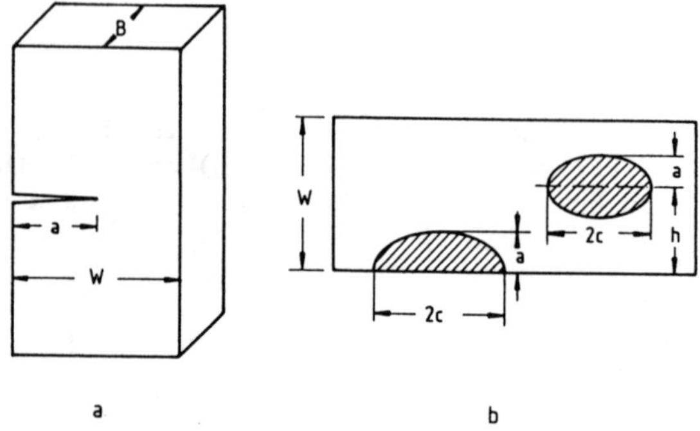

Fig. 3:   One dimensional and two-dimensional cracks

For two-dimensional problems such as through-the-wall cracks in plates
the size of a crack is characterised by its length a. Most of the
real cracks are two-dimensional thus leading to a three-dimensional
problem. Usually, they are described as semi-elliptical surface cracks
or elliptical internal cracks with the semi-axes a and c (Fig. 3).

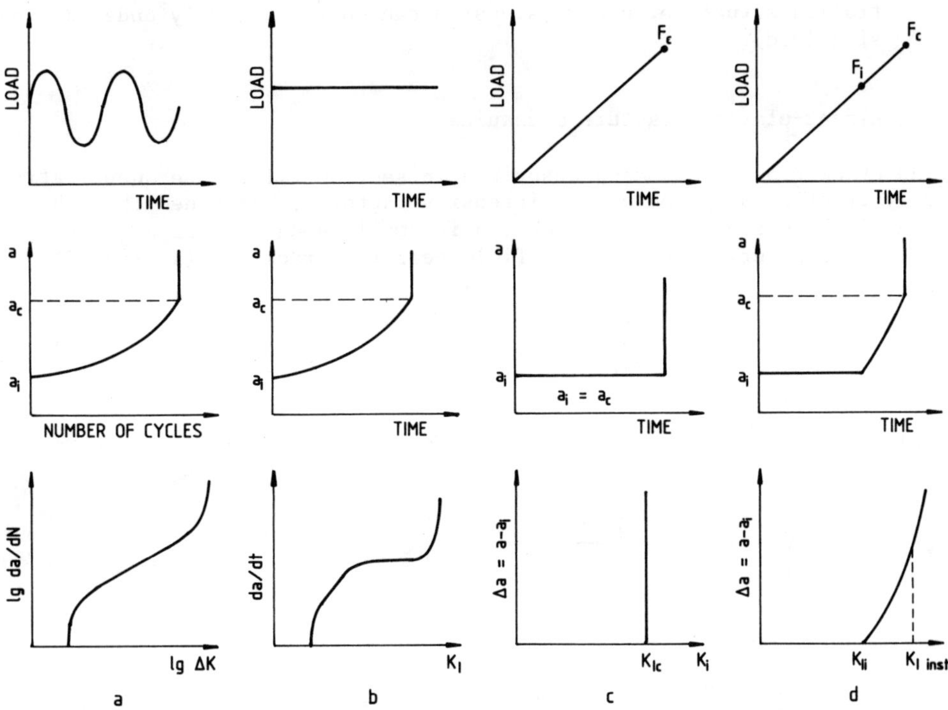

Fig. 4:   Different types of loading and material response

To describe the extension of these cracks, it is useful to distinguish between different types of loading (Fig. 4):

- Alternating loads lead to an increase in the crack size. This type of crack extension is called subcritical or stable. At a given crack size a_c, which depends on the maximum load, fast, unstable crack extension occurs (Fig. 4a).

- Constant loads may also lead to subcritical crack extension in corrosive environments (stress corrosion cracking) or at high temperatures (creep crack extension). Again the stable crack extension is terminated by unstable crack extension (Fig. 4b).

- Under an increasing load two types of material response are possible. Type A (Fig 4c) shows no crack extension until at a critical load unstable crack extension occurs. For Type B-behaviour (Fig. 4d) the crack starts to extend at a critical load F_i. For further crack

extension the load has to be increased further until unstable crack
extension occurs at a critical load F_c (in a displacement controlled situation, crack extension can continue stably under decreasing load).

2.2 Linear-elastic fracture mechanics

The crack growth phenomena described in section 2.1 can be quantitati-
vely treated using the stress intensity factor K_I, provided that the
plastic zone ahead of the crack tip is small enough. K_I is a quantity
which characterises the stress field near the crack typ (s. Fig. 5):

$$\sigma_{ij} = \frac{K_I}{\sqrt{\pi r}} \; f_{ij} \; (\Theta) \tag{5}$$

f_{ij} are functions of the angle Θ.

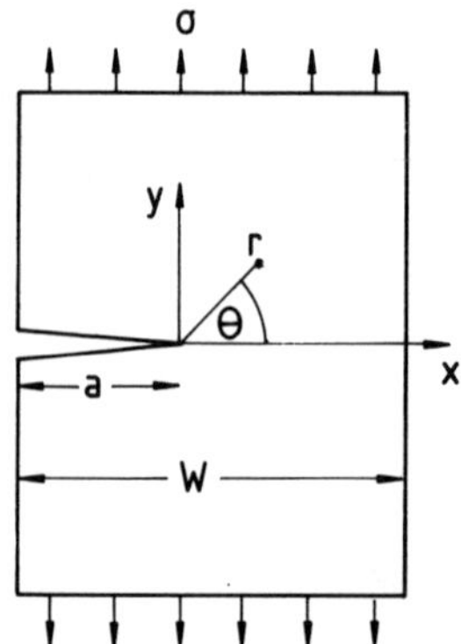

Fig. 5:   Geometric variables of a component with a crack

K_I is proportional to a stress parameter σ (e.g. tensile stress under
uniaxial tensile loading or outer fiber stress under bending loading)
and a function of the crack size and of the crack and component geometries. Usually, this function is written in the form

$$K_I = \sigma \sqrt{a} \cdot Y, \tag{6}$$

where a is a crack size parameter (e.g. crack depth for semi-elliptical
cracks) and Y is a dimensionless function of crack size and geometry
of the cracked component. Y can be calculated with numerical or analytical methods. The different crack propagation phenomena mentioned in
section 2.1 are described in the following way:

- For alternating loads the crack extension in one load cycle, desig-
  nated as da/dN (N = number of cycles) is a material specific function of the range of the stress intensity factor ΔK (Fig. 4a).

$$da/dN = f(\Delta K) \tag{7}$$

with

$$\Delta K = \Delta\sigma \cdot \sqrt{a}\, Y \tag{8}$$

For most materials the function $f(\Delta K)$ can be approximated in a
broad range of K by a power law:

$$\frac{da}{dN} = C(\Delta K)^n \tag{9}$$

where C and n are material parameters.

- For constant load in a corrosive environment the crack growth rate
  da/dt is a material/environment specific function of K_I (Fig. 4b)

$$\frac{da}{dt} = f(K_I) \tag{10}$$

- For a continuously increasing load, the fracture behaviour can be
  described by a material specific K_I-Δa-curve, where $\Delta a = a - a_i$ is
  the crack extension. As shown in Fig. 4c there can be unstable
  crack extension at $K_I = K_{Ic}$ without preceding stable crack extension (Type A); or as shown in Fig. 4d crack extension starts at
  $K_I = K_{Ii}$ and K_I increases during further crack extension until
  instability at $K_I = K_{Iinst}$, corresponding to $F = F_c$ (Typ B).

2.3 Elastic-plastic fracture mechanics

Linear-elastic fracture mechanics can only be applied if the plastic
zone ahead of the crack tip is small enough compared to the crack
length and to the remaining ligament. For the ligament width W-a this
requirement can be expressed by the relation

$$W-a > \beta \left(\frac{K_{Ic}}{\sigma_y}\right)^2 \tag{11}$$

with β on the order of 1. For a given material with the fracture
toughness K_{Ic} and the yield strength σ_y unstable fracture can be predicted with linear-elastic fracture mechanics for components large
enough to fulfill the requirements of eq. (11). For smaller components

the methods of elastic-plastic fracture mechanics have to be applied.
Different methods were developed in the past:

- Crack Opening Displacement (COD),

- J-Integral,

- Two Criteria Method (Fracture assessment Diagram).

These methods have reached different stages of development and all
are associated with problems, if applied to real structures. Here only
the J-Integral method and the Two-Criteria Method will be briefly
described. The J-Integral is a parameter describing the crack tip
singularity not only for elastic deformation but also for extended
plastic deformation at the crack tip region. The crack extension under
increasing load can be characterised by a material specific J_I-Δa-curve.
On the analogy of the linear-elastic behaviour a critical value at the
onset of stable crack extension J_{Iinst} can be defined. The main problem
for practical application of the J-Integral concept is the calculation
of J for a given cracked component.

The Two-Criteria Method |1| considers two basic failure modes:
failure in the linear-elastic regime, which is described by

$$K_{Ic} = \sigma_c \sqrt{a}\ Y \tag{12}$$

and failure by plastic collapse. The latter failure mode is observed
for very tough materials where the effect of the crack consists mainly
in a reduction in the cross section. The critical stress for this mode
σ_L is given by

$$\sigma_L = \sigma_f \cdot f(a/w), \tag{13}$$

where σ_f is the flow stress of the material (usually the average of
yield and ultimate tensile strength) and $f(a/w)$ is a material independent function of the crack length and the geometry of the component.

The Two-Criteria Method, which was developed by CEGB in England,
is a failure criterion covering the total range between linear elastic
fracture mechanics and plastic collapse.

The failure diagram shown in Fig. 6 is based on two parameters

$$K_r = \frac{K_I}{K_{Ic}} \quad \text{and} \quad S_r = \frac{\sigma}{\sigma_L}, \tag{14}$$

where K_I is calculated using eq. (6) and σ is the applied stress. K_r and S_r increase with increasing σ until the failure curve is reached. This is given by

$$K_r^* = S_r^* \left| \frac{8}{\pi^2} \ln \sec \left(\frac{\pi}{2} S_r^*\right)\right|^{-1/2} \tag{15}$$

Therefore, the loading parameters are K_r, S_r and the material resistance parameters are K_r^*, S_r^*.

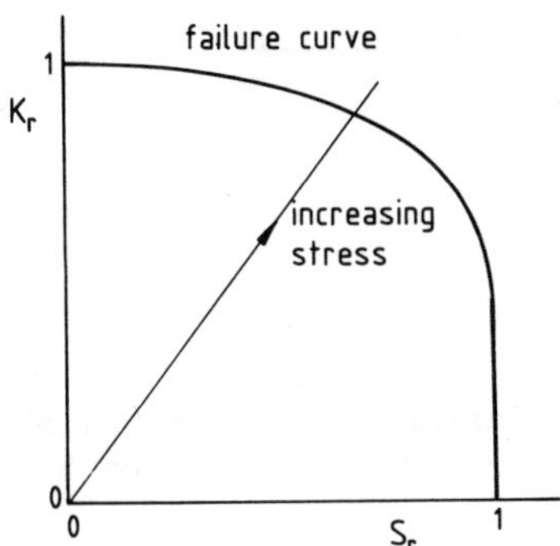

Fig. 6: Failure-assessment-diagram

2.4 Deterministic failure prediction

A deterministic failure prediction starts from a given flaw which is characterised by its size and shape (a and c for elliptical or semi-elliptical cracks) and by its location (depth h in Fig. 3). This crack extends stably due to the service loads of the component. For fatigue crack extension the stress-time variation has to be separated into different stress ranges $\Delta\sigma$ and the extension of the crack can be calculated cycle by cycle using eq. (9). This leads to a crack size a(t), c(t) increasing with time. If the crack length is sufficiently large, a high stress may cause unstable fracture. In lifetime calculations, which fix the admissible service life of a component, the maximum stress σ_{max} used for the calculation of the critical crack length a_c is very often not a normal service stress, but it is a stress resulting from an emergency situation (Fig. 7).

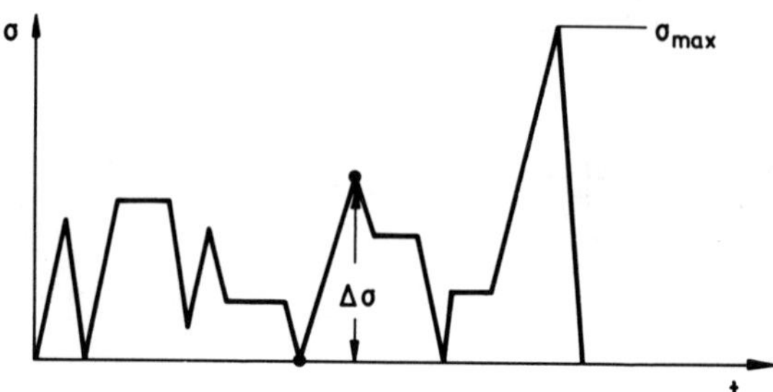

Fig. 7: Service load and maximum load σ_{max} for emergency situation

For reactor components these are for instance loss-of-coolant accidents or seismic events. Thus, the critical crack size is given by

$$a_c = f\ (\sigma_{max},\ \text{crack resistance}) \tag{16}$$

The instability criterion can be written as

$$a\ (t) > a_c \tag{17}$$

In case of the applicability of linear-elastic fracture mechanics

$$a_c = \left[\frac{K_{Ic}}{\sigma_{max}\ Y}\right]^2 \quad \text{or} \quad a_c = \left[\frac{K_{Iinst}}{\sigma_{max}\ Y}\right]^2 \tag{18}$$

The instability criterion can also be expressed as

$$K_I > K_{Ic} \quad \text{or} \quad K_I > K_{Iinst} \tag{19}$$

If the Two Criteria Method is applied the instability criteria is

$$(S_r,\ K_R) > K_R^*\ ,\ S_r^* \tag{20}$$

The number of cycles until failure is given by

$$N_f = \int_{a_i}^{a_c} \frac{1}{da/dN} \, da \tag{21}$$

where a_i is the initial crack size. For the simple case of a constant stress range $\Delta\sigma$ and the power law relation of eq. (9). N_f can be calculated by

$$N_f = \int_{a_i}^{a_c} \frac{1}{da/dN} \, da = \frac{1}{C(\Delta\sigma)^n} \int_{a_i}^{a_c} \frac{1}{a^{n/2} Y^n} \, da \tag{22}$$

3. THE PROBABILISTIC FRACTURE MECHANICS APPROACH

3.1 General approach

In probabilistic fracture mechanics failure probabilities are calculated from the

- flaw size distribution at the beginning of the service of the component (for semi-elliptical or elliptical cracks distribution of a and c);

- distribution of the flaw locations (for instance h in Fig. 3);

- distribution of the material parameters characterising the subcritical crack extension (for instance C and n in eq.9);

- distribution of the material parameters characterising instable crack extension (for instance K_{Ic} or K_r^*, S_r^*);

- distribution of the stress ranges $\Delta\sigma$;

- distribution of the stress under emergency conditions.

The scheme of calculation is shown in Fig. 8.

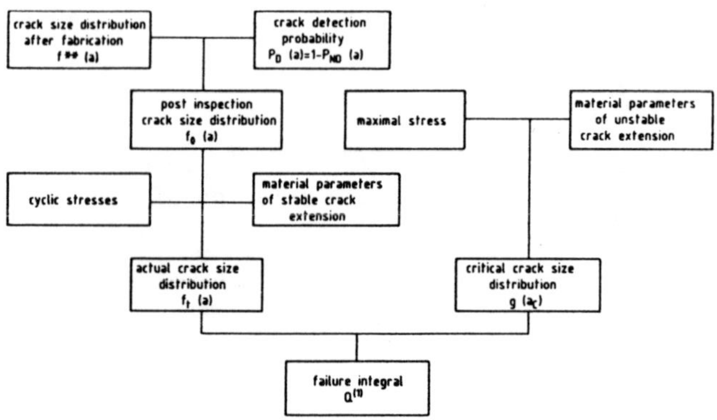

Fig. 8:   Calculation scheme for failure probability

Within the range of application of the linear-elastic fracture mechanics the loading parameter is the stress-intensity factor K_I, and
the crack-resistance parameter is the fracture toughness K_{Ic}. The
distribution of K_{Ic} follows from experimental results. The distribution
of K_I follows according to eq. (6) from the distribution of the crack
size and of the stress applied. Instead of applying the distributions
of K_I and K_{Ic} it is also possible to use the distribution of the crack
size f_o(a) and of the ciritical crack size $g_o(a_c)$ (Fig. 9).

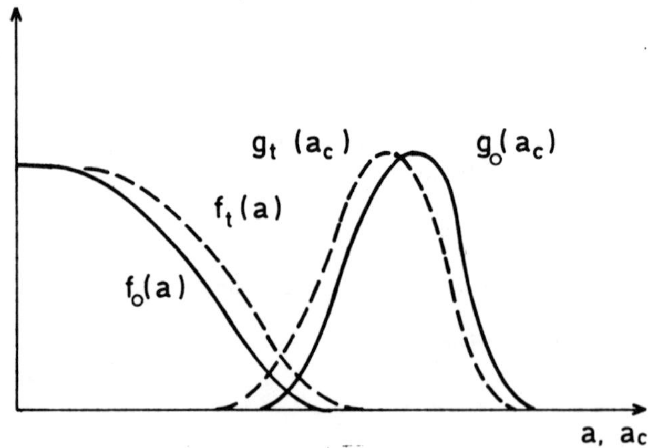

Fig. 9:   Density distribution of actual crack size a and
          critical crack size a_c (—— at the beginning of
          life, - - - after time t)

Then $g_o(a_c)$ follows from the distribution of the fracture toughness and of the stress. Outside the linear-elastic range the critical
crack size distribution can be obtained from the corresponding deterministic failure equation.

The distribution of the actual crack size changes with time because
of stable crack growth. The distribution of the critical crack size
may also change with time due to a change in the fracture resistance,
for instance by radiation effects or by microstructural changes.

The failure probability for a given crack of size a (for simplicity only one size parameter is used here) is given by

$$Q_1(a) = \int_o^a g_t(a_c)\, da_c, \tag{23}$$

where $g_t(a_c)$ is the time-dependent distribution density of the critical
crack size. For one randomly selected crack the failure probability
is given by

$$Q_1 = \int_o^\infty Q_1(a) \cdot f_t(a)\,da = \int_o^\infty \left(\int_o^a g_t(a_c)\,da_c \right) \cdot f_t(a)\,da, \tag{24}$$

where $f_t(a)$ is the time-dependent density distribution of the actual
crack size.

For n cracks in a component, the failure probability of this
component is the probability that at least one crack causes failure
and this is given by

$$Q_n = 1 - (1-Q_1)^n \tag{25}$$

If $\bar{n}$ is the average number of cracks in the component, the probability of having n cracks is

$$P(n,\bar{n}) = \frac{e^{-\bar{n}}\, \bar{n}^n}{n!} \tag{26}$$

Finally, the failure probability of a rancomly selected component
is

$$Q_{total} = \sum_{n=1}^\infty P(n,\bar{n}) \cdot Q_n = 1-e^{-\bar{n}Q_1} \tag{27}$$

For small values of Q_1 this equation can be approximated by

$$Q_{total} = n \cdot Q_1 \qquad (28)$$

3.2 Crack size distribution

One of the most important distributions entering probabilistic fracture mechanics evaluation is the initial crack size distribution. In welded components it is necessary to distinguish between the distribution density after fabrication $f^{**}(a)$ and the distribution density $f(a)$ after nondestructive inspection and repair of the detected cracks. A similar situation exists in components free of fabrication cracks, where fatigue cracks are initiated during service loading, for instance in aircraft structures. Again there are crack size distributions before inspection and after inspection.

The relation between pre-inspection and post-inspection crack size distributions after removing of the detected cracks is given by the probability of detecting a crack $P_D(a)$ which is a function of the crack size:

$$f(a) = C \left[1 - P_D(a) \right] f^{**}(a) \qquad (29)$$

If - as usual - $f^{**}(a)$ is not known, $f(a)$ can be derived from the distribution of the detected cracks $f^{*}(a)$:

$$f(a) = C_1 \frac{1 - P_D(a)}{P_D(a)} f^{*}(a) \qquad (30)$$

C and C_1 are normalisation constants.

The experimental data leading to $f^{*}(a)$ are available only in a limited range of defect size. Small defects cannot be detected and large cracks are not found because of limitations in the sample size. Therfore an incorrect crack size distribution is obtained from $f^{*}(a)$ as shown in Fig. 10.

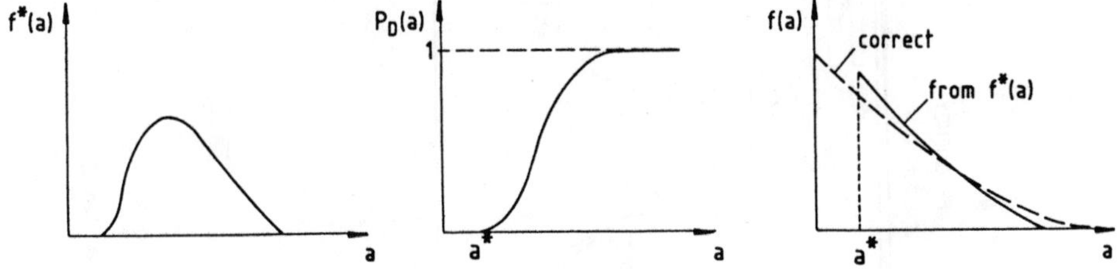

Fig. 10:   Experimental distribution of detected crack size
f*(a) and distribution of not detected cracks

Most of probabilistic fracture mechanics calculations have been
performed so far for welded structures, especially for pressure vessels
and pipes of nuclear power plants |2-5|. The distribution densities
used in these calculations are based on

- engineering judgement,

- results of nondestructive testing,

- model calculations.

Results of nondestructive evaluations first have to be transformed
into crack sizes. Then the crack size density has to be calculated
using eq. (30). An important step is the selection of a distribution
density for the results which are available in the form of a histogram.
It is not only important to select a density function whithin the
range of the experimental data. Very often, especially for components
with high reliability, the calculated failure probability is strongly
dependent on the crack density of large cracks outside the range where
measurements are available. The extrapolation to large crack sizes
cannot be performed unequivocally. This can be seen from Fig. 11,
where different distribution densities are fitted to experimental
data.

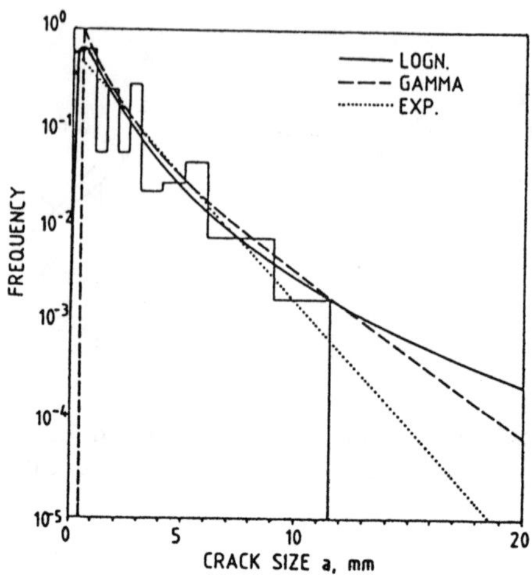

Fig. 11:   Crack size distribution measured by Becher
and Hansen |14| with different distribution
functions

Fig. 12 shows some of the distributions of the crack depths as
used so far for weldments in reactor components |6|.

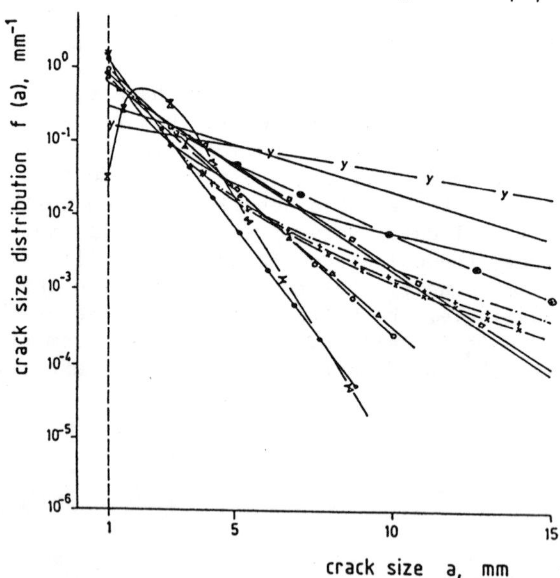

Fig. 12: Crack size distributions found in the literature
normalized with lower bound a_o^*

3.3 Scatter of material parameters

The scatter of the material parameters for stable and unstable crack
extensions have to be described by distribution functions.

If stable crack extension by fatigue loading can be described by
eq. (9) it is sufficient to use a fixed value of n and to treat only
C as a statistical variable. Wellein $|7|$ used a Beta-function bet-
ween the bounds $\bar{C}/10$ and $10\,\bar{C}$, where $\bar{C}$ is the mean value. Nilsson $|8|$
used a log-normal distribution.

For the fracture toughness normal as well as two- and three-
parameter Weibull distributions have been applied. The lower tails
of the distribution are important for the calculation because failure
of high-reliability components can only be expected for very low
values of fracture toughness.

If only mean and standard deviations are available, different
distribution functions can be selected. For conservative calculations
of failure probabilities an upper bound of the lower tail of the distribution has to be chosen. Analysing different distribution functions
Wirsching $|9|$ proposed the power-distribution

$$f(x) = \begin{cases} \alpha \, b^{\alpha} \, x^{\alpha-1} & x < b \\ o & x > b \end{cases} \tag{31}$$

$$\text{with} \quad \mu = \frac{b}{\alpha+1} \,, \qquad \sigma^2 = \frac{\alpha b^2}{(2+\alpha)\,(1+\alpha)^2}$$

as the upper bound curve for the lower tail.

However, by application of upper bounds the failure probability
is overestimated. It is therefore necessary to consider the lower tail
behaviour more realistically. The distribution functions usually
applied for material parameters (normal, lognormal, two-parameter
Weibull) have a lower bound of zero or $-\infty$, which is unrealistic. The
variations in the fabrication process (composition, heat treatment,
welding, mechanical treatment etc.) are within given limits. Quality
control is a further reason for the existence of a lower bound
(Fig. 13), because a component with properties not meeting a given
standard is rejected. It should, however, be taken into account that,
usually, material testing is destructive which means that one or
several test specimens have to be cut from a plate. Due to variations
whithin the plate there may be lower material properties than required. Therefore, distributions as shown in Fig. 14 can be expected.

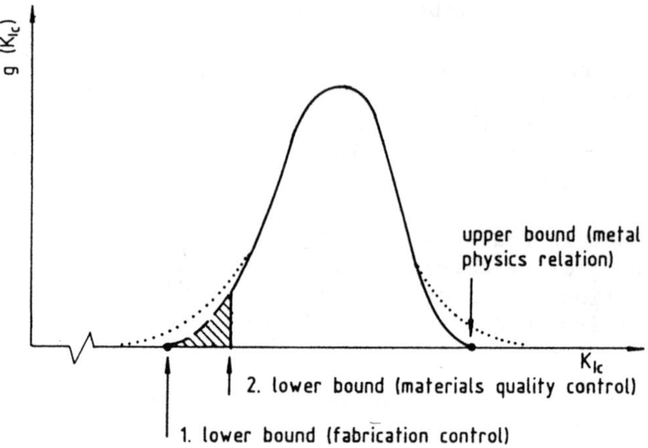

Fig. 13: Distribution of K_{Ic} (schematically) with upper and lower bound

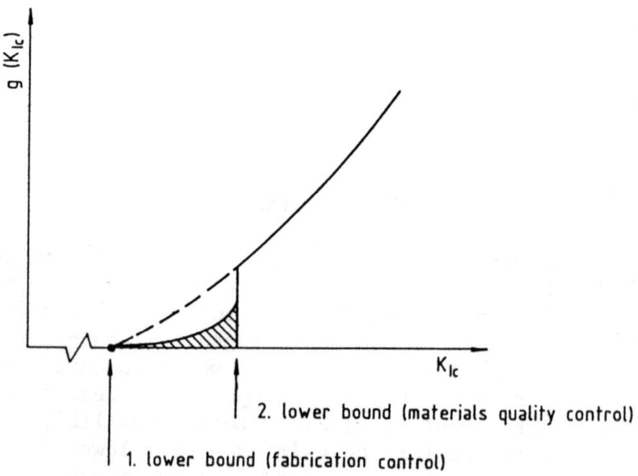

Fig. 14: Lower bound region of K_{Ic} distribution

3.4 Differential failure probabilities

In eq. (24) the failure probability Q_1 for one randomly selected crack
is calculated from the distribution of the actual and the critical
crack sizes. For two-dimensional surface cracks, which can be described
by a semi-ellipse with the crack depth a and the length 2c, the distribution f(c) or f(a/c) has to be considered in addition. For embedded
cracks the depth distribution f(h) is important. Generally, if we have
n additional distribution densities $f(x_1)$ ... $f(x_n)$ the failure equation (24) has to be replaced by

$$Q_1 = \int f(x_1) \cdot f(x_2) \ldots f(x_n) \int_o^\infty f_t(a) \int_o^a g_t(a_c) da_c \, da \, dx_n \ldots dx_n$$

$$(32)$$

Sometimes the relative or differential failure probabilities dQ_1/dx_i
are more important than the absolute failure probabilities Q_1. These
are the (non-normalised) distribution densities of the failed components.

For instance - starting from eq. (24) -

$$\frac{dQ_1}{da} = f_t(a) \int_o^a g_t(a_c) da_c \tag{33}$$

is the distribution of the crack size of the failed components.

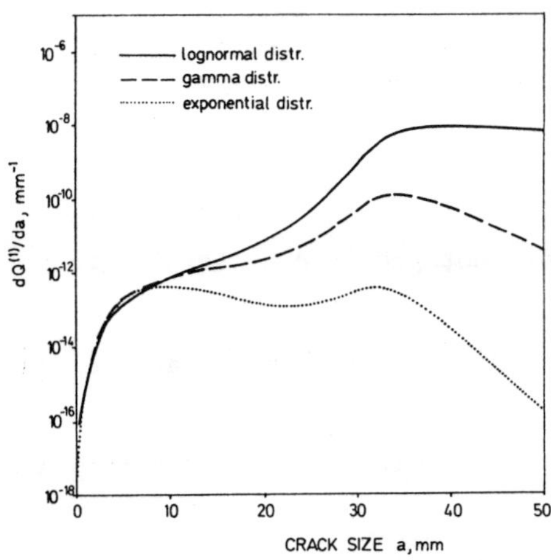

Fig. 15: Distribution of crack size for the failed
          components for crack size distribution of Fig. 10

In Fig. 15 dQ_1/da-a-curves are shown for the crack size distribution
of Fig. 11 and a two-parameter Weibull distribution for the fracture
toughness. It can be seen that in the range where crack size data are
available all crack size distributions lead to the same value of dQ_1/da.
However, great differences are obtained in the region of extrapolation.

4. SCATTER IN THE DUCTILE-BRITTLE TRANSITION REGION OF FERRITIC STEELS

Large scatter in fracture resistance is observed in the ductile-brittle
transition region of ferritic steels. Four different fracture modes for
components with fatigue cracks can be observed which are dependent on
temperature, loading rate, crack size and size of the component. These
fracture modes are shown schematically in Fig. 16 as load-displacement
curves of a specimen with a crack.

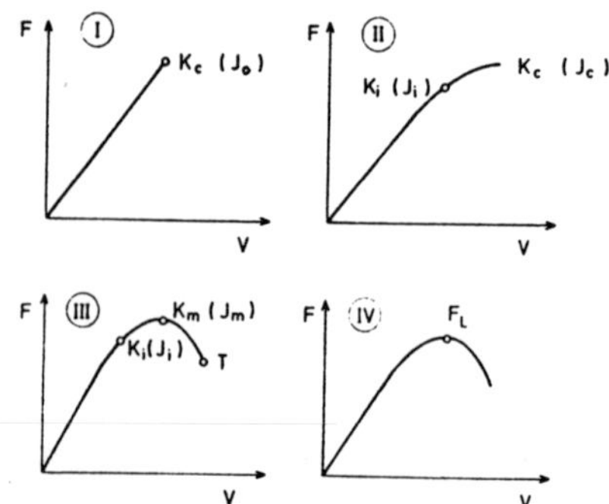

Fig. 16: Load displacement curves for the different
          fracture modes with fracture mechanic
          parameters

The fracture modes are:

I: Unstable cleavage fracture not preceded by stable crack
     extension

II: Stable ductile crack extension starting from fatigue crack and
     subsequent cleavage crack.

III: Stable crack extension beyond maximum load (in a displacement
     controlled test).

IV: No crack extension before maximum load.

In fracture modes I and II the maximum load is given by the critical load for cleavage crack, in mode III by the ductile crack growth resistance curve, and in mode IV by the plastic collapse load, which is governed by the plastic deformation behaviour of the material.

To describe the crack growth behaviour for the different fracture modes the following characteristic values are important. These values are given in terms of stresses σ, loads F, stress-intensity factors K, or the J-Integral J.

K_c or J_c at the onset of unstable cleavage crack extension,

F_i, K_i or J_i at the onset of ductile crack extension,

F_m, K_m or J_m at the maximum load for ductile crack extension,

F_L or σ_L for plastic instability.

There is a general trend from mode I to mode IV with increasing temperature. However, it is observed that at a given temperature some specimens may fail in mode I, others in mode II and in mode III. The critical values are also dependent on the specimen size. Fig. 17 shows the different effects of thickness on K_i and K_c in the ductile-brittle transition region $|15|$. In thick specimens there may be cleavage crack not preceded by ductile crack extension and in thin specimens ductile crack extension preceding unstable cleavage crack extension. Again Fig. 17 shows only the general trend. For a given thickness some specimens may fail in mode I, others in mode II.

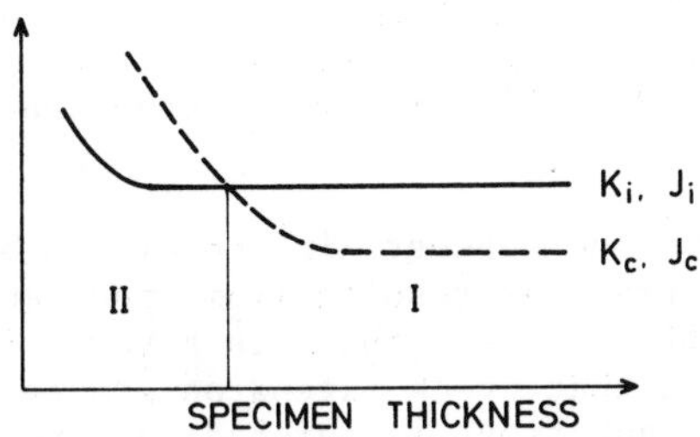

Fig. 17: Effect of specimen thickness on K_i or J_i at the onset of ductile crack extension and on K_c or J_c at the onset of cleavage crack extension

The problem in this ductile-brittle transition region is the conversion of the results from specimens tested in the laboratories to the behaviour of real components. A different scatter behaviour can be expected for the different fracture modes (s. Fig. 18).

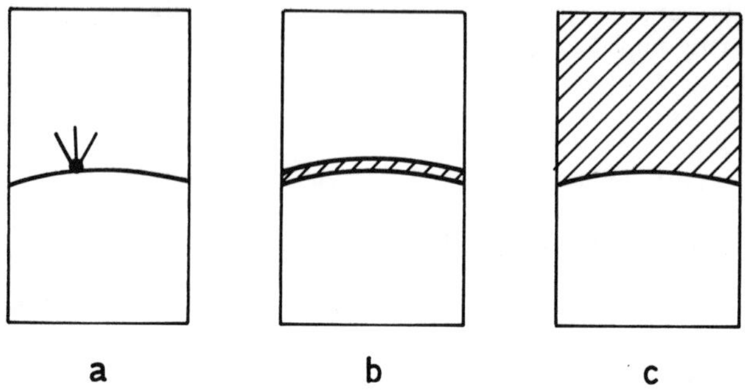

a   b   c

Fig. 18: Averaged region of material parameters
for onset of cleavage crack extension (a),
onset of ductil tearing (b),
plastic collapse (c)

For the onset of cleavage fracture one assumption is that there exists a distribution of K_{Ic} along the crack front and that rapid crack propagation starts, if at one point along the crack front the minimum value of K_{Ic} is reached. An alternative approach is that the applied K_I is equal to the averaged K_{Ic} along a given length of crack front. For the onset of ductile tearing usually a critical value is defined after some crack extension of for instance 0.1 mm. Then this value K_i or J_i is a value averaged along the crack front. For plastic collapse the plastic deformation behaviour is determined by averaging along the ligament area. From these considerations it is concluded that the amount of scatter should be higher for cleavage than for ductile tearing.

For cleavage crack extension the weakest link model can be applied to describe the scatter and to transfer results from specimens to components |11|. The basic assumption is that there is a variation of K_{Ic} for cleavage crack extension and that crack extension starts if the lowest value of K_{Ic} along the crack front is reached. A more realistic assumption is that a value K_{Ic} averaged over a given length ΔS has to be reached (Fig. 19).

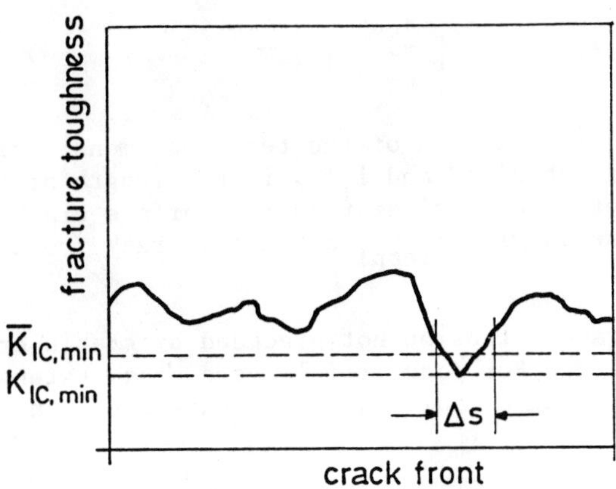

Fig. 19: Variation of K_{Ic} along the crack front

The scatter in fracture toughness measured with test specimens
with a crack front length 1 is described by the Weibull distribution.
The cumulative distribution is

$$F_1(K_{Ic}) = 1 - \exp\left| - (\frac{K_{Ic}}{b})^m \right| \qquad (34)$$

Then the cumulative distribution of K_{Ic} in a specimen or component
with a crack front length 1_1 is

$$F_{1_1}(K_{Ic}) = 1 - \left| 1 - F_1 \right|^{1_1/1} \qquad (35)$$

$$= 1 - \exp\left| - (\frac{K_{Ic}}{b_1})^m \right| \qquad (36)$$

with
$$b_1 = b \, (\frac{1_1}{1})^{1/m} \qquad (37)$$

This distribution again is a Weibull distribution with the parameters
m and b_1.

Eq. (36) can be interpreted as the failure probability of a component with one crack of length 1_1 for an applied stress intensity
factor $K_{Iappl.}$ For a randomly selected crack from a distribution f(a)
the failure probability is

$$Q_1 = \int\limits^{\infty} f(a) \left\{ 1 - \exp \left| - \left(\frac{K_{Iappl}}{b}\right)^m \frac{1}{l_1(a)} \right| \right\}$$

where l is the crack front length of the test specimens, for which the parameters b and m are obtained and $l_1(a)$ is the length of the crack of depth a. A more detailed consideration for surface cracks has also taken into account the varying K_{Iappl} along the crack front for surface cracks $|12|$.

So far cleavage crack extension not preceded by stable crack extension has been considered. For mode II (cleavage crack extension preceded by stable crack extension) a modification of this approach is possible $|13|$.

References

| 1| Harrison, R.P.; Loosemore, K., Milne, I., Dowling, A.R.:
Assessment of the integrity of structures containing defects,
Central Electricity Generating Board, U.K., Report R/H/R6-
Rev. 2 (1980).

| 2| Lidiard, A.G.: Applications of probabilistic fracture mechanics
to light water pressure vessels and piping. Nucl.Eng. Des. 60
(1980) 49-56.

| 3| Harris, D.O.; Lim, E.Y.; Dedhia, D.D.: Probability of pipe fracture
in the primary coolant loop of a PWR plant. Report NUREG-CR 2189,
Vol. 5, Lawrence Livermore Laboratory, 1981.

| 4| Dufresne, M.J. et al.: PWR Reactor vessel failure probabilities.
Proc. of the 4th Int. Conf. on pressure vessel technology.
I. Mech. E., London 1980, paper C 13/80.

| 5| Brückner, A.; Häberer, R.; Munz, D.; Wellein, R.: Reliability
study for the steel containment of a nuclear power plant using
probabilistic fracture mechanics. ASME conf. on pressure vessels
and piping, 1983, paper 83-PVP-86.

| 6| Häberer, R.; Munz, D.; Brückner A.: Fehlergrößenverteilungen und
Fehlerauffindwahrscheinlichkeiten in geschweißten Bauteilen,
Primärbericht des Kernforschungszentrums Karlsruhe 10.02.12P
02P, 1981.

| 7| Wellein, R.: Bericht der Kraftwerk Union Erlangen, KWU/R212 (1981).

| 8| Nilsson, F.: A model for fracture mechanical estimation of the
failure probability of reactor pressure vessels. Proceedings
of the Third International Conference on Pressure Vessel Technology, Part II, pp. 593-601.

| 9| Wirsching, P.H.: On the behavior of statistical models used for
design, Trans. ASME, J. of Engineering for Industry (1976),
pp. 601 - 606.

|10| Munz, D.; Keller, H.D.: Effect of specimen size on fracture
toughness in the ductile brittle transition region of steel,
Fracture and Fatigue, Proceeding of the Third European Conference
on Fracture, Pergamon Press 1980, pp. 105 - 117.

|11| Landes, J.D.; Shaffer, D.H.: Statistical characterisation of
fracture in the transition region, in Fracture Mechanics, 12th
Conference, ASTM STP700, 1980 pp. 368 - 382.

|12| Brückner, A; Munz, D.: Prediction of failure probabilities for
cleavage fracture from the scatter of crack geometry and of
fracture toughness using the weakest link model,
Eng. Fract. Mech. 18 (1983), 359 - 375.

|13| Brückner, A.; Munz, D.: Scatter of fracture toughness in the
brittle ductile transition region of a ferritic steel,
ASME publication PVP - Vol. 92, pp 105-111.

|14| Becher, P.E.; Hansen, B.: Statistical evaluation of defects in
welds and design implications, Report ot the Danish Atomic
Energy Comission, 1974.

ADAPTIVE DYNAMIC MODELLING, RISK ASSESSMENT, FAILURE DETECTION AND CORRECTION

Demetrios G. Lainiotis[*]
Department of Informatics and Computer Engineering
University of Patras - Greece

ABSTRACT. System failure can be, in many instances, very catastrophic, e.g. nuclear reactor failure, structural failure of large, multi-story buildings due to earthquakes, aircraft structural failures, chemical plant malfunctions, etc. Present-day system are, moreover, very complex, large-scale, and consequently sensitive to failure. Their models - to reflect accurately their complexity - must be realistically only partially known. Thus, the complexity, large-scale dynamic and partially known nature of their models, combined with the disastrous, socially and economically, consequences of system malfunction or failure, necessitate the use of on-line, real-time, automatic, adaptive/learning methodologies and related algorithms for modelling, monitoring, risk assessment, reliability forecasting, failure detection and isolation (recognition), and for effective system stabilization and malfunction correction. Such real-time, adaptive/learning algorithms, based on Lainiotis' multi-model partitioning framework, are presented. It is shown that they are remarkable in several respects, namely they have a decoupled, parallel-processing, easily implementable structure, they are robust to numerical errors, and subsystem failure, and exhibit quick learning/adaptation even in the absence of prior information on the range of unknown models.

1. INTRODUCTION

System failure can be, in many instances, very catastrophic, e.g. nuclear reactor failure, structural failure of large, multi-story buildings due to earthquakes, aircraft structural failures, chemical plant malfunctions, etc. Present-day systems are, moreover, very complex, large-scale and, consequently, sensitive to failure. Their models - to reflect accurately their complexity - must be large-scale (high

---

[*] Formerly Professor and Chairman, Department of Electrical Engineering, State University of New York.

A. C. Lucia (ed.), *Advances in Structural Reliability, 165–193.*
© 1987 by ECSC, EEC, EAEC, Brussels and Luxembourg.

state-vector dimensionality), non-linear, perhaps distributed, stochastic, time varying - either parametrically or structurally or both - and realistically only partially known, this ignorance being of a parametric or structural nature, e.g. unknown state-vector dimensionality. For example, the model of a satellite may depend on its position in a given trajectory, and the dynamics of a missile or perhaps of a pressure vessel may have to change to represent it with the same accuracy, as the liquid in the missile or the pressure vessel changes, e.g. the state-vector dimensionality may have to change, and/or the model parameter values. Moreover, these changes cannot be known a-priori. Indeed, the complexity, time-varying nature and sensitivity to failure of present-day systems makes their complete a-priori modelling practically unfeasible and economically and socially very expensive. The approximate and time-invariant or static modelling of these systems is not adequate for later use, especially in view of the catastrophic consequences of erroneous decisions based on incomplete, a-priori models, and the related static risk (reliability) assessments. In addition, in many practical applications of great current interest, it may also be unfeasible to account or predict all possible malfunctions or system failures.

Thus, the complexity, large-scale dynamic nature, time-varying, stochastic and unknown nature of their models, combined with the disastrous, socially and economically, consequences of system malfunction or failure, necessitate the use of on-line, real-time, automatic, adaptive/learning methodologies and related algorithms for modelling, monitoring, risk assessment, reliability forecasting, failure detection and isolation (recognition), and for preventive maintenance, effective system stabilization and malfunction correction. The use of such real-time, on-line, adaptive/learning algorithms presupposes the following:

a) The existence of realistic and efficient design procedures which include the criteria to be met and, most importantly, the algorithms to be implemented;
b) The existence of the relevant software and hardware technology for their cost-effective realization.

Fortunately, all of the above prerequisites can be met because of recent developments. Specifically, the methodology of multi-model partitioning and the related fast, parallel-processing and robust algorithms, Lainiotis [1,15], as well as the advent of relatively inexpensive, reliable and very fast computers (e.g. microprocessor-based, parallel-processing, etc.) make possible the ready and cost-effective implementation of the multi-model partitioning algorithms for automatic, real-time, adaptive/learning modelling, monitoring, risk assessment, failure detection and isolation, and malfunction correction.

2. STATE-VARIABLE DYNAMIC MODELS

The state-variable methodology lends itself to the realistic modelling of dynamic systems. It constitutes a canonical framework and it readily accommodates continuous or discrete (sampled data) models, lumped or distributed models, linear or non-linear models, etc. Because of space and time limitations, this presentation will be limited to lumped models. The canonical continuous linear, state-variable model is given by:

$$\frac{dx(t)}{dt} = F(t)x(t) + G(t)u(t) + w(t) \tag{1}$$

$$z(t) = H(t)x(t) + J(t)u(t) + v(t) \tag{2}$$

where $x(t)$ is the nx1 state-vector process, fully representing the "state" of the system at any given time instant t, and $z(t)$ is the mx1 vector measurement process. The $w(t)$, $v(t)$ are the nx1 and mx1 "plant" and measurement vector noise random processes, representing the stochastic disturbances present in the system. These noise processes may be gaussian or non-gaussian, e.g. Poisson. However, because of time and space limitations, this presentation will concentrate on gaussian noise processes. The $u(t)$ is a known px1 vector input, representing known "control" inputs and/or biases. The matrix $F(t)$ is the system dynamic matrix incorporating the "memory" (eigenvalues) of the system, and the nxp, and mxp, respectively, matrices $G(t)$ and $J(t)$ constitute the input and output system matrices (transducers) that represent the coupling of the known input $u(t)$ with the state and measurement process. In general, $F(t)$, $G(t)$ and $J(t)$ may be time-varying or not.

The corresponding discrete, linear, state-variable model is given by:

$$x(k+1) = \phi(k+1,k)x(k) + G(k)u(k) + w(k) \tag{3}$$

$$z(k+1) = H(k+1)x(k+1) + J(k+1)u(k+1) + v(k+1) \tag{4}$$

where $\phi(k+1,k)$ is the so-called system transition matrix incorporating the "memory" of the system, and the other quantities are as defined previously. Moreover, in both models, the stochastic processes $w(k)$ and $V(k+1)$ are gaussian and zeromean, uncorrelated simply for convenience. They are also independent of the initial state $x(0)$ which is also gaussian with a-priori mean $\hat{x}_0$ and covariance P_0.

The canonical non-linear continuous and discrete state-variable models are given by:

$$\frac{dx(t)}{dt} = f[x(t),t] + G(t)u(t) + w(t) \tag{5}$$

$$z(t) = h[x(t),t] + J(t)u(t) + v(t) \tag{6}$$

and

$$x(k+1) = f[x(k),k] + G(k)u(k) + w(k) \tag{7}$$

$$z(k+1) = h[x(k+1,k+1)] + J(k+1)u(k+1) + v(k+1) \tag{8}$$

where the f and h functions are non-linear functions of the state $x(t)$. All other quantities are as described previously.

    The above non-linear state-variable models can be linearized to
yield linear models of the type given above. Moreover, the continuous
linear and non-linear models may be discretized to yield the corresponding discrete linear and non-linear models.

    As indicated previously, the model of the plant or structure under
consideration constitutes the cardinal part in the design of performance monitoring, failure detection/isolation and malfunction correction
systems to be used for the plant or structure, since the plant model is
the vital link between the "physical" problem in which the designed
system will operate and the mathematical realm in which they will be
designed. The efficacy and usefulness of the performance monitoring,
failure detection/isolation and malfunction correction systems, depends
strongly on the realism with which the model represents the underlying
physical situation. Unfortunately, the more realistic the model, the
greater its complexity and, consequently, the greater the difficulty of
the associated design problem, and the greater the difficulty in implementing the designed system. The difficulties are compounded further
by the fact that, as the realism of the model increases, so does the
lack of parametric or structural knowledge of the model. In most physical situations, complete knowledge of complex models is neither available nor readily forthcoming and one is confronted with the design of
optimal monitoring, failure detection, etc. systems in the face of incomplete model knowledge. If, moreover, the failure detection/monito-
ring system design is to be done in real-time or while data are still
being acquired, it constitues an adaptive/learning problem.

    To account for model ignorance, whether parametric or structural,
a vector parameter θ of dimension rx1 is introduced, each dimension of
which corresponds to an unknown scalar model parameter. As such, the
partially unknown models take the following modified forms for the
linear models:

$$\frac{dx(t)}{dt} = F(t,\theta)x(t) + G(t,\theta)u(t) + w(t) \tag{9}$$

$$z(t) = H(t,\theta)x(t) + J(t,\theta)u(t) + v(t) \tag{10}$$

and

$$x(k+1) = \phi(k+1,k,\theta)x(k) + G(k,\theta)u(k) + w(k) \tag{11}$$

$$z(k+1) = H(k+1,\theta) + J(k+1,\theta)u(k+1) + v(k+1) \tag{12}$$

The partially unknown non-linear models are similarly given by:

$$\frac{dx(t)}{dt} = f[x(t),t,\theta] + G(t,\theta)u(t) + w(t) \tag{13}$$

$$z(t) = h[x(t),t,\theta] + J(t,\theta)u(t) + v(t) \tag{14}$$

and

$$x(k+1) = f[x(k),k,\theta] + G(k,\theta)u(k) + w(k) \tag{15}$$

$$z(k+1) = h[x(k+1),k+1,\theta] + J(k+1,\theta)u(k+1) + v(k+1) \qquad (16)$$

At this point the concept of system structure may be conveniently introduced, namely system "structure" is defined as the dimensionality n of the state-vector. In general, structural and/or parametric ignorance of the model may be present. However, we may imbed structural uncertainty into parametric uncertainty by assuming a model structure less than a given fixed number n. This assumption may be justified on the basis of physical considerations, or in order to limit the complexity of the resulting monitoring, failure detection, or failure correction system. In the latter case, the problem may be viewed as on-line, adaptive system approximation, or adaptive constrained modelling. The assumption of an upper bound n to the system dimensionality permits us to imbed structure adaptation into parameter adaptation by choosing the model structure as n (upper bound) and, subsequently, determining adaptively the elements of the system matrices F,F,J,H,ϕ structure is less than n.

The above, partially-known models are specified up to the unknown parameter vector θ, which may be time-invariant or time-varying. The time-invariance of θ may be justified on the basis of physical considerations or as an approximation to slowly time-varying θ. Indeed, the latter approximation may be adequate since the physical distinction between the state $x(t)$ and the parameter θ of a given process or structure lies in the fact that a parameter has much slower time-constants than the state; otherwise it should have been included in the state. As such, θ is approximately time-invariant compared to the state-vector $x(t)$. The time-invariance of θ is given mathematically by:

$$\frac{d\theta(t)}{dt} = 0 \qquad \text{or} \qquad \theta(k+1) = \theta(k) \qquad (17)$$

for continuous and discrete models, respectively.

At this point, the subject of modelling system malfunction or failure may be faced. One may consider, roughly, two broad classes of malfunction, namely malfunctions due to gradual changes of parameter values with time (soft or gradual failures), and those due to abrupt changes of parameter values with time (hard failures). The first may be due to physical ageing or deterioration of the system, and the second may be due to "breaking" of some subsystem or component of the system. Mathematically, the first may be modelled as a slowly time-varying parameter θ or, more conveniently, as piece-wise time-invariant parameter θ. The hard failure situation may be modelled as an abrupt jump in the value of the parameter θ or of one of its dimensions.

The failure model as discussed above may account for any combination of subsystem failures or malfunctions by having failures in the scalar parameters associated with these subsystems. In this way, sensor failures, transducer or actuator failures, subsystem failures, biases, etc. may be accounted for. For example, for a continuous non-linear system, a "hard" sensor failure may be simply modelled mathematically as follows:

$$z(t) = \theta h[x(t),t] + J(t)u(t) + v(t) \tag{18}$$

$$\frac{dx(t)}{dt} = f[x(t),t] + G(t)u(t) + w(t) \tag{19}$$

where θ takes one of two values 1 or 0, the first corresponding to non-failed sensors, while the second corresponding to failed or completely non-operational sensors, when the measurements consist of noise only. On the other hand, "soft" or gradual sensor failures that may be due to sensor degradation in the form of bias or increased inaccuracies, may be modelled as an increase in the measurement noise covariance $R(t,\theta)$ only. In the case of hard sensor failure, we may want to detect this failure and sound an alarm, while in the case of soft failure, estimates of the bias or the increase in noise will permit continued use of the sensor, albeit in degraded mode. Obviously, the particular application will dictate the tasks to perform, e.g. estimation, failure detection and isolation, etc. If a human operator is present, we may only need to generate a failure alarm (failure detection) that tells him to take certain action. If, however, back-up sensors are available, failure isolation is needed, without estimation. On the other hand, in the absence of hardware redundancy, the degraded instrument may still have to be used, necessitating estimation information.

In a similar manner, particular subsystem failures may be modelled for a linear system as follows:

$$z(t) = H(t)x(t) + J(t)u(t) + v(t) \tag{20}$$

$$\frac{dx(t)}{dt} = \begin{bmatrix} F_{11}(t) & F_{12}(t,\theta) \\ F_{12}(t,\theta) & F_{22}(t,\theta) \end{bmatrix} x(t) + \begin{bmatrix} G_1(t) \\ G_2(t,\theta) \end{bmatrix} u(t) + \begin{bmatrix} w_1(t) \\ w_2(t) \end{bmatrix} \tag{21}$$

where $x(t) = \begin{bmatrix} x_1(t) \\ x_2(t) \end{bmatrix}$, and the plant noise covariances are $Q_1(t)$ of

$w_1(t)$, and $Q_2(t,\theta)$ of $w_2(t)$. The partial state vector $x_2(t)$ pertains to the subsystem whose failure we are interested in, and $x_1(t)$ is the remaining part of the system state-vector. The parameter vector θ takes values appropriate to "hard" and "soft" subsystem failures.

In a similar manner, other possible malfunctions could be easily handled. For example, actuator malfunctions could be handled through $G(t,\theta)$, sensor biases could be handled via a non-zero mean value for $v(t)$, etc. In general, for realistic modelling, in addition to parameters corresponding to malfunctions, there will be other parameters in θ, corresponding to model ignorance. However, using the multi-model partitioning framework, the treatment of all these cases is unified. This is presented in the following section.

3. MULTI-MODEL PARTITIONING - RATIONALE AND METHODOLOGY

Given the above models representing mathematically a wide class of
physical situations of interest concerning structures and plants, it is
appropriate to also describe mathematically the goals and objectives of
monitoring, modelling (identification), probabilistic risk assessment,
failure detection, failure isolation (recognition), and finally of system stabilization. System monitoring has the objective of monitoring
the "state" of the system as closely as possible. This may be stated
mathematically as a state-estimation problem, namely estimate the state
of the system $x(\ell)$, by an estimate $\hat{x}(\ell/k+1)$ based on all the a-priori
information $\lambda(0)$, and all the operating data available at the present
instant k+1, namely on $\lambda(k+1) \equiv \{\lambda(0),z(1),z(2),\ldots,z(k+1)\}$, where
$\ell \geq k+1$. The optimal choice of $\hat{x}(\ell/k+1)$ in the mse-sense - which minimizes the distance of the estimate to the true state - is the conditional mean

$$x(\ell/k+1) = E\{x(\ell)/\lambda(k+1)\} \tag{22}$$

If this estimation is done on-line and in the presence of model uncertainty, it constitutes adaptive estimation. By adaptive modelling or
on-line system identification is meant the estimation of the uncertainty θ of the system given $\lambda(k+1)$, while the system is operating, and its
state also monitored.

Probabilistic risk assessment (pra) involves the calculation of
the probabilities of the various possible failure modes, based on the
current operating data and/or the a-priori knowledge (prior to the current operation of the system), namely the calculation of $p[\alpha_j/\lambda(k+1)]$
or of $p[\alpha_j/\lambda(0)]$, where α_j is the j-th failure mode. If $p[\alpha_j/\lambda(0)]$ is
the one used for pra, it is an a-priori or static pra, while if
$p[\alpha_j/\lambda(k+1)]$ is used, it constitutes on-line dynamic pra, and if, moreover, there exists model uncertainty, it constitutes adaptive pra.
Failure detection and isolation has the objective of deciding whether
and which failure has occurred with minimum error probability. The per-
tinent decision rule is: decide failure α_i has occurred if $p[\alpha_i/\lambda(k+1)]$
is the $\max\{p[\alpha_i/\lambda(k+1)]$, $j=1,2,\ldots,M\}$, and, moreover, if $p[\alpha_j/\lambda(k+1)]$
$> p[\bar{\alpha}/\lambda(k+1)]$, where $j=1,2,\ldots,M$ are all the possible failure modes,
and $\bar{\alpha}$ is the no-failure case. If failure detection and isolation is
done on-line and in the presence of model uncertainty, then it constitutes adaptive failure detection/isolation. It must be noted that there
are numerous other decision rules that can be used, based on different
criteria. However, common to all the decision rules is the need for
obtaining the a-posteriori failure probabilities $p[\alpha_j/\lambda(k+1)]$ or
$p[\alpha_j/\lambda(0)]$.

Finally, system stabilization has the objective of obtaining the
optimal input u(k) so as to maintain the performance of the system as
closely as possible to a desired or acceptable system operation. If,
moreover, the design of u(k) is done on-line and in the presence of
model uncertainty, it constitutes adaptive stabilization.

It is well known [1-17] that the solution of all of the above problems involves the following cardinal elements, namely the model of the

plant or structure and the calculation of the mse estimate $\hat{x}(k+1/k+1)$.
The powerful and unifying nature of the multi-model partitioning framework comes to fore especially in the treatment and calculation of the
above quantities.

As was pointed out earlier, the model realism constitutes the most
vital element. However, the more realistic the model, the greater its
complexity, consequently, the greater is the difficulty in designing
the estimators and the greater the difficulty in implementing these estimators. The difficulties are compounded by the fact that, as the
realism of the model increases, so does the model uncertainty. To fully
account for this necessitates increasing even further the model complexity. Specifically in the classical approach, the state-vector
$x(k+1)$ is augmented with the unknown parameter vector θ, thus sub-
stantially increasing the state-vector dimensionality. For example, for
a completely unknown linear system, namely for unknown $\phi,J,Q,H,G,R,$
$\hat{x}_O$ and P_O, the dimensionality r of the unknown parameter vector θ is

575, for $n = m = p = 10$! Thus, $x_\alpha(k+1) \equiv \begin{bmatrix} x(k+1) \\ \theta \end{bmatrix}$ will have a vastly increased dimensionality, namely 585, in comparison to that of the state-
vector $x(k+1)$, which is $n = 10$.

Moreover, state augmentation leads from an originally linear system to essentially a non-linear one. Thus, following the classical approach of state augmentation leads us from a linear small-scale model
$(n = 10)$, to a non-linear large-scale model $(n_a = 585)$!

The classical approaches to the problem are:

a) To admit the more complex but realistic model, resulting in state
   augmentation, thus bearing the consequences of difficult estimation design problems, practically unimplementable estimators and
   ad-hoc approximations of optimal estimators, whose efficacy cannot
   be assessed without actual implementation;
b) To approximate the complex model by a less complex and hence less
   realistic model. The consequences are suboptimal estimators whose
   performance cannot be assessed without actual implementation.

Multi-Model Partitioning (MMP), Lainiotis [1-15], does not con-
front the original complex model directly nor does it approximate it by
a simpler but unrealistic model of reduced complexity. Instead, it de-
composes (partitions) the original estimation problem into a set of
estimation subproblems, the models of which are of considerably reduced
complexity. Namely, the MMP approach replaces the large-scale and complex model by a set of simpler and smaller-scale models. Since the sub-
problem models are simpler, the corresponding estimators are far easier
to derive and to implement. Specifically the MMP approach consists of
the selection of a parameter vector v of the model, each possible value
of which specifies a particular realization of the model. Thus, the
possible values of α describe a set of submodels. This gives rise to
the name MMP for the approach. Following the Bayesian viewpoint, the
choice of a particular model realization (value of α) is random (by
nature or an adversary) with a-priori pdf consistent with the nature of
the original complex model. Most importantly, if α is chosen to be a

pivotal parameter vector for the original model, then each possible
value of α corresponds to a simpler model, namely conditioning on α
(in the statistical sense) the original complex model reduces to a far
simpler one. There exist large and important classes of estimation
problems, both linear and non-linear, gaussian and non-gaussian, lumped
and distributed, where such a natural partitioning parameter exists.
In the problems under consideration in this paper the natural partitioning parameter vector is the parameter θ of Section 2.

In the following sections, the MMP algorithms for adaptive modelling, monitoring, failure detection and stabilization will be presented.

4. ADAPTIVE ESTIMATION

Because of time limitations, this presentation will be restricted
to discrete, linear, lumped, gaussian models, such as those given by
Eqs.(11)-(12) of Section 2. Moreover, to simplify the presentation, the
matrices G and I will be assumed zero. The corresponding optimal adaptive estimation algorithm is given, Lainiotis [1,4,9,16,17], by the
following:

<u>Algorithm I:</u>

The optimal estimate (forecasting) of the future state $x(\ell)$, where
$\ell > k+1$, is given by:

$$\hat{x}(\ell/k+1) = \Sigma\phi(\ell,k+1,\theta_i)\hat{x}(k+1/k+1,\theta_i)P(\theta_i/k+1) \tag{23}$$

and the optimal estimate of the current state $x(k+1)$ (filtering) is
given by:

$$\hat{x}(k+1/k+1) = \Sigma\hat{x}(k+1/k+1,\theta_i)P(\theta_i/k+1) \tag{24}$$

where the parameter-conditional estimate $\hat{x}(k+1/k+1,\theta_i)$ is given by the
well-known Kalman filter [20] or correspondingly by the Lainiotis filter [1,4,18,19]. The Kalman filter algorithm is given by:

$$\hat{x}(k+1/k+1,\theta_i) = \phi(k+1,k,\theta_i)\hat{x}(k/k,\theta_i) + K(k+1,\theta_i)[z(k+1)$$
$$- H(k+1,\theta_i)\phi(k+1,\theta_i)\hat{x}(k/k,\theta_i)] \tag{25}$$

and the gain matrix $K(k+1)$ is given by the following recursive equations:

$$K(k+1,\theta_i) = P(k+1/k,\theta_i)H^T(k+1,\theta_i)[H(k+1,\theta_i)P(k+1/k,\theta_i)H^T(k+1,\theta_i) +$$
$$+ R(k+1,\theta_i)] \tag{26}$$

$$P(k+1/k,\theta_i) = \phi(k+1,k,\theta_i)P(k,0/\theta_i)\phi^T(k+1,k,\theta_i) + Q(k,\theta_i) \tag{27}$$

$$P(k+1,0/\theta_i) = [I-K(k+1,\theta_i)H(k+1,\theta_i)] \tag{28}$$

The corresponding a-posteriori probabilities $p[\theta_i/\lambda(k+1)] \equiv p(\theta_i/k+1)$ are given by:

$$p(\theta_i/k+1) = \frac{L(k+1/k+1,\theta_i)}{\displaystyle\sum_{j=1}^{M} L(k+1/k+1,\theta_j)p(\theta_j/k+1)} \, p(\theta_i/k) \tag{29}$$

where the likelihood ratios $L(k+1/k+1,\theta_i)$ are:

$$L(k+1/k+1,\theta_i) \equiv \left| P_{\underset{z}{\sim}}(k+1/k,\theta_i) \right|^{-1/2} \cdot \exp\left\{ -\frac{1}{2}\tilde{z}(k+1/\theta_i) \cdot \right.$$

$$\left. \cdot P_{\underset{z}{\sim}}^{-1}(k+1/k,\theta_i)\tilde{z}(k+1/\theta_i) \right\} \tag{30}$$

and

$$\tilde{z}(k+1/\theta_i) \equiv z(k+1) - H(k+1,\theta_i)\phi(k+1,k,\theta_i)\hat{x}(k/k,\theta_i) \tag{31}$$

$$P_{\underset{z}{\sim}}(k+1/k,\theta_i) \equiv H(k+1,\theta_i)P(k+1,0/\theta_i)H^T(k+1,\theta_i) + R(k,\theta_i) \tag{32}$$

The above recursive algorithm is to be used recursively for $k = 0,1,..$, and with initial conditions $\hat{x}(0/0,\theta_i)$ and $P(0,0/\theta_i) = P_o(\theta_i)$ specified a-priori. The model parameter values θ_i, their number M and the corresponding a-priori probabilities $p(\theta_i)$, are obtained from the physical problem and a-priori knowledge, or are obtained by a suitable quantization of a continuous parameter space and $p(\theta)$. Such quantization procedures are given in Lainiotis [1,4,17].

The optimal mse estimate $\hat{\theta}_s(k+1)$ of the unknown model parameter vector θ is given by:

$$\hat{\theta}_s(k+1) = \sum_{i=1}^{M} \theta_i p(\theta_i/k+1) \tag{33}$$

and the maximum a-posteriori likelihood estimate $\hat{\theta}_1(k+1)$ of θ is $\hat{\theta}_s(k+1) = \theta_i$, where θ_i corresponds to $p(\theta_i/k+1)$ such that

$$p(\theta_i/k+1) = \max\{p(\theta_j/k+1) \quad j = 1,2,\ldots,M\} \tag{34}$$

<u>Remarks</u>

The following remarks on Lainiotis' MMP adaptive estimator shed considerable light on its interesting structure, its implementational advantages, its learning capacity and its robust nature.

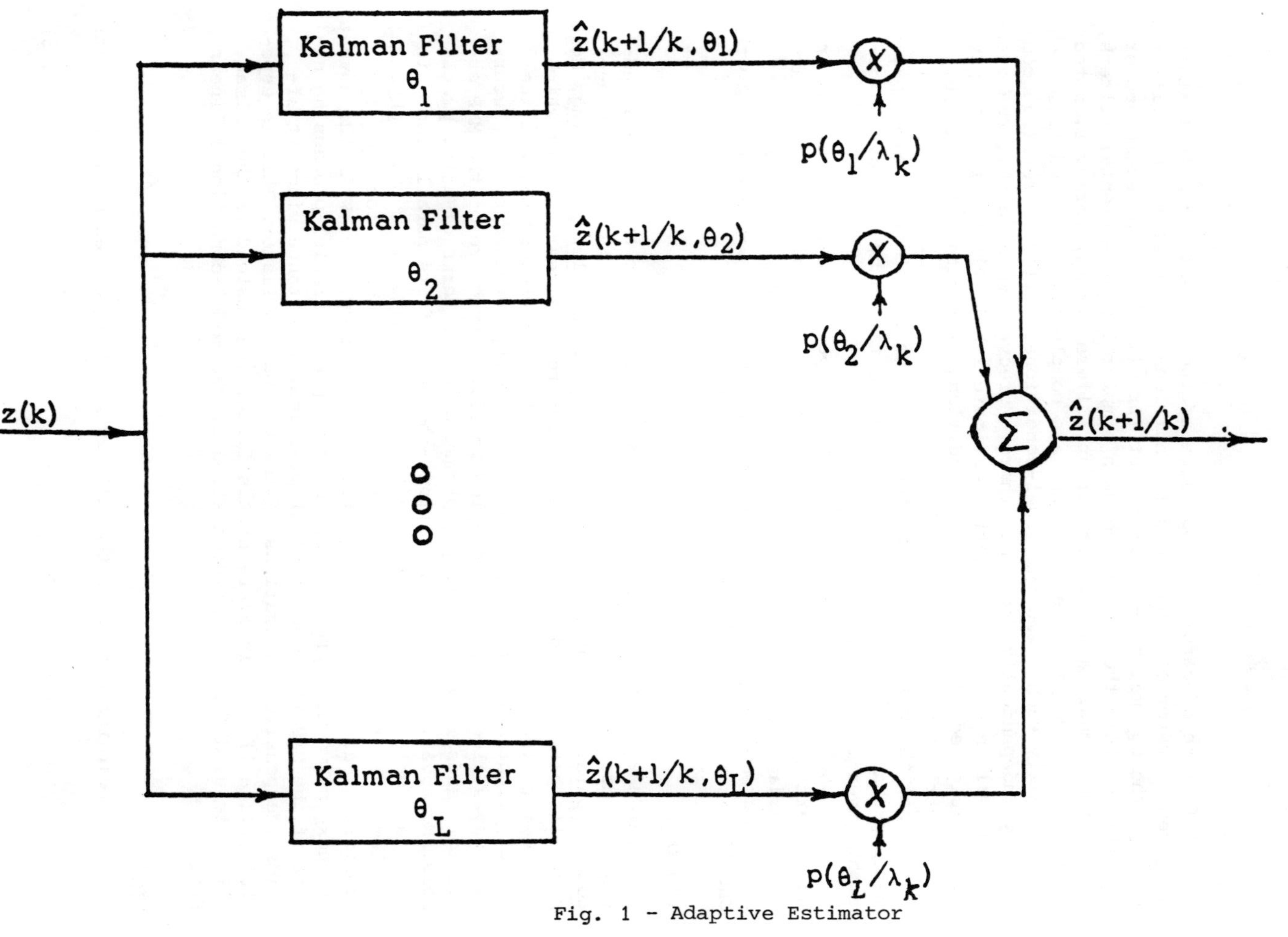

Fig. 1 – Adaptive Estimator

<u>Remark 1.</u>

A feature of cardinal importance of the MMP estimator is that it is given in terms of <u>linear</u> model-conditional or model-matched filters. Namely, the optimal non-linear filter for $\hat{x}(k+1/k+1)$ is decomposed or partitioned into the set of the much simpler <u>linear</u> elemental filters, namely the linear Lainiotis or Kalman filters. It is, moreover, noted that these elemental linear filters are complexely <u>decoupled</u> from each other, namely each elemental model-conditional filter is realized completely independently from the others. Moreover, we note that the overall optimal, non-linear, adaptive estimate $\hat{x}(k+1/k+1)$ is given simply as the weighted sum of the model-conditional estimates $\hat{x}(k+1/k+1,\theta_i)$. This constitutes a natural parallel realization form. Finally, it is noted that the weighting coefficients are the a-posteriori model parameter probabilities which are recursively obtained by the Bayes algorithm given in Eq.(29).

In conclusion, it is seen that the adaptive non-linear estimator has a partitioned or decomposed structure which consists of a <u>linear</u> and non-adaptive part, composed of a set of linear estimators and a non-linear and adaptive part composed of the set of a-posteriori probabilities acting as weighting coefficients.

<u>Remark 2.</u>

The MMP adaptive estimator has considerable implementational advantages due to its partitioned structure. Namely, its <u>naturally decoupled parallel structure</u> lends admirably to <u>parallel processing</u>. Moreover, its parts consist of the set of recursive linear filters which are easily implementable and do all of the data processing, and the set of weighting coefficients which are easily obtained by the recursive Bayes algorithm. The decoupled parallel structure of the MMP estimator lends it well to easy implementation in terms of array processors. Moreover, this implementation - because of its parallelism - is <u>very fast</u> which enables us to use the MMP estimator in complex physical problems that require large amounts of data processing for <u>real-time</u> monitoring, adaptive modelling, failure detection, etc. With the advent of inexpensive microprocessors, the decoupled, parallel-processing nature of the MMP is of substantial practical importance since it affords inexpensive realizations of the optimal estimator. This is especially true if the linear Lainiotis filter [8] instead of the Kalman filter is used for the implementation of the model-conditional linear filters.

<u>Remark 3.</u>

The learning nature of the MMP can be readily seen by noting that:

$$\lim_{k\to\infty} P(\theta_i/k+1) \to 1 \qquad (35)$$

where θ_i is the true model parameter value or the <u>closest to the true</u> if the true value is not included, inadvertedly, in the set of values of θ. This clearly demonstrated in the following examples. Moreover, the learning or adaptive nature of the MMP estimator can be seen further by noting that the MMP converges to the linear filter matched to true model that generated the processed data or to the model closest to it, namely:

$$\lim_{k \to \infty} \hat{x}(k+1/k+1) \to \hat{x}(k+1/k+1, \theta_i) \tag{36}$$

where $\hat{x}(k+1/k+1, \theta_i)$ is the linear filter matched to the true parameter value or the closest to it.

Remark 4.

    The MMP has also robust nature. This can be demonstrated by noting that the partitioned realization has a natural "fault detection" and "correction" mechanism built into its weighted-sum parallel structure. Specifically, if a model-conditional filter fails, its estimate $\hat{x}(k+1/k+1, \theta_j)$ will not provide good state estimates, and consequently, the predicted estimates of $z(k+1)$ given $\lambda(k)$ will also be inferior to those of the other filters. Thus, the corresponding one-step prediction error $\tilde{z}(k+1/\theta_j)$ for the failing j-th filter will be larger than those of the other filters, namely:

$$\tilde{z}(k+1/\theta_j) = z(k+1) - \hat{z}(k+1/k, \theta_j)$$

$$= z(k+1) - H(k+1, \theta_j) \phi(k+1, k, \theta_j) \hat{x}(k/k, \theta_j) \tag{37}$$

will be larger than the $z(k+1/\theta_i)$ for $i = 1, 2, \ldots, M$ but $i \neq j$. This, in turn, will make the corresponding $p(\theta_j/k+1)$ tend to 0, as can be seen from Eqs.(29)-(31), which will cut off the diverging conditional estimate $\hat{x}(k+1/k+1, \theta_j)$ from the weighed-sum, thus, in essence, correcting this filter malfunction.

Remark 5.

    Finally, a theoretical remark. It is seen that the multi-model partitioning framework has expanded the optimal non-linear estimate $\hat{x}(k+1/k+1)$ into a set of basis functions, namely the linear, model-conditional estimates. Correspondingly, the MMP realization of the non-linear estimator constitutes a partitioning of the original non-linear mse estimation problem into a set of elemental linear mse estimation subproblems, which are, moreover, completely decoupled from each other. In this sense, the non-linear orthogonal projection necessary to yield the non-linear mse estimate $\hat{x}(k+1/k+1)$ has been decomposed into the set of linear orthogonal projections yielding the model-conditional estimates $\hat{x}(k+1/k+1, \theta_i)$.

Remark 6.

    The MMP computational burden depends on the dimensionality of the state and on whether θ is discrete or continuous. There exist important

applications where θ is naturally discrete such as in the case of
failure detection. In other applications, θ is a continuous rv, resulting in a continuous a-posteriori pdf. This requires a non-denume-
rable infinity of linear filters for the exact realization of the optimal adaptive estimator. Thus, there is an apparent penalty to be
paid in using the partitioning approach. However, this penalty is only
apparent since it is very simple to obtain parameter-space quantiza-
tions yielding the algirithm I. These quantization Lainiotis [1,4,17]
are very effective and yield excellent estimates of the state as well
as of the unknown parameters. This is especially true in view of the
remarkable property of the MMP, namely that of its convergence to the
true value of θ or to its closest if the true value is not included.
This will be demonstrated in a following example.

Example 1 -   Structure and parameter uncertainty

This example illustrates the use and effectiveness of the algorithm in adaptive estimation with structural and parametric uncertainty.
The model is given by:

$$z(k+1) = \begin{bmatrix} 1 & 1 \end{bmatrix} x(k+1) + v(k+1) \tag{38}$$

$$x(k+1) = \begin{bmatrix} e^{-\alpha_0} & 0 \\ 0 & e^{-\beta} \end{bmatrix} x(k) + \begin{bmatrix} 1 & 0 \\ 0 & 1 \end{bmatrix} u(k) \tag{39}$$

where $R(k,\theta) = r$, and

$$Q(k,\theta) = \begin{bmatrix} g_s(1-e^{-2\alpha}) & 0 \\ 0 & g_n(1-e^{-2\beta}) \end{bmatrix} \tag{40}$$

The unknown parameter vector is defined as $\theta^T \equiv [\alpha \ g_s \ \beta \ g_n \ r]^T$. It is
thus seen that the uncertainty is both structural as well as parametric
since it is not known whether the state is 2-dimensional or 1-dimen-
sional (as it would be if $\beta = 0$ and $g_n = 0$), and the value of the parameter vector is unknown. In the simulation, the actual data was generated by a 1-dimensional signal, with actual θ-value
$\theta^T = [0.2 \ 2.0 \ 0 \ 0 \ 1.0]^T$.

To implement the algorithm, ten Kalman filters were used (ten
quantization levels for θ were chosen), one of which matched the data
generation model. The first two filters have the same state-vector
dimensionality as the data generation model, and filter 1 in the bank
of Kalman filters has the true parameter values (as given by θ^*) while
filter 2 has erroneous parameter values. Filters 3-10 were matched to
a 2-dimensional state vector, with various values for θ as given in
Table I.

The results after 300 samples are summarized in Table I, namely
the a-posteriori probability $p(\theta_i/k)$ after $k = 300$ samples for each
filter (for each $\theta = \theta_i$) is given in Table I.

Fig. 2 illustrates the results. It is seen from Fig. 1 that the

TABLE I

Filter	$\theta^T = [\ \alpha$	g_α	β	g_β	$r\ ]^T$	$p(\theta/300)$
1*	$\theta_1^T = [0.2$	2.0	0	0	$1.0]^T$	1.000
2	$\theta_2^T = [0.2$	2.0	0	0	$2.0]^T$	0.000
3	$\theta_3^T = [0.2$	2.0	0.02	4.0	$1.0]^T$	0.000
4	$\theta_4^T = [0.2$	2.0	0.02	4.0	$2.0]^T$	0.000
5	$\theta_5^T = [0.2$	2.0	0.02	8.0	$1.0]^T$	0.000
6	$\theta_6^T = [0.2$	2.0	0.02	8.0	$2.0]^T$	0.000
7	$\theta_7^T = [0.2$	2.0	0.04	4.0	$1.0]^T$	0.000
8	$\theta_8^T = [0.2$	2.0	0.04	4.0	$2.0]^T$	0.000
9	$\theta_9^T = [0.2$	2.0	0.04	8.0	$1.0]^T$	0.000
10	$\theta_{10}^T = [0.2$	2.0	0.04	8.0	$2.0]^T$	0.000

correct values of the true model as well as the correct state-vector
dimensionality were identified in less than 20 samples, and convergence
was achieved in less than 250 samples.

Example 2 - Response of tall buildings to random ground motion

Knowledge of the response of tall buildings to random ground motion
is important in the study of earthquakes on such structures. Eringen
[23] and Lin [24] have obtained a simplified model of a tall building
in which they assume concentrated masses at the floor levels, with the
masses connected by springs, and each mass subjected to viscous damping.
The system is driven by a random ground displacement.

The ground displacement g is assumed to be generated by a second
order dynamical system that is driven by white gaussian noise,

$$\ddot{g} + 2v\,\alpha\,\dot{g} + \alpha^2 = \sigma w \tag{41}$$

where v is the damping rate, α is the undamped natural frequency, and
σ is the standard deviation of the driving noise. The structural differential equations are written in terms of damping ratios n_i and undamped natural frequencies β_i, where $i = 1,2,\ldots,N$

$$\ddot{x}_i + 2n_i\beta_i x_i + \beta_i^2(x_i - x_{i-1}) + \beta_{i+1}^2(x_i - x_{i+1}) = 0 \tag{42}$$

$$i = 1,2,\ldots,N-1$$

$$\ddot{x}_n + 2n_N\beta_N\dot{x}_N + \beta_N^2(x_N - x_{N-1}) = 0 \tag{43}$$

$$x_0 = g \tag{44}$$

By letting $n_i = n$ and $\beta_i = \beta$ for $i = 1,2,\ldots,N-1$, these equations reduce
to the system discussed by Lin.

The response of a three story building to random ground motion was
simulated. The data processed was generated by solving the system of
Eqs. (36)-(39) using a Runge-Kutta-Gill algorithm. The damping ratios
$n_i = n = 0.8$, and the undamped natural frequencies $\beta_i = \beta = 2$ radians/
second. The damping rate $v = 0.5$, and the undamped natural frequency

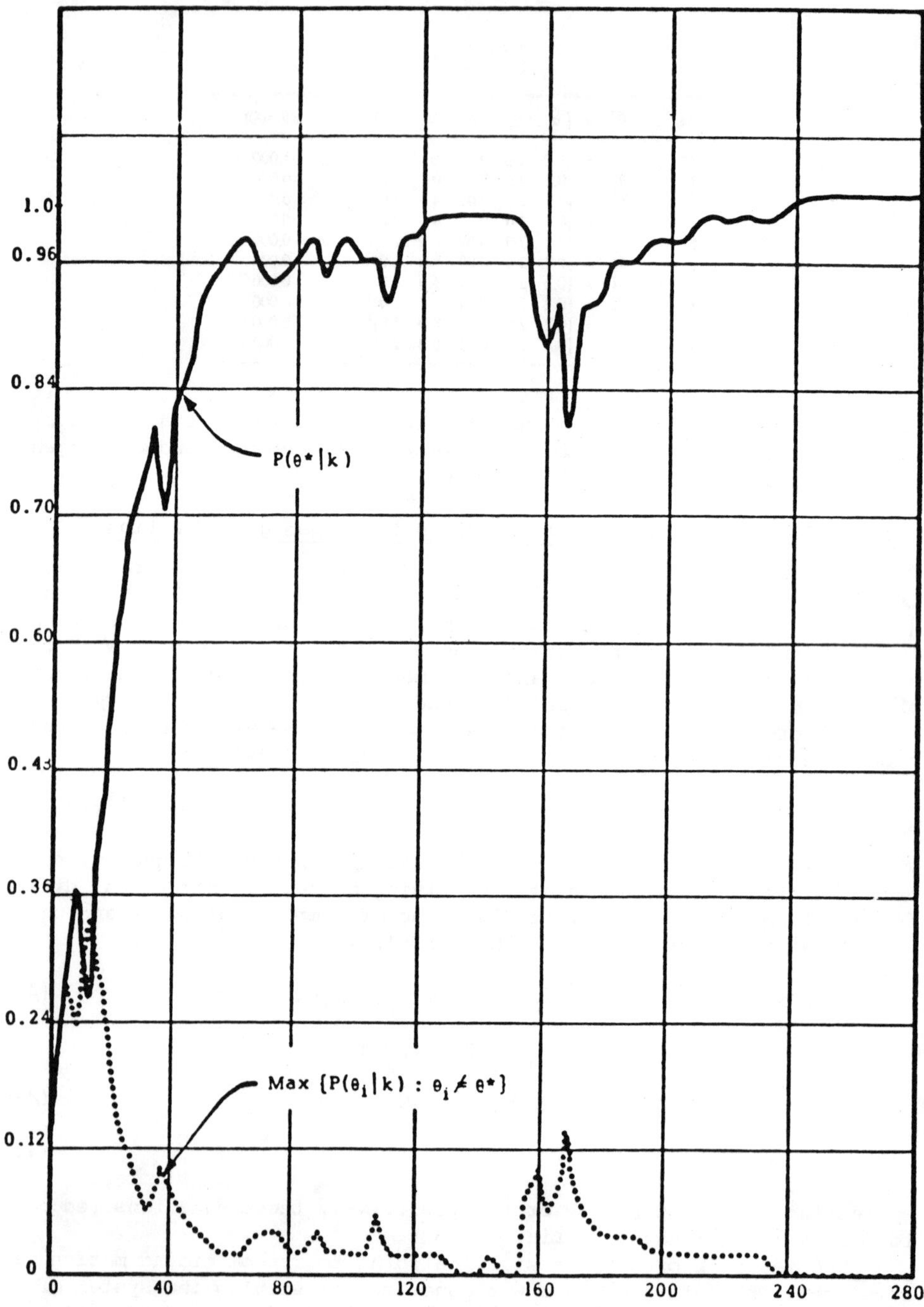

Fig. 2 - Vertical scale: a-posteriori probability $P(\theta/k)$. Horizontal scale: number of samples.

$\alpha = 1$ radians/sec. The standard deviation σ of the driving noise is
0.05. The time sampling interval Δ is set equal to $\pi/5$ which gives
10 samples per period of the higher undamped natural frequency α. It
was assumed that measurements of the velocity of each floor and the
ground velocity are available at each sampling instant. For each measurement, the measurement error is considered to be 20% of the true
velocity.

In the simulation, the unknown parameter vector consisted of
$\theta^T \equiv [\eta \quad \beta]$, namely the damping ratio and the undamped natural frequencies of the three-story building. The results of the simulation
are shown in Figs. 3-6. The true values are shown with solid lines in
the figures. It is seen that after a transient period at the beginning,
the estimates track the true values very well. Moreover, the unknown
parameters were learned quickly.

Example 3 - Parametric uncertainty

This example illustrates the effectiveness of the Lainiotis MMP
algorithm whether or not the true parameter value is included in the
set of possible θ values. The model for this problem is given by:

$$x(k+1) = \theta x(k) + x(k) \tag{45}$$

$$z(k+1) = x(k+1) + v(k+1) \tag{46}$$

where $x(0)$ is gaussian, zero-mean, and with unit variance, and the
true parameter value of θ is 0.4. There were two simulations run. In
the first, the data generated by the above model (namely the measurements z) were processed by an MMP estimator with seven model-conditional filters matched to the following possible values of θ {0.1, 0.2,
0.3, 0.4, 0.5, 0.6, 0.7}, all with equal a-priori probability
$p(\theta_i) = 1/7$, $i = 1,2,\ldots,7$. It is noted that in this simulation the
true value 0.3 was included in the set of θ-values. The results of this
simulation are shown in Fig. 7, which demonstrates the learning property of the MMP, namely the convergence to the true θ value, when it
is included in the set.

In the second simulation, the data generated by the above model
was processed by an MMP estimator with five model-conditional filters
matched to following values of θ {0.1, 0.2, ..., 0.5}, all having equal
a-priori probabilities $p(\theta_i) = 1/5$, $i = 1,2,\ldots,5$. It is noted that in
this simulation the true value of θ was not included in the included
in the set of possible values. Moreover, the 0.3 value was also omitted
making the 0.5 value the closest to the true value. The results of this
simulation are shown in Fig. 8, which demonstrates the remarkable
learning property of the MMP, namely its convergence to the value closest to the true value, in the case where the true value is not included. Moreover, we note that this convergence occurs earlier than in
the previous simulation where 7 possible values were included. This
shows that with fewer choices the convergence is faster.

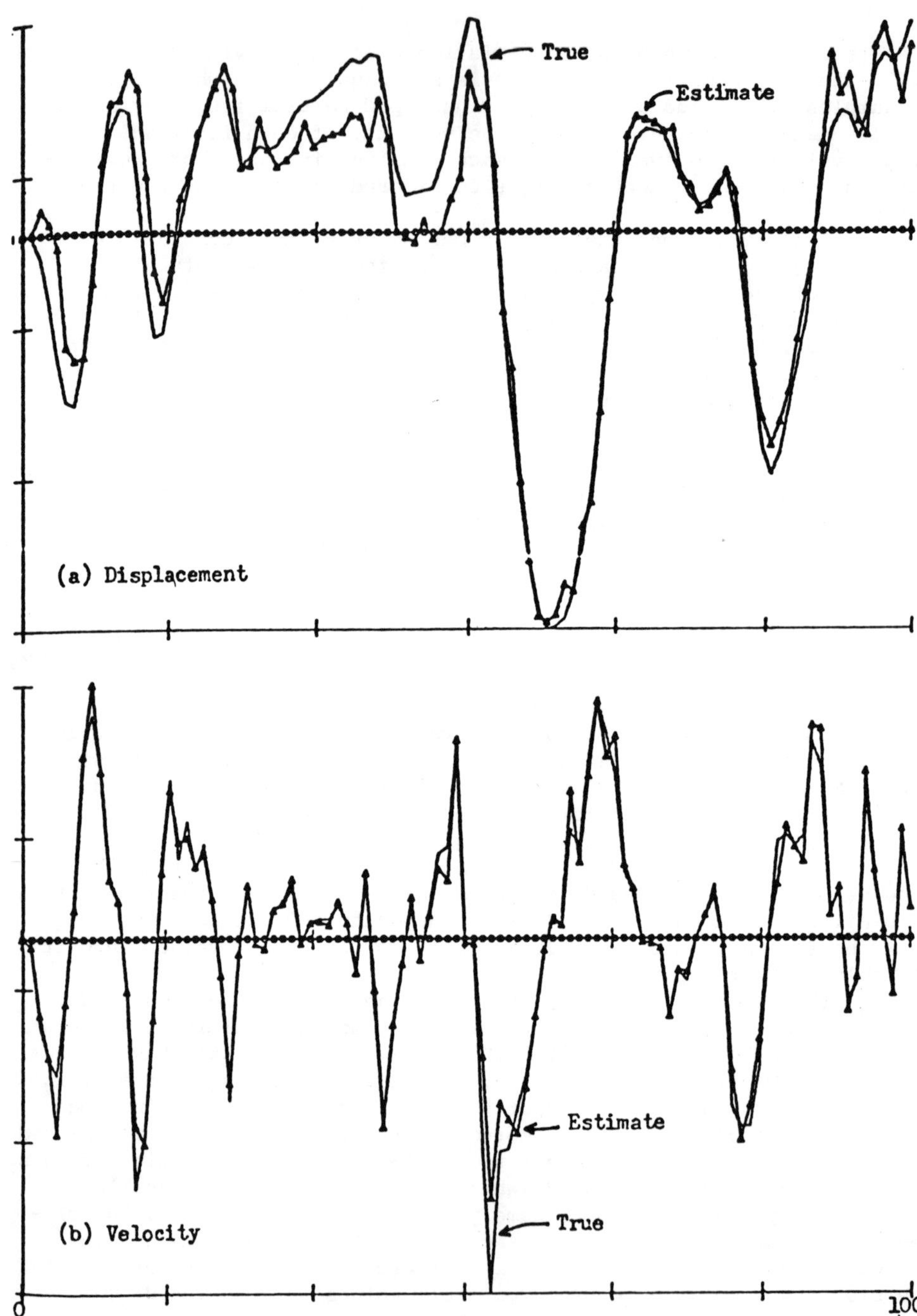

Fig.   3 –   Random ground motion

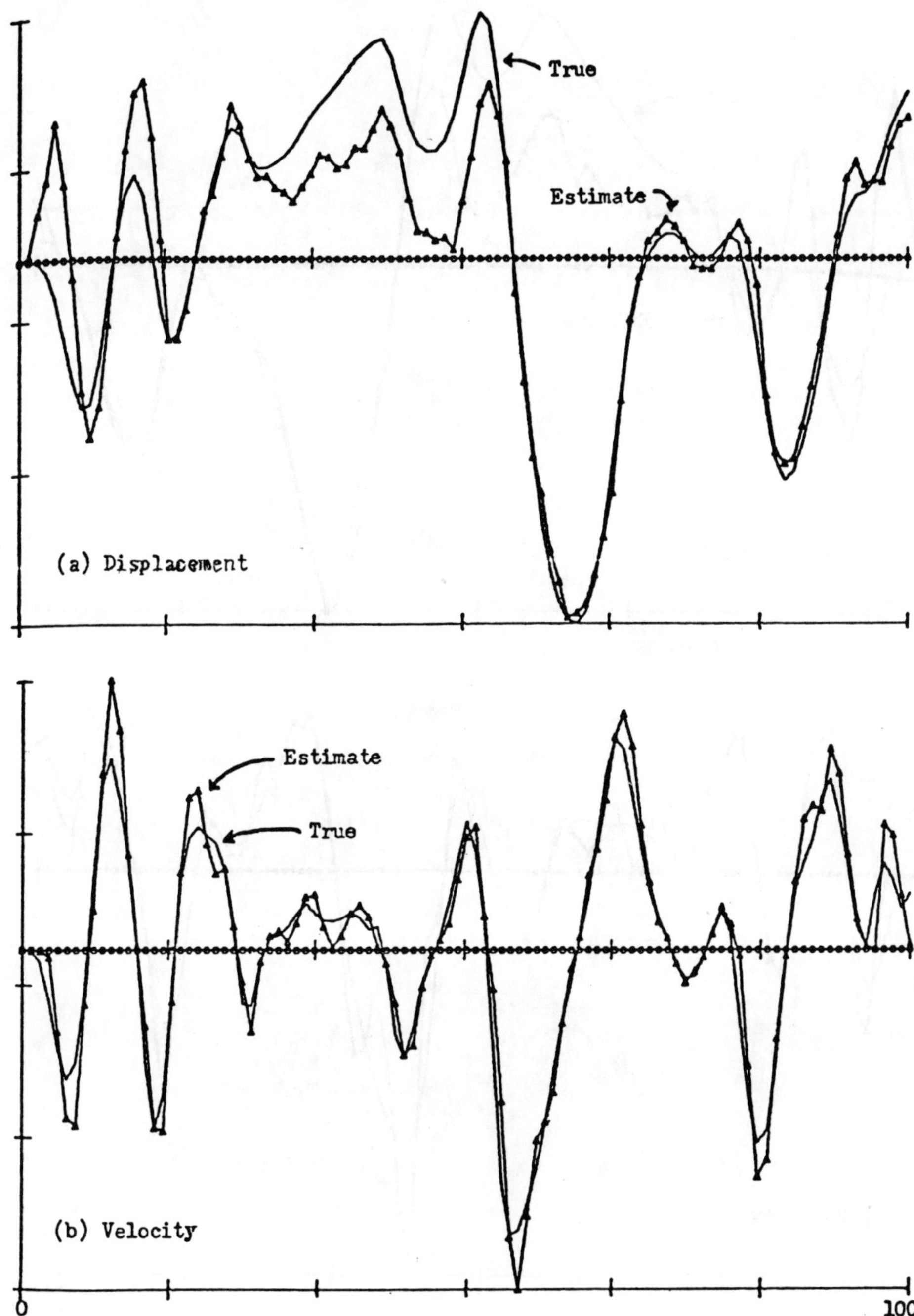

Fig. 4   —   First floor motion

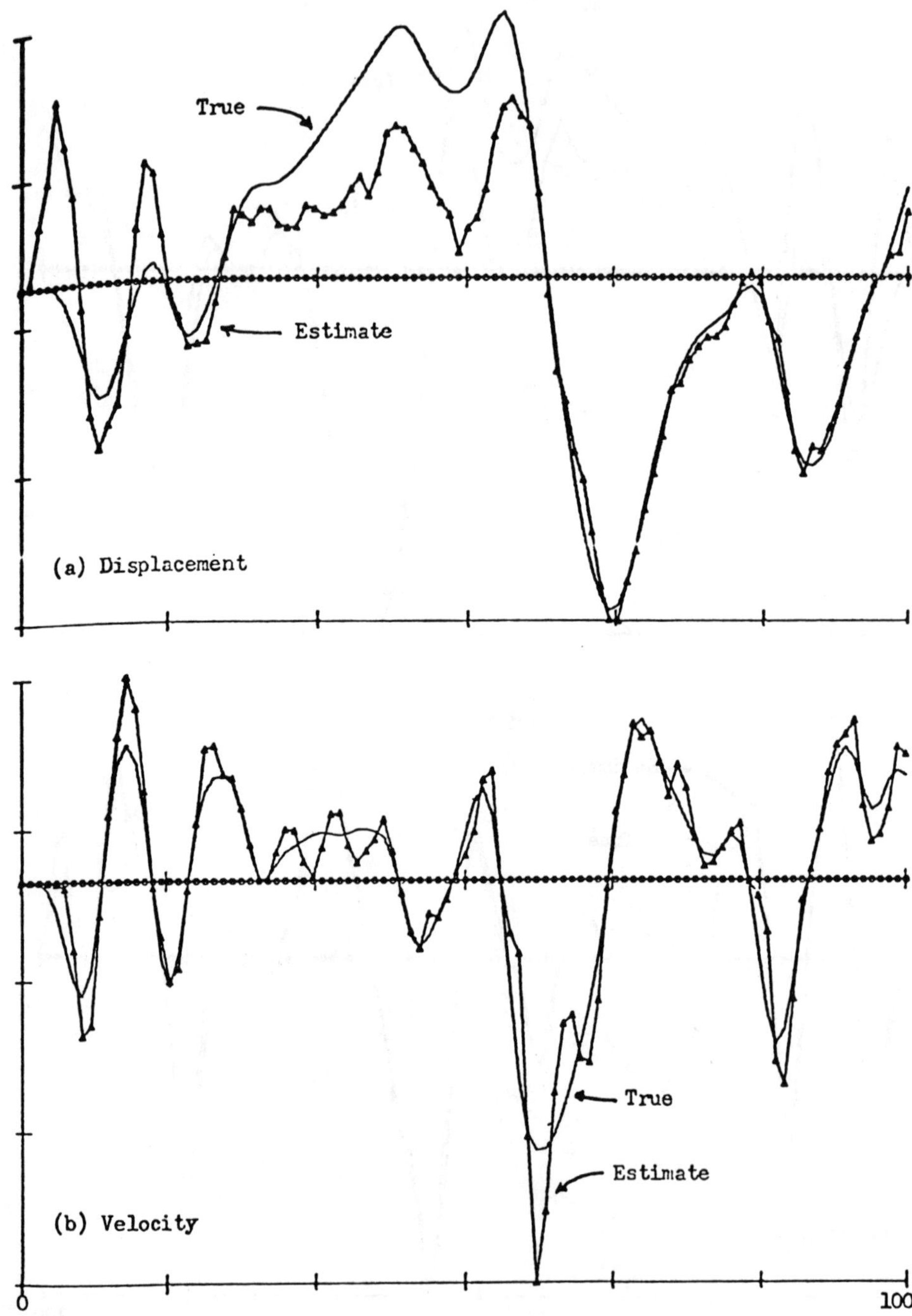

Fig.   5 -   Second floor motion

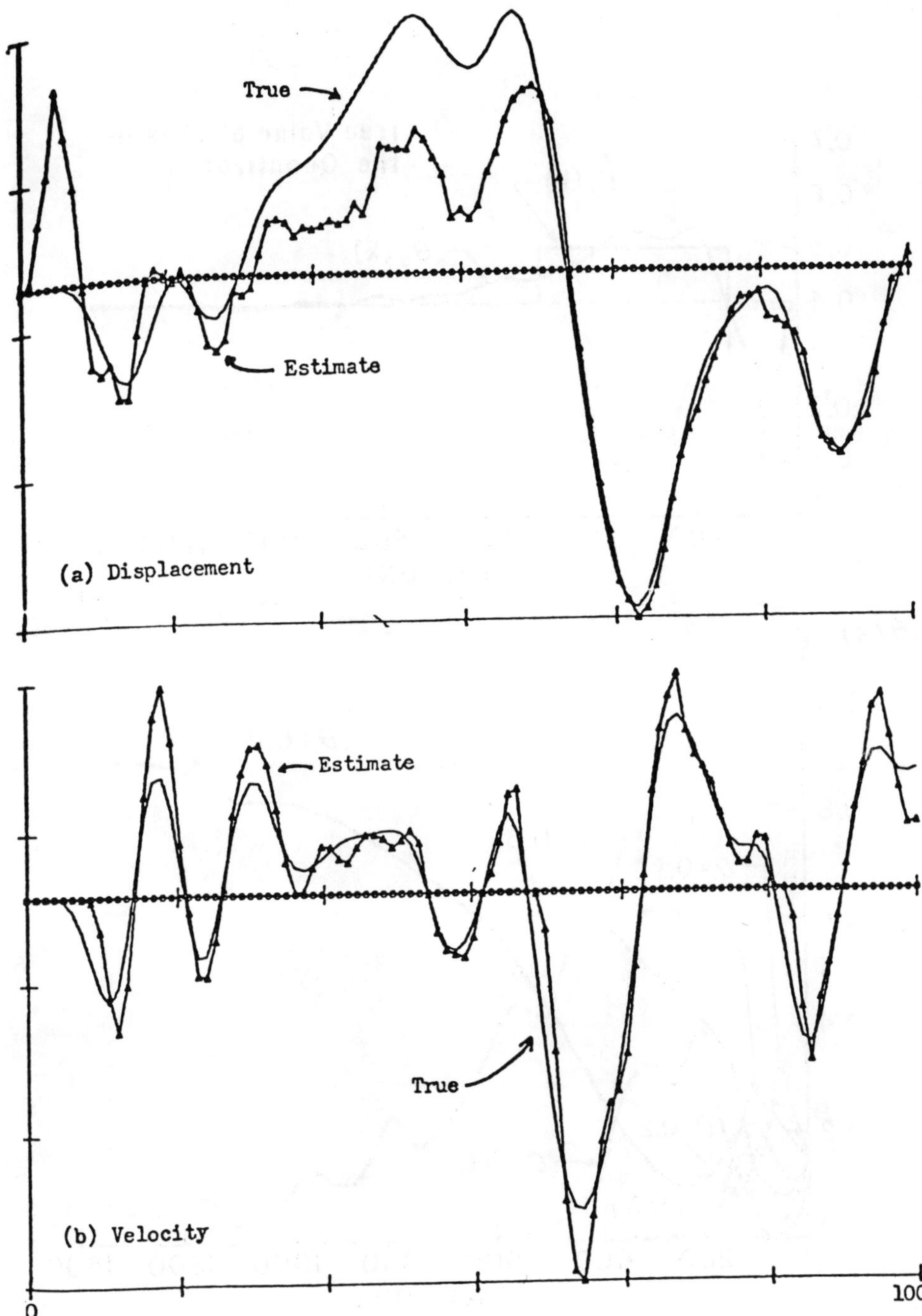

Fig.   6 -   Third floor motion

True Value of θ is in the Quantization
0.7
0.6
0.5
0.4
0.3
0.2
0.1
0
$\hat{\theta}_\ell(k)$
$\hat{\theta}_S(k)$
0   200   400   600   800   1000   1200   1400
TIME UNIT

p(θ/k)
1.0
0.8
0.6
0.4
0.2
0
θ = 0.4
θ = 0.5
θ = 0.1
θ = 0.3
θ = 0.2
θ = 0.6
θ = 0.7
0   200   400   600   800   1000   1200   1400
TIME UNIT

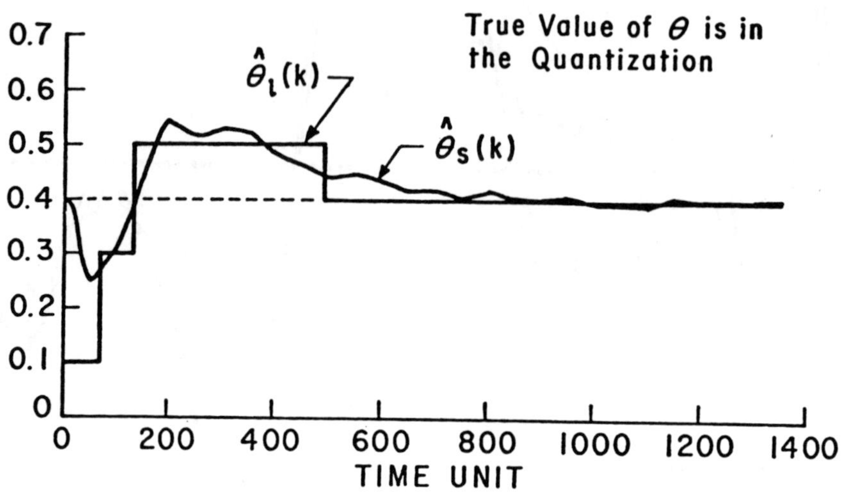

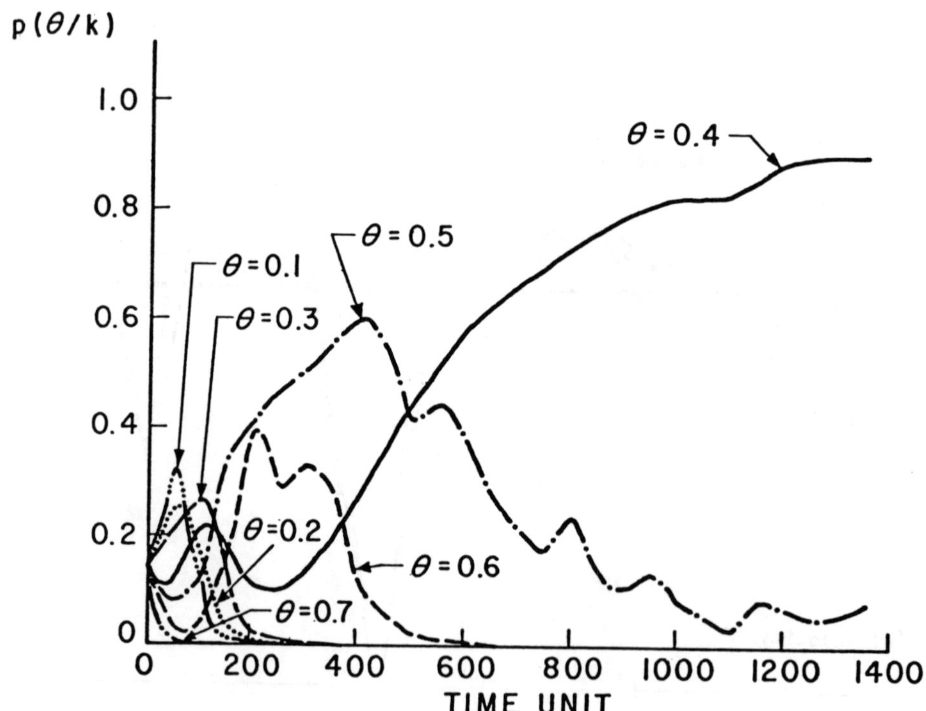

Fig. 7

True Value of θ is not
in the Quantization
0.7
0.6
0.5
0.4
0.3
0.2
0.1
0
θ̂s
θ̂l
0   200   400   600   800   1000   1200   1400
TIME UNIT

p(θ/k)
1.0
0.8
0.6
0.4
0.2
0
θ = 0.5
θ = 0.1
θ = 0.2
θ = 0.7
θ = 0.6
0   200   400   600   800   1000   1200   1400
TIME UNIT

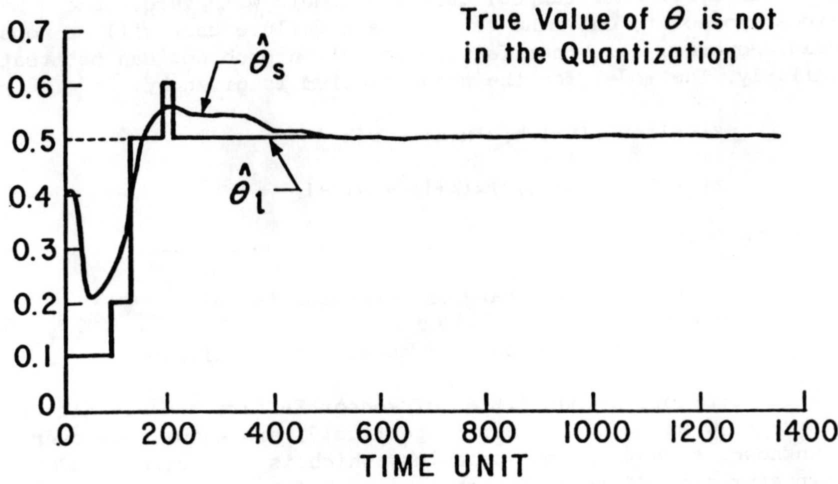

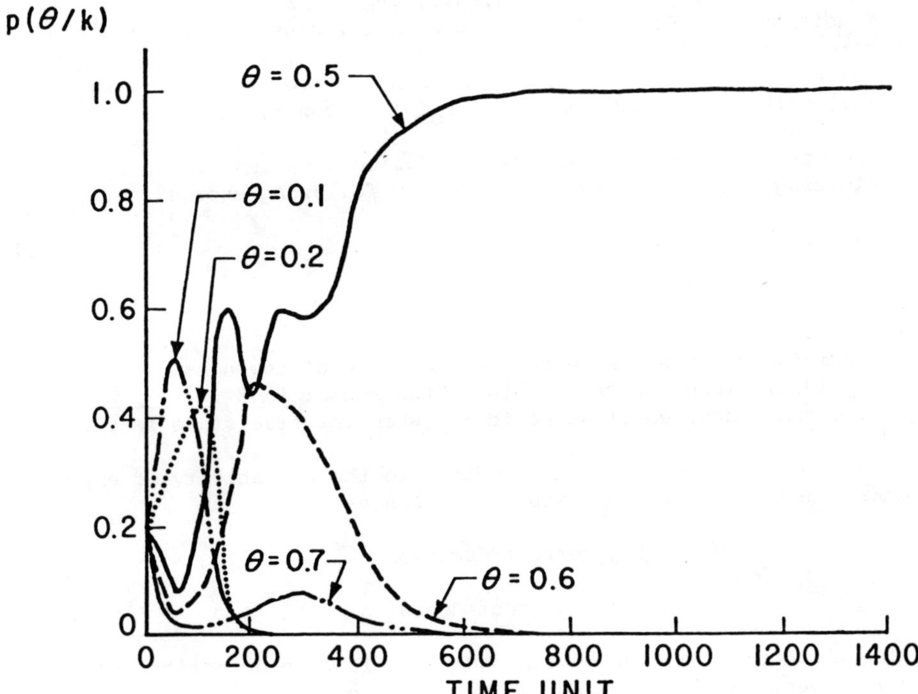

Fig. 8

5. ADAPTIVE FAILURE DETECTION

Because of time limitations this presentation will be restricted to discrete, linear, lumped, gaussian models with zero G and J matrices. Moreover, only the total sensor failure case will be considered. However, all other failure detection problems can be treated similarly. The model for the above problem is given by:

$$x(k+1) = \phi(k+1,k,\theta)x(k) + w(k) \tag{47}$$

$$z(k+1) = \beta H(k+1,\theta)x(k+1) + v(k+1) \tag{48}$$

where the indicator variable β takes values as follows:

$$\beta = \begin{cases} 0 & \text{for the case of no sensor failure} \\ 1 & \text{for the case of total sensor failure} \end{cases}$$

The a-priori probabilities of sensor failure and no-failure $p(\beta = 1) = P_1$, and $p(\beta = 0) = P_o$, respectively, are known. Moreover, θ is an unknown, time-invariant paramter, which is considered random with known a-priori pdf $p(\theta)$. For convenience of the presentation, θ takes discrete values only, namely $\theta \epsilon \{\theta_1, \theta_2, \ldots, \theta_M\}$.

Conditioned on θ, $v(k+1)$, $w(k)$ are white, gaussian, uncorrelated, with zero means, and covariances $R(k+1,\theta)$ and $Q(k,\theta)$, respectively. The $x(0)$ conditioned on θ is also gaussian uncorrelated with $v(k+1)$, and $w(k)$, with $\hat{x}_o(\theta)$, and covariance $P_o(\theta)$. Given θ, the model is completely known.

The optimal decision algorithm is desired, optimal in the sense of minimizing the risk as expressed by the following expression:

$$\text{Risk} = \sum_{i,j} C_{ij} P(\beta_j, \delta_i) \tag{49}$$

where
β_j = state β_j of the sensor, $\beta_j \epsilon (0,1)$;
δ_i = decision d_i, that is we decide the state of the sensor is $\delta_i \epsilon (0,1)$, when the true state of the sensor is β_j;
C_{ij} = the cost incurred if we decide δ_i when the true state is β_j.

We note that the above risk reduces to the probability of error in making decisions if the costs are chosen as

$$C_{ij} = \begin{cases} 0 & \text{if } i=j \text{ correct decision} \\ 1 & \text{if } i \neq j \text{ incorrect decision} \end{cases}$$

For the above problem, the optimal adaptive failure-detection procedure is given, Lainiotis [1-4] by the following recursive algorithm.

Algorithm II:

Decide according to whether $\Lambda(N)$ is greater or smaller than the threshold T, namely

$$\text{decide} \begin{cases} \text{no sensor failure if } \Lambda(N) > T \\ \text{sensor failure if } \Lambda(N) < T \end{cases}$$

where:

$$T = \frac{P_o(C_{10}-C_{00})}{P_1(C_{01}-C_{11})} \tag{50}$$

N = the total number of measurements used for detection and $\Lambda(N)$ is obtained recursively using the following algorithm:

$$\Lambda(k+1) = L(k+1/k)\Lambda(k) \tag{51}$$

for $k = 0,1,\ldots,N$ and with initial value $\Lambda(0) = 1$.

The quantity $L(k+1/k)$ is defined and obtained recursively by the Bayes algorithms given in Eqs. (52)-(54) in the following page. The quantities $P_{\tilde{z}}(k+1/k,\theta_\mu)\,\hat{z}(k+1/\theta_\mu)$ are obtained also recursively by a linear filter, e.g. Kalman, matched to the model with parameter value θ_μ and with no sensor failure. That is they are defined by Eqs. (31)-(32), and obtained recursively by Eqs. (25)-(28). Since θ takes M possible values, M linear filters are required, each one matched to a possible value θ may take.

Remarks

The remarks made in connection with Algorithm I pertain here as well. Namely, Algorithm II exhibits a similar parallel-processing, robust, failure-detection and correction nature. It has an adaptive/learning capacity and it is also easily implementable.

Example 4 – Adaptive detection

This example illustrates the performance of the adaptive detection algorithm. The model is given by:

$$x(k+1) = \frac{1}{4} x(k) + w(k) \tag{55}$$

$$z(k+1) = \beta \begin{bmatrix} 2 \\ 1 \end{bmatrix} x(k+1) + \begin{bmatrix} v_1(k+1) \\ v_2(k+1) \end{bmatrix} \tag{56}$$

where β (0,1), with a-priori probabilities $P_o = P_1 = \frac{1}{2}$ and costs $C_{ij} = \begin{matrix} 0 \ i = j \\ 1 \ i \neq j \end{matrix}$. The $P_o(\theta) = 10$, $\hat{x}_o(\theta) = 0$, $R_1(k+1,\theta) = 1$, $R_2(k+1,\theta) = 10$ and $Q(k,\theta) = \theta$. That is the plant noise covariance $Q(k,\theta)$ is unknown from considerations of the physical situation that θ can take either of two

$$L(k+1/k) = \frac{\sum_{\mu=1}^{M} |P_z^{\sim}(k+1/k,\theta_\mu)|^{-1/2} \exp\left\{-\frac{1}{2} \tilde{z}^T(k+1/\theta_\mu) P_z^{\sim^{-1}}(k+1/k,\theta_\mu) \tilde{z}(k+1/\theta_\mu)\right\} p(\theta_\mu/k,H_1)}{\sum_{v=1}^{M} |R(k+1,\theta_v)|^{-1/2} \exp\left\{-\frac{1}{2} z^T(k+1) R^{-1}(k+1,\theta_v) z(k+1)\right\} p(\theta_v/k,H_o)} \tag{52}$$

$$p(\theta_\mu/k,H_1) = \frac{|P_z^{\sim}(k/k-1,\theta_\mu)|^{-1/2} \exp\left\{-\frac{1}{2} \tilde{z}^T(k/\theta_\mu) P_z^{\sim^{-1}}(k/k-1,\theta_\mu) \tilde{z}(k/\theta_\mu)\right\} p(\theta_\mu/k-1,H_1)}{\sum_{i=1}^{M} |P_z^{\sim}(k/k-1,\theta_i)|^{-1/2} \exp\left\{-\frac{1}{2} \tilde{z}^T(k/\theta_i) P_z^{\sim^{-1}}(k/k-1,\theta_i) \tilde{z}(k/\theta_i)\right\} p(\theta_i/k-1,H_1)} \tag{53}$$

$$p(\theta_v/k,H) = \frac{|R(k,\theta_v)|^{-1/2} \exp\left\{-\frac{1}{2} z^T(k) R^{-1}(k,\theta_v) z(k)\right\} p(\theta_v/k-1,H_o)}{\sum_{\rho=1}^{M} |R(k,\theta_\rho)|^{-1/2} \exp\left\{-\frac{1}{2} z^T(k) R^{-1}(k,\theta_\rho) z(k)\right\} p(\theta_\rho/k-1,H_o)} \tag{54}$$

TABLE II

$P(\theta_1/k,H_1)$	
01	.047890
02	.001140
03	.000071
04	.000000
05	.000000
06	.000000
07	.000000
08	.000000
09	.000000
10	.000000

$P(\theta_1/k,H_o)$	
01	.333333
02	.500000
03	.400000
04	.454545
05	.423077
06	.440678
07	.430657
08	.436306
09	.433103
10	.434913

$P(\theta_2/k,H_1)$	
01	.952110
02	.998860
03	.999929
04	1.000000
05	1.000000
06	1.000000
07	1.000000
08	1.000000
09	1.000000
10	1.000000

$P(\theta_2/k,H_o)$	
01	.666667
02	.500000
03	.600000
04	.545455
05	.576923
06	.559322
07	.569343
08	.563694
09	.566897
10	.565087

k	$L(k+1/k)$
01	11.145E+001
02	51.827E+004
03	11.860E+008
04	64.228E+052
05	22.305E+020
06	98.819E+007
07	69.311E+013
08	25.250E-001
09	16.741E+003
10	11.745E+006

values, namely θ (0,5) with a-priori probabilities $P(\theta = 0) = \frac{1}{3}$ and $P(\theta = 5) = \frac{2}{3}$. Thus, the initial conditions for the problem are:

$$\hat{x}(0/0,\theta_1) = \hat{x}(0/0,\theta_2) = 0 \tag{57}$$

$$P(0,0/\theta_1) = P(0,0/\theta_2) = 10 \tag{58}$$

$$P(\theta_1/0,H_1) = p(\theta_1/0,H_o) = 1/3 \tag{59}$$

$$P(\theta_2/0,H_1) = p(\theta_2/0,H_o) = 2/3 \tag{60}$$

We are given 10 measurements $(N = 10)$ which we are to use to decide whether there is failure or not. The decision rule is

$$\Lambda(10) \underset{H_o}{\overset{H_1}{\underset{<}{>}}} 1$$

where $T = 1$, since $\qquad T = \dfrac{P_o(C_{10} - C_{00})}{P_1(C_{01} - C_{11})} = 1$

$\Lambda(10)$ is obtained recursively from Eq. (51), with initial value $\Lambda(10) = 1$.

The simulation was made under no-sensor failure conditions and with $Q(k,\theta) = 5$. Of course, this fact is not known to the adaptive detection system. Given the 10 measurements and processing them recursively yields: $\Lambda(10) = 3.3 \cdot 10^{127} > 1$ and, hence, we decide that there is no sensor failure. Moreover, we note that the correct value of the parameter θ was learned essentially after using the first measurement only. Similarly, the correct decision concerning failure would have been reached using only one measurement. This can be readily deduced from Table II consisting of the computer data.

REFERENCES

[1] LAINIOTIS, D.G., "Optimal adaptive estimation: structure and parameter adaptation", IEEE Transactions Automatic Control, Vol. AC-16, pp.160-170, April 1971.
[2] LAINIOTIS, D.G. and PARK, S.K., "On joint detection, estimation and system identification: discrete data case", Int. J. Control, Vol.17, No.3, pp.609-633, March 1973.
[3] LAINIOTIS, D.G., "Supervised learning sequential structure and parameter adaptive pattern recognition: discrete data case", IEEE Trans. Inform. Theory, Vol.IT-17, pp.106-110, Jan. 1971.
[4] LAINIOTIS, D.G., "Partitioning: a unifying framework for adaptive systems. Part I: estimation", Proc. IEEE, pp.1026-1043, Aug. 1976.
[5] LAINIOTIS, D.G., DESHPANDE, J.G. and UPADHYAY, T.N., "Optimal adaptive control: a non-linear separation theorem", Int. J. Control, Vol.15, No.5, pp.877-888, May 1972.

[6] DESHPANDE, J.G., UPADYAY, T.N. and LAINIOTIS, D.G., "Adaptive
     control of linear stochastic systems", Automatica, Vol.9, pp.107-
     115, Jan. 1973.

[7] LAINIOTIS, D.G., "Partitioning: a unifying framework for adaptive
     systems, II: control", Proc. of the IEEE, Vol.64, No.8, pp.1182-
     1198, Aug. 1976.

[8] LAINIOTIS, D.G., "Partitioning filters", J. Inform. Sci., Vol.
     17, pp.177-192, Jan. 1979.

[9] LAINIOTIS, D.G., "Partitioning: the multi-model framework for es-
     timation and control, I: estimation", Proc. Int. Symp. on Systems
     Optimization and Analysis, A. Bensoussan and J.L. Lions (eds.),
     Springer-Verlag, 1979.

[10] EULRICH, B.J., ANDRISANI II, D. and LAINIOTIS, D.G., "Partitioning
     identification algorithms", IEEE Trans. on Automatic Control,
     Vol.AC-25, No.3, pp.521-528, June 1980.

[11] GOVINDARAJ, K.S. and LAINIOTIS, D.G., "A unifying framework for
     discrete linear estimation: generalized partitioned algorithms",
     Int. J. Control, 1978.

[12] ASHER, R.B. and LAINIOTIS, D.G., "Adaptive estimation of doubly
     stochastic Poisson processes with applications to adaptive optics",
     J. Inform. Sciences, Vol.12, Oct. 1977.

[13] LAINIOTIS, D.G. and PARK, S.K. and KRISHNAIAH, R., "Optimal state-
     vector estimation for non-gaussian initial state-vector", IEEE
     Trans. Automat. Contr., Vol.AC-16, pp.197-198, April 1971.

[14] PARK, S.K. and LAINIOTIS, D.G., "Monte-Carlo study of the opti-
     mal nonlinear estimator: linear systems with non-gaussian initial
     state", Int. J. Control, Vol.16, No.6, pp.1029-1040, 1972.

[15] LAINIOTIS, D.G. (ed.), Estimation Theory, American Elsevier,
     New York, 1974.

[16] LAINIOTIS, D.G., "Optimal nonlinear estimation", Int. J. Control,
     Vol.14, No.6, pp.1137-1148, 1971.

[17] LAINIOTIS, D.G., "Partitioned estimation algorithms I: nonlinear
     estimation", J. Inform. Sciences, Vol.7, No.3, pp.203-255, 1974.

[18] COSMIDIS, E.T., "Comparison of the Kalman and Lainiotis filters",
     Proc. IEEE Mediterranean Electrotechnical Conference (MELECON),
     Col.II, May 1983.

[19] BAPTISTA, E.A. and COSMIDIS, E.T., "Kalman-Bucy and Lainiotis
     filters: comparisons and interconnections", Proc. 1979 Conf. on
     Modelling and Simulation, pp.285-291, Instrument Society of
     America, 1979.

[20] KALMAN, R.E., "A new approach to linear filtering and prediction
     problems", J. Basic Engin. 82: 34-45 çMarch 1960).

[21] SAGE, A.P. and MELSA, J.L., Estimation Theory, McGraw-Hill Book
     Company, New York, 1971.

[22] AOKI, M., Optimization of Stochastic Systems, New York, Academic
     Press, 1967.

[23] ERINGEN, A.C. "Response of tall buildings to random earthquakes",
     Proc. 3rd US Nat. Cong. Applied Mech., pp.141, ASME, 1958.

[24] LIN, Y.K., Probabilistic Theory of Structural Dynamics, McGraw-
     Hill, 1967.

DEVELOPMENT OF STRUCTURAL DAMAGE SUPERVISING SYSTEMS

W. Wedig, Karlsruhe, FRG

Summary:

For elastic structures of a high demand of security it is necessary to observe and to control the stiffness parameters
and their rate of changement. This is possible by measuring
and analysing the structural response when the system is in
motion driven by its natural excitations such as turbulent
flows or gusty winds without any knowledge on their intensities, mean values and statistical dependences or where they
are in action.

The applied parameter estimation methods are based on the Maximum Likelihood theory and derivable by means of the principle of minimal excitation energy. Problems of interest are
crack identification in bending oscillators, stiffness monitoring in vibration chains and nonlinear estimation procedures
in case of restricted measurements or for systems under band-
limited stochastic excitations. The methods are verified in
numerical simulations using time-discrete models.

1. Introduction

A supervising or monitoring of elastic structures is of increasing importance if a high demand of security is required. To
fulfill the requirement we permanently have to control the
stiffness parameters of the elastic structure when it is in
motion or in action effected by its natural excitations such
as gusty winds, turbulent flows, traffic loads or harmonic mass
forces of rotating machines.

Provided that we know a mathematical model of these excitation
processes as well as of the elastic structure, we are able to

A. C. Lucia (ed.), Advances in Structural Reliability, 195–215.
© *1987 by ECSC, EEC, EAEC, Brussels and Luxembourg.*

estimate the parameters of both models by measuring only the
response of the structure without any further knowledge on the
excitations, particularly, on their intensities, mean values,
statistical dependences, frequency distributions or where
they are applied.

The basis of such parameter estimation methods is the principle
of minimal excitation energy which determines the unknown parameters from the measured response of the system in such a
way that the energy of the non-measurable excitation becomes
minimal. In case of stationary white noise excitations this
principle leads to estimators which coincide with those derived
by means of the Maximum Likelihood theory. However, the principle is also valid for nonstationary processes with arbitrary
frequency distributions and in the deterministic case of impact excitations, as well.

To show the efficiency of the applied estimation methods we investigate some interesting problems such as the measurement of
cracks in bending oscillators and the stiffness supervising in
one-dimensional structures. Furthermore, we give extensions to
nonlinear estimation procedures in case of coloured noise excitations or if from the total structural response only one degree of freedom is measurable. Such estimations may be adapted
for a permanent control of an existing structure of a high demand of security.

2. Parameter estimations in dynamic systems

To introduce the basics of parameter estimation methods, we
first consider the minimal model of such dynamic problems given
by the following scalar differential equation of a low-pass system driven by a white noise excitation.

$$\dot{X}_t + \omega_g X_t = \sigma \dot{W}_t, \qquad E(\dot{W}_t \dot{W}_s) = \delta(t-s). \qquad (2.1)$$

Herein, X_t is the measurable response process from which we have
to analyse the unknown frequency parameter ω_g of the system.
Subscript t denotes the time dependence and dots are derivatives
with respect to the time t. The system excitation is given by

zero mean white noise $\dot{W}_t$ with an unknown intensity σ and the normed delta-correlation, noted in (2.1).

The measurements in (2.1) are simulated by corresponding Arma (autoregressive moving average) models. They are time-discrete representations of the form:

$$\Delta x(k) = - \omega_g \Delta t\ x(k) + \sigma \Delta t\ z(k), \quad IC:\ x(0) = 0,$$

$$\Delta x(k) = x(k+1) - x(k), \qquad \text{for } k = 0,1,2,\ldots \qquad (2.2)$$

This is a recurrence formula which may be started under the initial condition $x(0) = 0$. Then, it allows to calculate the next states $x(1)$, $x(2),\ldots$ for each further time step Δt and for any given excitation process $z(k)$. To approximate white noise $\dot{W}_t$, we apply a digital generator producing pseudo-random numbers $z(k)$ which are uncorrelated and uniformly distributed with zero mean.

For a vanishing intensity $\sigma = 0$, the homogeneous difference equation (2.2) possesses the well-known solution

$$x(k+1) = \rho\ x(k), \qquad\qquad \rho = 1 - \omega_g \Delta t, \qquad\qquad (2.3)$$

$$|\rho| < 1, \qquad\qquad 0 < \Delta t < 2/\omega_g,$$

where ρ is the eigenvalue determining the stability of the numerical simulation in (2.2). Obviously, $x(k)$ is asymptotically stable if the absolute value of ρ is smaller than one. From this it follows that the applied scan rate Δt has to be smaller than $2/\omega_g$.

If we are now performing measurements of the states $x(k)$ and their increments $\Delta x(k)$ e.g. for the first three steps $k = 0$, 1 and 2, we obtain three equations from (2.2) with the unknown parameter ω_g and the non-measurable excitations $z(0)$, $z(1)$ and $z(2)$. Obviously, this corresponds to the well-known situation of the purely statical investigation of a beam with four support reactions. Since the three equilibrium equations are not sufficient to determine the four unknowns, we need one additional equation which follows from the minimal principles of mechanics.

In our situation of parameter estimation we analogously formulate a further equation or requirement that we have to determine the unknown parameter ω_g in such a way that the measured response $x(k)$ of the system is effected by a minimal excitation energy, denoted by

$$\sum_k z^2(k) = \text{Min!}, \qquad \sum_k [\Delta x + \omega_g \Delta t\, x]^2 = \text{Min!} \qquad (2.4)$$

Inserting the recurrence formula (2.2) and differentiating with respect to ω_g the necessary condition for the existence of the minimum is calculated to

$$\hat{\omega}_g = -\frac{\sum x(k)\, \Delta x(k)}{\Delta t\, \sum x^2(k)} \qquad (2.5)$$

This is the well-known frequency estimator of a low-pass system. It is the time-discrete representation of the corresponding Maximum Likelihood estimator (ML). Some more details on the ML theory and its application to dynamic systems are given in [1,2,3].

The estimator (2.5) is unbiased and consistent. To verify the first property we make use of the independence of Markow processes denoted by

$$\sum_k x(k)\, z(k) = 0. \qquad (2.6)$$

Clearly, the initial state $x(0)$ can arbitrarily be chosen and is therefore not influenced by the excitation $z(0)$. The same is true for the following situations $k = 1,2,\ldots$. Consequently, both processes are uncorrelated. This corresponds to the relation (2.6). For its application we multiply the recurrence formula (2.2) by $x(k)$ and sum up over all time steps k

$$\sum_k [\Delta x(k) + \omega_g \Delta t\, x(k)]x(k) = \sigma \Delta t \sum_k x(k)z(k) \qquad (2.7)$$

in arriving to the same estimator (2.5) if we take into account the statistical independence property (2.6). Some numerical evaluations of (2.5) are given in [4].

It is now quite easy to extend such considerations to the general case of linear or nonlinear dynamic systems subjected to several white noise processes $z_i(k)$ (i=1,2,...n) which are uncorrelated in space and in time. As shown in [5], we require that the unknown parameters are estimated from the measured system response in such a way that the mean square sum of all excitations becomes minimal.

$$\sum_{i,k} \sum z_i^2(k) = \text{Min} ! \qquad (2.8)$$

It can be evaluated by inserting the associated system equations and by differentiating with respect to the unknown parameters. Following [5], the derived estimators are time-discrete versions of the Maximum Likelihood results given in [2]. In the following we show some further applications which are interesting for the development of a structural damage supervising system.

3. Crack identification in bending oscillators

As a first application of the minimal principle (2.8) let us consider the mechanical model of a cracked bending oscillator shown in figure 1. We assume a bilinear spring characteristic

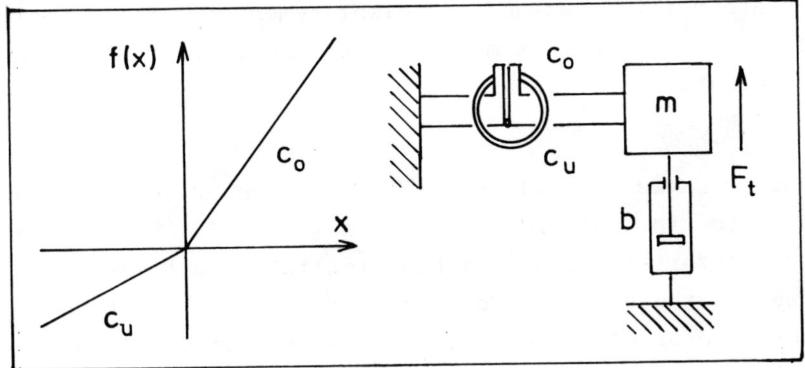

Fig. 1: Model of a cracked bending oscillator

with two stiffness parameters c_o and c_u in correspondence to the momentary displacement situation that the crack is closed for positive deflections or **opened**, in the inverse case. Furthermore, b denotes the coefficient of a viscous external damping and m is the mass of the oscillator under the force excitation F_t which may be given by zero mean white noise $\dot{W}_t$.

Acoordingly, the equation of motion of the cracked bending oscillator is given by

$$\ddot{X}_t + 2\delta\dot{X}_t + [\omega_o^2 s_o(X_t) + \omega_u^2 s_u(X_t)]\,X_t = \sigma\dot{W}_t, \qquad (3.1)$$

$$s_o(X_t) = \begin{cases} 1 \text{ for } X_t \geq 0, \\ 0 \text{ for } X_t < 0, \end{cases} \qquad s_u(X_t) = \begin{cases} 0 \text{ for } X_t \geq 0, \\ 1 \text{ for } X_t < 0. \end{cases}$$

Herein, s_o and s_u are unit step functions, ω_o^2 and ω_u^2 are the crack frequencies associated to the stiffness parameters c_o and c_u, δ is the damping coefficient related to the mass and σ is the intensity of the white noise excitation $\dot{W}_t$. The equation (3.1) is simulated by its discrete version of the form:

$$\Delta x_1(k) = \Delta t\, x_2(k), \qquad \Delta x_i(k) = x_i(k+1) - x_i(k), \qquad (3.2)$$
$$\Delta x_2(k) = -2\delta\Delta t x_2(k) - [\omega_o^2 s_o + \omega_u^2 s_u]\Delta t x_1(k) + \sigma\Delta t z(k),$$
$$\text{IC: } x_1(0) = 0, \quad x_2(0) = 0, \qquad \text{for } k = 0,1,2,\ldots$$

where Δt is the scan rate and Δx_i are corresponding increments of the state realizations of the displacement and the velocity response at the discrete time points k. The simulation of the random excitation z(k) is already described in the chapter before.

The figure 2 shows typical realizations of the recurrence formula (3.2) for the data $\omega_o = 10\text{ s}^{-1}$, $\omega_u = 5\text{ s}^{-1}$, $\delta = 1\text{ s}^{-1}$ and $\Delta t = 0.01$ s starting with vanishing initial conditions (IC). Above, we see the broad-banded excitation, below the response of the oscillator displacement. A comparison of both realizations shows clearly the well-known filter effect of the oscillator as well as the shifted mean value of its deflections as a consequence of the asymmetric restoring mechanism of the bilinear spring characteristic. According to (2.3) we check the numerical stability of the simulation formula (3.2) by the set
-up

$$x_i(k+1) = \rho \, x_i(k), \qquad \text{for } i = 1,2$$

$$\rho_{1,2} = 1 - \delta \Delta t + i \Delta t \sqrt{\omega_{o,u}^2 - \delta^2} \tag{3.3}$$

leading to the above noted eigenvalue ρ for $\delta < \omega_{o,u}$. Consequently, the homogeneous solutions of (3.2) are asymptotically stable if the condition

$$|\rho_{1,2}| < 1, \qquad 0 < \Delta t < 2\delta/\omega_{o,u}^2 \tag{3.4}$$

is satisfied for each branch of the bilinear spring characteristic of the oscillator.

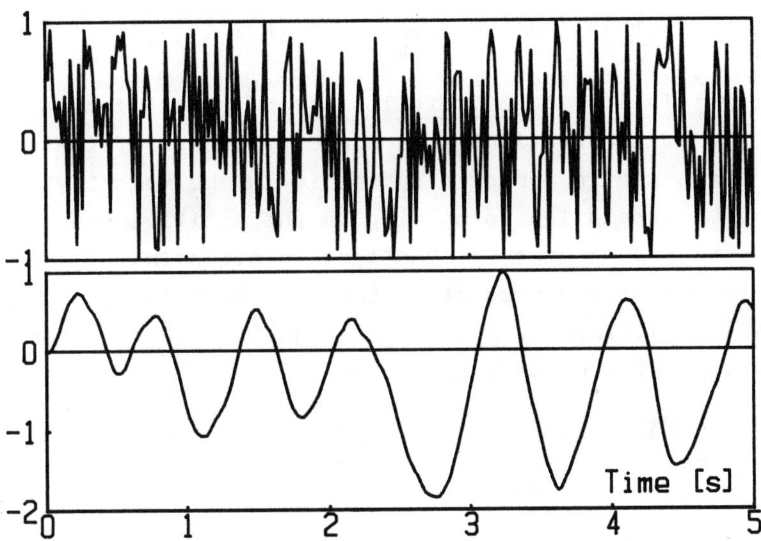

Fig. 2: Excitation and response of the oscillator

Now, we are in the position to discuss the estimation problem. Since the excitations $z(k)$ are not measurable, we calculate them from the measured response $x_1(k)$ and $x_2(k)$ in such a way that the excitation energy becomes minimal. In terms of the system data this minimal requirement reads as

$$\sum_k [\Delta x_2 + 2\delta\Delta t x_2 + (\omega_o^2 s_o + \omega_u^2 s_u)\Delta t x_1]^2 = \text{Min!} \tag{3.5}$$

It is evaluated by partial differentiations with respect to the three unknown parameters of the crack oscillator leading to the following necessary conditions

$$
\begin{bmatrix}
\Sigma s_o^2 x_1^2 & \Sigma s_o s_u x_1^2 & \Sigma s_o x_1 x_2 \\
\Sigma s_o s_u x_1^2 & \Sigma s_u^2 x_1^2 & \Sigma s_u x_1 x_2 \\
\Sigma s_o x_1 x_2 & \Sigma s_u x_1 x_2 & \Sigma x_2^2
\end{bmatrix}
\begin{bmatrix}
\hat{\omega}_o^2 \Delta t \\
\hat{\omega}_u^2 \Delta t \\
2\hat{\delta}\Delta t
\end{bmatrix}
= -
\begin{bmatrix}
\Sigma s_o x_1 \Delta x_2 \\
\Sigma s_u x_1 \Delta x_2 \\
\Sigma x_2 \Delta x_2
\end{bmatrix}
\tag{3.6}
$$

for the existence of the energy minimum (2.8).

The estimation equations (3.6) can considerably be improved using some a-priori knowledges which are physically obvious.

$$
\sum_k x_1(k)x_2(k) = 0, \qquad \sum_k s_u(k)x_1(k)x_2(k) = 0, \tag{3.7}
$$

$$
\sum_k s_o(k)x_1(k)x_2(k) = 0, \qquad \sum_k s_o(k)s_u(k)x_1(k)x_2(k) = 0.
$$

The first relation in (3.7) follows from the statistical independence of the stationary response processes of the oscillator. Since the response is symmetric in the velocity range, the same uncorrelation is holding in the positive and in the negative range, separately. The last relation in (3.7) is obvious because of the product of the unit step functions defined in (3.1). Inserting (3.7) the equation (3.6) decouples to the simplier frequency estimators

$$
\hat{\omega}_{o,u}^2 = - \frac{\sum_k x_1(k)\, s_{o,u}(k)\, \Delta x_2(k)}{\Delta t \sum_k x_1^2(k)\, s_{o,u}^2(k)} \tag{3.8}
$$

meanwhile the damping estimator coincides with that of the linear oscillator.

In the figure 3 we give a simultaneously performed evaluation of the estimators (3.8). The crack oscillator was simulated by the recurrence formula (3.2) starting with vanishing initial conditions. After a simulation period of 20 sec the frequency estimators (3.8) were applied and then stopped at the end of the measuring time T = 20 sec. The figure 3 shows two statistically independent estimation realizations simply performed by restarting the measurements. The applied scan rate is $\Delta t = 0.01$ sec, the damping coefficient is $\delta = 1.2 \ s^{-1}$ and the bilinear

frequencies are $\omega_o = 12\ \text{s}^{-1}$ and $\omega_u = 10\ \text{s}^{-1}$. We observe that
the frequency estimates converge to the true values ω_o and ω_u.
Already after 2 sec the relative errors of both estimates are
in the range of 10 %. At the end of the measuring time T the
relative errors are smaller than 2 %.

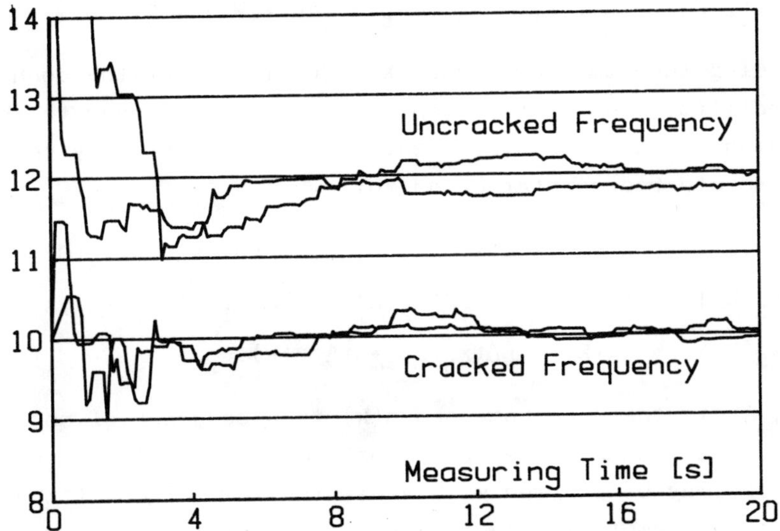

Fig. 3: Estimates of the bilinear frequencies

4. Nonlinear estimations for coloured excitations

The above applied minimal principle of estimation theory is
holding if the system excitation is characterized by a broad
frequency distribution so that the excitation and the response
of the system are statistically independent. To maintain this
Markow property in more realistic cases of band-limited excitations we reduce them to white noise processes by means of
shaping filters. In general, therefore, we have to introduce
mathematical models for both, the mechanical structure and for
the filtered excitations and to estimate all parameters of the
entire system from one or several measured signals of the
elastic structure.

As a simple example let us consider the linear oscillator with
the natural frequency ω_1 and the damping ratio D excited by a

force process F_t with the limiting frequency ω_g, which is generated from white noise by means of a low-pass filter. Hence, the entire problem is described by the following two differential equations:

$$\ddot{X}_t + 2D\omega_1\dot{X}_t + \omega_1^2 X_t = F_t, \qquad \dot{F}_t + \omega_g F_t = \sigma\dot{W}_t, \qquad (4.1)$$

$$X_{1,t} = X_t, \quad X_{2,t} = \dot{X}_t, \quad X_{3,t} = F_t, \qquad E(\dot{W}_t\dot{W}_s) = \delta(t-s).$$

Introducing the state processes $X_{i,t}$ (i=1,2,3) of the mechanical oscillator and of the shaping filter, we go over to the associated first order system which is approximated by the time-discrete representation as follows:

$$x_1(k+1) = x_1(k) + \Delta t x_2(k), \qquad \text{IC: } x_1(0)=x_2(0)=x_3(0)=0,$$

$$x_2(k+1) = x_2(k) - 2D\omega_1\Delta t x_2(k) - \omega_1^2\Delta t x_1(k) + \Delta t x_3(k),$$

$$x_3(k+1) = x_3(k) - \omega_g\Delta t x_3(k) + \sigma\Delta t z(k). \qquad (4.2)$$

Herein, Δt is the scan rate and $z(k)$ is a sequence of uncorrelated random numbers, as previously mentioned. The simulation (4.2) may be started with vanishing initial conditions (IC). It is asymptotically stable, if the applied scan rate satisfies the inequalities

$$0 < \Delta t < 2/\omega_g, \qquad 0 < \Delta t < 2D/\omega_1, \qquad (D<1) \qquad (4.3)$$

in correspondence to (2.3) and (3.5).

We assume that the processes F_t and $\dot{X}_t$ are not measurable. Only, a finite set of measured displacements $x_1(k)$ at the discrete time points $k=0,1,2,\ldots$ is available to estimate the parameters ω_1^2 and $2D\omega_1$ of the mechanical system as well as the limit frequency ω_g of its driving force. To prepare this estimation problem and also, to speed up the corresponding simulation, we eliminate in (4.3) the nonmeasurable realizations $x_2(k)$ and $x_3(k)$ in arriving finally to the following discrete model for the measurable oscillator deflections $x_1(k) = x(k)$:

$$x(k+3) - a_1 x(k) - a_2 x(k+1) - a_3 x(k+2) = \sigma\Delta t^3 z(k),$$

$$\text{IC: } x(0) = x(1) = x(2) = 0, \qquad k = 0,1,2,\ldots \qquad (4.4)$$

The introduced abbreviations a_i (i=1,2,3) are nonlinear combinations of the system parameters as follows:

$$a_1 = 1-q_3+q_2-q_1, \quad q_1 = p_3 p_1, \quad p_1 = (\omega_1 \Delta t)^2,$$
$$a_2 = -3+2q_3-q_2, \quad q_2 = p_1+p_2 p_3, \quad p_2 = 2D\omega_1 \Delta t, \quad (4.5)$$
$$a_3 = 3-q_3, \quad q_3 = p_2+p_3, \quad p_3 = \omega_g \Delta t.$$

The realizations $x(k)$ coincide exactly with the original simulation results (4.2) if the recurrence formula (4.4) is started with the initial conditions (IC), noted above.

Applying now the minimal principle (2.8) to the dynamic system (4.4) we obtain

$$\sum_k \left[x(k+3) - a_1 x(k) - a_2 x(k+1) - a_3 x(k+2) \right]^2 = \text{Min!} \quad (4.6)$$

This minimal requirement can be evaluated by differentiating with respect to the unknown parameters a_i (i=1,2,3).

$$(4.7)$$

$$\begin{bmatrix} \sum x^2(k) & \sum x(k)x(k+1) & \sum x(k)x(k+2) \\ \sum x(k)x(k+1) & \sum x^2(k+1) & \sum x(k+1)x(k+2) \\ \sum x(k)x(k+2) & \sum x(k+1)x(k+2) & \sum x^2(k+2) \end{bmatrix} \begin{bmatrix} \hat{a}_1 \\ \hat{a}_2 \\ \hat{a}_3 \end{bmatrix} = \begin{bmatrix} \sum x(k)x(k+3) \\ \sum x(k+1)x(k+3) \\ \sum x(k+2)x(k+3) \end{bmatrix}$$

These are linear equations from which we are able to estimate the parameters a_i simply by measuring the discrete displacements of the oscillator at four different time points, multiplying them and summing up over all k. The estimates $\hat{a}_i$ are unbiased since the equations (4.7) can be derived from the simulation formula (4.4) directly by multiplying it with $x(k)$, $x(k+1)$ and $x(k+2)$, respectively and summing up over all time points k. Because of the Markow properties

$$\sum_k x(k)z(k) = \sum_k x(k+1)z(k) = \sum_k x(k+2)z(k) = 0 \quad (4.8)$$

the right-hand sides of these three equations are vanishing and their left sides coincide exactly with (4.7). The statistical independence (4.8) is physically obvious since the initial state $x(0)$, $x(1)$, $x(2)$ of the dynamic system (4.4) is not influenced by $z(0)$. The same situation is observable for the further time steps $k = 1,2,..$, as well.

The estimation equation (4.7) can considerably be improved by using further a-priori knowledges on the stationary system response. To show this we first insert the parameters q_i (4.5) into the minimal requirement (4.6) yielding the form

$$\sum_k [\Delta^3 x(k) + q_1 x(k) + q_2 \Delta x(k) + q_3 \Delta^2 \ddot{x}(k)]^2 = \text{Min!} \qquad (4.9)$$

$$\Delta x(k) = x(k+1)-x(k), \quad \Delta^2 x(k) = x(k+2)-2x(k+1)+x(k),$$

$$\Delta^3 x(k) = x(k+3)-3x(k+2)+3x(k+1)-x(k), \quad k=0,1,2,\ldots$$

Herein, $\Delta x(k)$ is the increment of the displacement $x(k)$. Accordingly, $\Delta^2 x(k)$ and $\Delta^3 \ddot{x}(k)$ are higher order increments. The differention with respect to the parameters q_i results in

$$\begin{bmatrix} \sum x^2 & \sum x\,\Delta x & \sum x\,\Delta^2 x \\ \sum x\,\Delta x & \sum (\Delta x)^2 & \sum \Delta x\,\Delta^2 x \\ \sum x\,\Delta^2 x & \sum \Delta x\,\Delta^2 x & \sum (\Delta^2 x)^2 \end{bmatrix} \begin{bmatrix} \hat{q}_1 \\ \hat{q}_2 \\ \hat{q}_3 \end{bmatrix} = \begin{bmatrix} -\sum x\,\Delta^3 x \\ -\sum \Delta x\,\Delta^3 x \\ -\sum \Delta^2 x\,\Delta^3 x \end{bmatrix} \qquad (4.10)$$

which is clearly a linear combination of the original estimation equation (4.7). In this form, we are now able to apply the a-priori knowledges

$$\sum_k x(k)\Delta x(k) = \sum_k \Delta x(k)\Delta^2 x(k) = \sum_k x(k)\Delta^3 x(k) = 0, \qquad (4.11)$$

$$\sum_k x(k)\Delta^2 x(k) = -\sum_k (\Delta x)^2, \qquad \sum_k \Delta x(k)\Delta^3 x(k) = -\sum_k (\Delta^2 x)^2.$$

The first relations follow from the statistical independence of differentiated processes, the second can be derived by means of partial integration. Therewith, the estimation prescription (4.10) is simplified to

$$\hat{q}_2 \sum_k [\Delta x(k)]^2 = \sum_k [\Delta^2 x(k)]^2,$$

$$\begin{bmatrix} \sum x^2(k) & -\sum [\Delta x(k)]^2 \\ -\sum [\Delta x(k)]^2 & \sum [\Delta^2 x(k)]^2 \end{bmatrix} \begin{bmatrix} \hat{q}_1 \\ \hat{q}_2 \end{bmatrix} = \begin{bmatrix} 0 \\ -\sum \Delta^2 x(k)\,\Delta^3 x(k) \end{bmatrix} \qquad (4.12)$$

and is performable by measuring the oscillator displacements x(k) and their increments taking only four different summations.

Knowing the estimates $\hat{q}_i$ (i=1,2,3) we are able to calculate the parameters p_i of interest which contain the original data ω_1^2, $2D\omega_1$ and ω_g of the mechanical system and the coloured excitation. According to (4.5) both sets of parameters are connected by three nonlinear equations. Resolving them we find e.g. for the third parameter $p_3 = p$ the cubic equation

$$p = p_3 : \qquad p^3 - \hat{q}_3 p^2 + \hat{q}_2 p - \hat{q}_1 = 0, \qquad (4.13)$$

$$p_{v+1} = p_v + \frac{\hat{q}_1 - \hat{q}_2 p_v + \hat{q}_3 p_v^2 - p_v^3}{\hat{q}_2 - 2\hat{q}_3 p_v + 3p_v^2} , \quad v = 0,1,2,\ldots$$

At least, there is always one real-valued root $p = e$ which is calculated from the estimates $\hat{q}_i$ by means of the iteration procedure, noted above, starting with any arbitrary initial value e.g. $p_0 = 0$. If there still exist two other real-valued roots, they are calculable from the first result $p = e$ by reducing (4.13) to a quadratic equation with the solutions

$$p_{2,3} = \tfrac{1}{2}(e-\hat{q}_3) \pm \sqrt{\tfrac{1}{4}(e-\hat{q}_3)^2 - (e^2-\hat{q}_3 e+\hat{q}_2)}, \quad p_1 = e,$$

$$2D\omega_1 \Delta t = \hat{q}_3 - p, \quad (\omega_1 \Delta t)^2 = \hat{q}_2 - \hat{q}_3 p + p^2, \quad \omega_g \Delta t = p \qquad (4.14)$$

so that finally we obtain three triples of system parameters of interest. If the estimates $\hat{q}_i$ are sufficiently correct, there exists only one real-valued root of (4.13) for $D < 1$. However, during the initial measuring period the estimates $\hat{q}_i$ can be so far away from their true values that we find sometimes three real-valued roots. Then, we have to know a sufficiently good a-priori knowledge on ω_1, D or ω_g to decide which of them has to be selected. This corresponds to the common difficulties of a nonlinear theory.

In the figures 4 and 5 we show some numerical evaluations of the linear estimations (4.12) and the nonlinear equations (4.13) and (4.14). The dynamic system of the oscillator and the low-pass filter of its coloured excitation are simulated by the discrete model (4.4) with the true values $\dot{\omega}_1 = 10$ s^{-1}, $\omega_g = 100$ s^{-1}, $D = 0.1$ and $\Delta t = 0.005$ s.

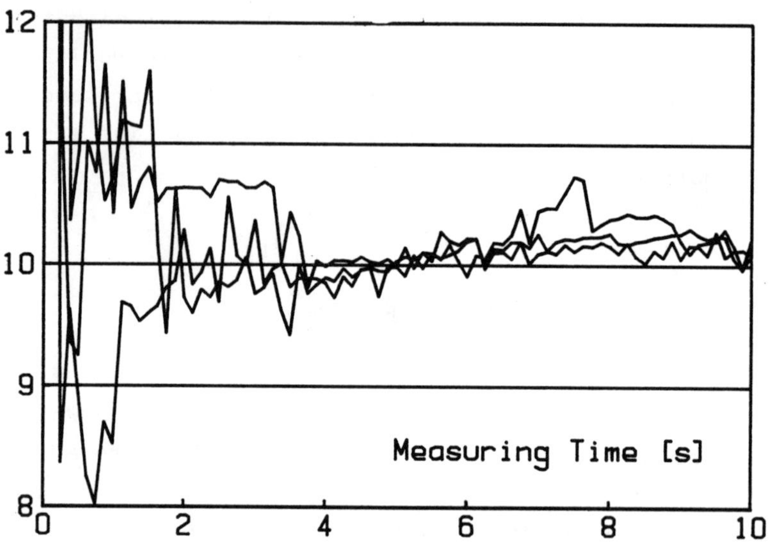

Fig. 4: Estimates of the system frequency $\omega_1 = 10$ s^{-1}

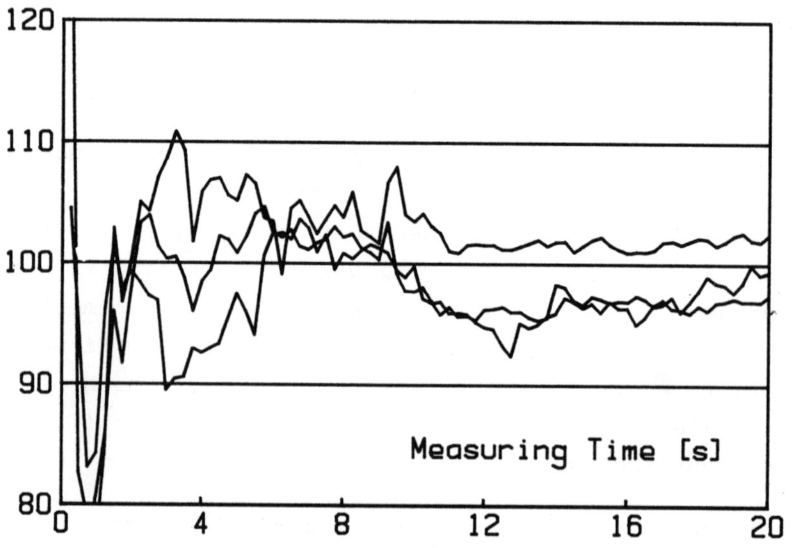

Fig. 5: Estimates of the excitation frequency ω_g

In each figure we show three statistically independent reali-
zations of the estimates $\hat{\omega}_1$ and $\hat{\omega}_g$ in a 20 % error range clearly indicating the consistency and the convergence rate of both
estimators.

5. Stiffness supervising by local measurements

It is physically obvious that the parameter estimation becomes
bader the less system informations or state measurements are
performable. Inversely, we need only short measuring times if
the number of state measurements is greater than that of the
stiffness parameters to be estimated. In this case, moreover,
we can restrict our modelling of the mechanical structure to
local finite elements if all state processes of these elements
are measurable. Particularly, we need no further knowledge or
modelling outside the local elements.

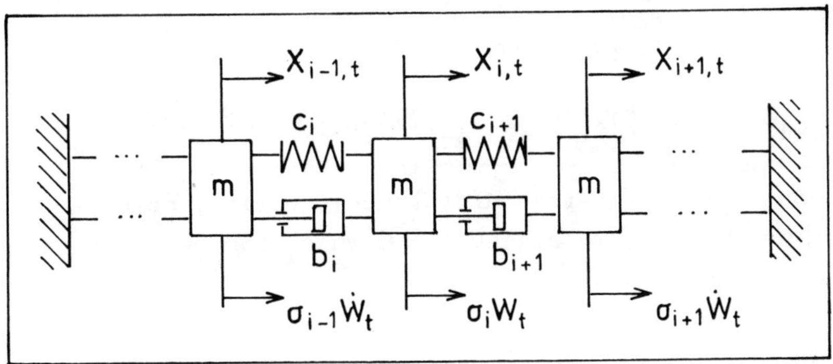

Fig. 6: Model of vibrating finite elements

As an example let us consider the vibration chain with n degrees of freedom, shown in figure 6. Assuming a unit mass distribution, different stiffness parameters c_i and internal damping coefficients b_i, we derive the equations of motion of the
following form:

$$\ddot{X}_{i,t} + \beta_i(\dot{X}_{i,t}-\dot{X}_{i-1,t}) + \beta_{i+1}(\dot{X}_{i,t}-\dot{X}_{i+1,t}) +$$
$$+ \gamma_i(X_{i,t}-X_{i-1,t}) + \gamma_{i+1}(X_{i,t}-X_{i+1,t}) = \sigma_i\dot{W}_t, \quad (5.1)$$
$$\text{BC:} \quad X_{0,t} = X_{n+1,t} = 0, \quad \gamma_i = c_i/m, \quad \beta_i = b_i/m.$$

Each mass may be excited by uncorrelated white noise processes
with vanishing mean values and the intensities σ_i. If the chain
is fixed at both ends, the boundary conditions (BC), noted in
(5.1), are holding, in addition. We simulate the differential
equations (5.1) by means of the time-discrete equations

$$\Delta^2 x_i + \Delta t \beta_i (\Delta x_i - \Delta x_{i-1}) + \Delta t \beta_{i+1} (\Delta x_i - \Delta x_{i+1}) +$$
$$+ \Delta t^2 \gamma_i (x_i - x_{i-1}) + \Delta t^2 \gamma_{i+1} (x_i - x_{i+1}) = \Delta t^2 \sigma_i z, \quad (5.2)$$

$$\Delta x_i = x_i(k+1) - x_i(k), \quad \Delta^2 x_i = x_i(k+2) - 2x_i(k+1) + x_i(k)$$

for all time points $k = 0,1,2,..$ and for each degree of freedom $i = 1,2,...\ n$. The excitations $\sigma_i z(k)$ are statistically
independent random numbers, as previously mentioned.

For the special case of a homogeneous vibration chain we analyse the numerical stability of the scanning system (5.2) by
means of the set-up

$$x_i(k+1) = \rho\, x_i(k), \quad \text{for } \gamma_i = \gamma ,\ \beta_i = \beta , \quad (i=1,..n) \quad (5.3)$$

$$x_{i+1}(k) - (2-\zeta)x_i(k) + x_{i-1}(k) = 0, \quad \zeta = \frac{\rho^2 - 2\rho + 1}{\Delta t \beta (1-\rho) - \gamma \Delta t^2} .$$

This leads to the time-free difference equation, noted above.
It contains the unknown eigenvalue ζ which can be calculated
via the well-known set-up

$$x_i(k) = C \sin i\varphi, \qquad \zeta = 2(1 - \cos\varphi), \qquad (5.4)$$

$$\text{BC:} \quad \sin(n+1)\varphi = 0, \qquad \varphi_\nu = \pi\nu/(n+1), \quad \nu = 1,2,,..n.$$

The set-up has to satisfy the vanishing boundary conditions
(BC) yielding all φ_ν values of interest. Therewith, we are able
to determine the eigenvalues ζ_ν and finally all ρ values by
means of the quadratic equation, already noted in (5.3).

$$\rho^2 - (2 - \Delta t \beta \zeta)\rho + 1 - \Delta t \beta \zeta + \Delta t^2 \gamma \zeta = 0,$$
$$|\rho| < 1 \implies 0 < \Delta t < \beta/\gamma, \quad \text{for } \gamma < \zeta \beta^2/4. \qquad (5.5)$$

Obviously, the absolute value of ρ is smaller than one if the
positive scan rate Δt is smaller than β/γ provided that the
internal damping ratio β is sufficiently small. For the sys-

tem data $\beta = 6$ s^{-1} and $\gamma = 600$ s^{-2} e.g., the applied scan rate has to be $\Delta t < 0.01$ sec to ensure a numerically stable simulation of the vibration chain.

In the inhomogeneous case of the discrete model (5.2) we are now interested to estimate the two different stiffness parameters γ_i, γ_{i+1} and the damping coefficients β_i, β_{i+1}. Assuming that local measurements of the three vibration displacements $x_{i-1}(k)$, $x_i(k)$ and $x_{i+1}(k)$ are performable, we are able to apply the minimal principle to the equation (5.2)

$$\sum_k [\Delta^2 x_i + \Delta t \beta_i (\Delta x_i - \Delta x_{i-1}) + \Delta t \beta_{i+1} (\Delta x_i - \Delta x_{i+1}) + \Delta t^2 \gamma_i (x_i - x_{i-1}) + \Delta t^2 \gamma_{i+1} (x_i - x_{i+1})]^2 = \text{Min!} \qquad (5.6)$$

The necessary conditions for the existence of a minimum are obtained by differentiating (5.6) with respect to the four parameters γ_i, γ_{i+1}, β_i, β_{i+1}. This results in the following matrix equation:

$$\begin{bmatrix} \sum(x_i-x_{i-1})^2 & \sum(x_i-x_{i+1})(x_i-x_{i-1}) & 0 & \sum(\Delta x_i-\Delta x_{i+1})(x_i-x_{i-1}) \\ \sum(x_i-x_{i-1})(x_i-x_{i+1}) & \sum(x_i-x_{i+1})^2 & \sum(\Delta x_i-\Delta x_{i-1})(x_i-x_{i+1}) & 0 \\ 0 & \sum(x_i-x_{i+1})(\Delta x_i-\Delta x_{i-1}) & \sum(\Delta x_i-\Delta x_{i-1})^2 & \sum(\Delta x_i-\Delta x_{i+1})(\Delta x_i-\Delta x_{i-1}) \\ \sum(x_i-x_{i-1})(\Delta x_i-\Delta x_{i+1}) & 0 & \sum(\Delta x_i-\Delta x_{i-1})(\Delta x_i-\Delta x_{i+1}) & \sum(\Delta x_i-\Delta x_{i+1})^2 \end{bmatrix} \begin{bmatrix} \Delta t^2 \hat{\gamma}_i \\ \Delta t^2 \hat{\gamma}_{i+1} \\ \Delta t \hat{\beta}_i \\ \Delta t \hat{\beta}_{i+1} \end{bmatrix} = \begin{bmatrix} -\sum(x_i-x_{i-1})\Delta^2 x_i \\ -\sum(x_i-x_{i+1})\Delta^2 x_i \\ -\sum(\Delta x_i-\Delta x_{i-1})\Delta^2 x_i \\ -\sum(\Delta x_i-\Delta x_{i+1})\Delta^2 x_i \end{bmatrix}$$

$$(5.7)$$

if the a-priori knowledge

$$\sum_k (\Delta x_i-\Delta x_{i-1})(x_i-x_{i-1}) = \sum_k (\Delta x_i-\Delta x_{i+1})(x_i-x_{i+1}) = 0 \qquad (5.8)$$

of vanishing correlations between relative displacement and velocity processes is already taken into account. Accordingly, the four structural parameters of interest are estimable by measurements of three associated state signals without any mea-

surement of the realizations and the intensities of the white
noise excitations and without any knowledge on the modelling
of the structure outside the three finite elements i, i+1 and
i-1.

The estimations (5.7) can once more considerably be simplified
if we assume that both pairs of parameters have the same values,
i.e. $\gamma_i = \gamma_{i+1} = \gamma$ and $\beta_i = \beta_{i+1} = \beta$. In this case the minimal requirement (5.6) reduces to

$$\sum_k [\Delta^2 x_i + \Delta t \beta(-\Delta x_{i-1} + 2\Delta x_i - \Delta x_{i+1}) +$$
$$+ \Delta t^2 \gamma(-x_{i-1} + 2x_i - x_{i+1})]^2 = \text{Min!} \qquad (5.9)$$

Differentiating it with respect to γ and taking into account
the a-priori knowledge of the statistical independence

$$\sum_k (-x_{i-1} + 2x_i - x_{i+1})(-\Delta x_{i-1} + 2\Delta x_i - \Delta x_{i+1}) = 0 \qquad (5.10)$$

we finally arrive at the estimator

$$\hat{\gamma} = - \frac{\sum_k [-x_{i-1}(k)+2x_i(k)-x_{i+1}(k)]\Delta^2 x_i(k)}{\Delta t^2 \sum_k [-x_{i-1}(k)+2x_i(k)-x_{i+1}(k)]^2} \qquad (5.11)$$

for the stiffness parameter γ. In a similar way, the damping
estimator $\hat{\beta}$ can be derived.

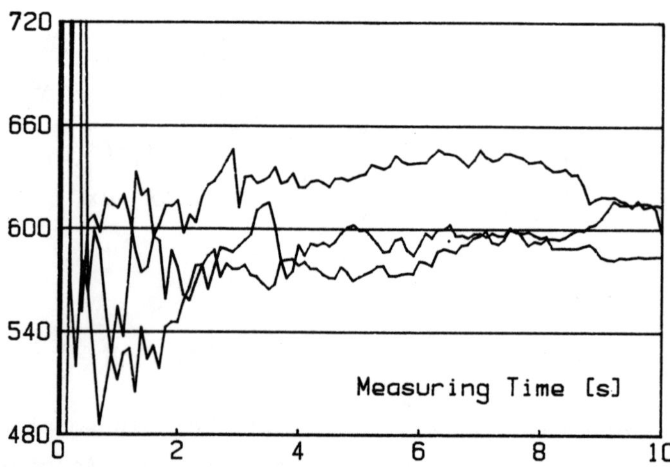

Fig. 7: Stiffness estimates in a vibrating chain

The figure 7 shows three typical results of the stiffness estimator (5.11) for the special case of a homogeneous vibration chain of five degrees of freedom and for the data $\gamma_i =$
$\gamma = 600$ s^{-2}, $\beta_i = \beta = 6$ s^{-1}, $\Delta t = 0.002$ s and $\sigma_i = \sigma$ (i=1,2,
3,4,5). Applying the recurrence formula (5.2) we performed the
simulation of this mechanical system. From this we measured
the deflections $x_2(k)$, $x_3(k)$ and $x_4(k)$ of the second, third
and fourth mass and we inserted the resulting simulations values into the estimation formula (5.11). In figure 7 we observe
that the 10 % error range is reached already after 2 sec. At
the end of the total measuring time T = 10 sec, the relative
error of the estimates related to the true stiffness parameter γ is smaller than 2.5 %.

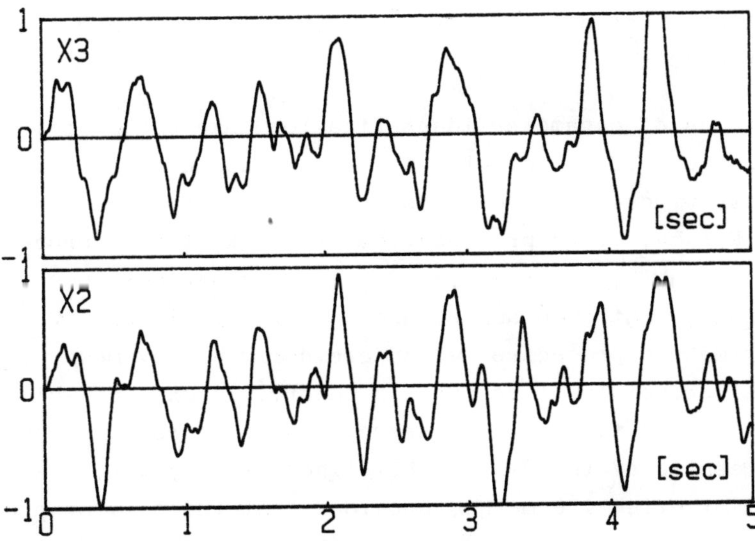

Fig. 8: Deflection processes of the chain

The figure 8 contains one vibration realization $x_3(k)$ of the
center mass of the chain, and below, one finds the simultaneously measured deflections $x_2(k)$ of the neighboured mass both
starting with vanishing initial conditions. For the case of
the given data, the lowest natural frequency of the chain is
$f_1 = 1.009$ Hz and the highest natural frequency $f_5 = 7.531$ Hz.

6. Conclusion

Elastic structures of a high demand of security as e.g. pipelines in nuclear power plants, highway bridges, aircraft wings,
rotating machines or vehicle-wheel systems require today an
early recognizing of a possible failure. It is obvious that
this can only be achieved by means of a permanent control or
measurement of the stiffness parameters of the elastic structure. The well-known methods of parameter identification, up
to now applied, require artificial excitations of the existing
structure e.g. by means of impacts (hammer) or unbalanced ro-
tors (shaker) in order to calculate the stiffness-, damping-
and massmatrices of multi degree of freedom systems using the
linear input-output relations and measuring the corresponding
realizations. Naturally, these classical procedures are either
nonrealistic or not permanently applicable for the important
goal, mentioned above.

Modern methods of parameter identification make use of the natural vibrations of the elastic structure caused by turbulent
flows, gusty winds, traffic loads or randomly profiled road
surfaces. Hereby, it is provided that such excitation processes
are not measurable, in practice. Since the classical input-output relations, mentioned above, are no more applicable, we make
use of estimation procedures which are derivable by means of
the minimal principle of estimation theory. In case of white
noise excitations the derived parameter estimators coincide
with the results of the Maximum Likelihood theory which is always the most effective method provided that the driving processes are gaussian.

Summarizing we can state that the minimal principle applied
gives principally the possibility to estimate several parameters of the structure and of the excitation from one measured
response process. The minimal principle is valid in case of
Markow excitations characterized by broad banded power spectra
or deterministic impulse spectra. In both cases a more realistic modelling is achieved by means of shaping filters leading

to an extended parametrization of the nonmeasurable excitation
process. Consequently, we have to develop mathematical models
of the elastic structure as well as of the realistic excitation. Then, the estimation of all parameters are immediately
derivable from the total system equations without any knowledge on their analytical solutions.

References:

1   Lee, T.S.; Kozin, F.: Almost Sure Asymptotic Likelihood
    Theory for Diffusion Processes. J. Appl. Prob., Vol. 14
    (1977) 527-537

2   Kozin, F.: Estimation of Parameters for Systems Driven by
    White Noise Excitations. Proceedings of the IUTAM-Sympo-
    sium Frankfurt/O 1982 on Random Vibrations and Reliability
    (ed. by K.Hennig). Berlin: Akademie Verlag 1983, 163-172

3   Bellach, B.: Parameter Estimation in Linear Stochastic
    Differential Equations and Their Properties. Proceedings
    of the IUTAM-Symposium 1982 on Random Vibrations and Re-
    liability (ed. by K.Hennig). Berlin: Akademie-Verlag 1983,
    137 - 144

4   Wedig, W.: Development of Structural Supervising Systems.
    Proceedings of the International Conference on Structural
    Failure, Product Liability and Technical Insurance, Wien
    1983 (ed. by H.P.Rossmanith). Amsterdam: Elsevier Science
    Publishers B.V. (North-Holland) 1984, 197 -203

5   Wedig, W.: Parameterschätzungen dynamischer Systeme nach
    dem Prinzip der minimalen Anregungsenergie, to appear in
    Ingenieur-Archiv.

RELIABILITY ASSESSMENT AND INSPECTION OPTIMIZATION OF AN AMMONIA TANK
WITH STRESS CORROSION

Sergio F. Garribba
Cesnef, Politecnico di Milano
Via G. Ponzio 34/3
Milano, Italy 20133

ABSTRACT. The survival probability of an ammonia tank is computed by
considering the mechanical properties of the construction materials,
and the distribution of defects resulting from loads experienced in a
service period. Essentially, the computation of the rupture probability
starts from the analysis of the behavior of a single crack in the tank
under stress corrosion. According to the circumstances, this crack may
lead the structure to break or to leak, provided that it propagates,
until its depth equals the thickness of the wall. Cracks are revealed
during inspections effected by the use of magnetoscopic detection
techniques. Once revealed, cracks undergo repair. It can be assumed
that after each maintenance operation the tank is brought to about the
same initial state. As a consequence of the analysis, recommendations
can be given to optimize inspection and repair actions.

1.   INTRODUCTION

The integrity of a tank for ammonia storage and transportation can
be assessed accurately only if one considers the real mechanical properties of construction materials, the actual load experienced and the
distribution of defects in the tank during its useful life. This task,
however, can be almost impossible to achieve in its most general terms,
therefore a number of simplifications is introduced by the adoption of
proper models. Precisely, the interest lies in the assessment of
interval reliability, or survival probability referred to a specific
operations sequence. Safety assessment requires indeed to estimate the
probability of absence of ruptures between serial inspection and repair
actions.

Let us then start with a formal definition and refer to a rupture
hypothesis and model, expressed by a set of failure patterns or modes
$H = \{H_1, \ldots, H_\alpha, \ldots, H_A\}$. Given a generic pattern H_α, the H_α - conditional absence of failure will be represented in terms of a (scalar or
vector) relation of the type

$$^\alpha h(^\alpha q_1, \ldots, ^\alpha q_i, \ldots, ^\alpha q_I; t, x) < {}^\alpha Q, \qquad\qquad (1.1)$$

217

A. C. Lucia (ed.), Advances in Structural Reliability, 217–232.
© *1987 by ECSC, EEC, EAEC, Brussels and Luxembourg.*

which is assumed to hold for all time instants t and all points x belonging to the operational cycle T and space domain X of interest for the tank, respectively. $^{\alpha}h$ denotes an analytic function on the set of parameters or characteristics $^{\alpha}q_i$ ($i = 1$, ..., I) that uniquely determine the behavior of the structure. Whereas $^{\alpha}Q$ denotes a level that must not be surpassed for sake of safety.

In a probabilistic frame, if reference is made to the failure mode H_{α} and the corresponding set of experimental data K_{β} , relation (1.1) allows to express the interval reliability of the tank as

$$R(T, X \mid H_{\alpha}, K_{\beta}) = \underset{\text{all } t \in T, x \in X}{\text{Prob}} \{^{\alpha}\tilde{h} < {}^{\alpha}Q\}, \qquad (1.2)$$

which signifies the conditional probability of absence of $^{\alpha}Q$-upcrossings by the stochastic process $^{\alpha}\tilde{h} = \{h; t \in T, x \in X \mid K_{\beta}\}$ in T and X. There is a notable difference in the treatment of time and space coordinates, however. When concern is with the problem of computation of rupture probability all upcrossings for $x \in X$ are of interest. Conversely, in time one should look for the first upcrossing along t, where $t \in T$. Would rupture patterns be mutually exclusive and exhaustive, each with probability of occurrence $P\{H_{\alpha}\} = P_{\alpha}$, unconditional reliability results

$$R(T,X \mid \tilde{H}_{\alpha},K_{\beta}) = \sum_{\alpha}^{1,\dots,A} R(T, X \mid H_{\alpha}, K_{\beta}) \cdot P_{\alpha} . \qquad (1.3)$$

Clearly, given the time interval $T = [t_0, t_F]$ and a generic time instant t' with $[t_0, t') \subset T$, it will be possible to compute the t'-interval reliability function $R(t', X \mid H, K_{\beta})$, the corresponding failure intensity function

$$\lambda(t', X \mid H, K_{\beta}) = -\partial \ln R(t', X \mid H, K_{\beta})/\partial t', \qquad (1.4)$$

and failure pdf R λ /1/.

In the problem under consideration, failure patterns H_{α} which are referred to, assume failure of the tank due to the propagazion of cracks or defects under stress corrosion. Essentially, the model for computing rupture probability 1-R starts from the consideration of a single crack in the tank. It is admitted that this crack may lead the structure either to break or to leak provided that it propagates, until its depth a(x,t) equals the thickness d of the wall (Fig. 1).

Cracks are revealed during inspections effected at times $t_j^{(m)}$ ($j = 1,\dots,J$) through magnetoscopic (and in some cases ultrasonic) detection techniques. Once revealed, cracks are repaired. It can be assumed that after each maintenance cycle, the tank is brought to the same initial condition. If reference is made to a generic operational cycle limited by the time interval $\Delta t_j^{(m)} = (t_{j+1}^{(m)} - t_j^{(m)})$ that corresponds to two sequential maintenance actions, one has for the T-interval

$$R(T,X \mid H, K_{\beta}) = \prod_{j}^{0,\dots,J} R(\Delta t_j^{(m)}, X \mid H, K_{\beta}), \qquad (1.5)$$

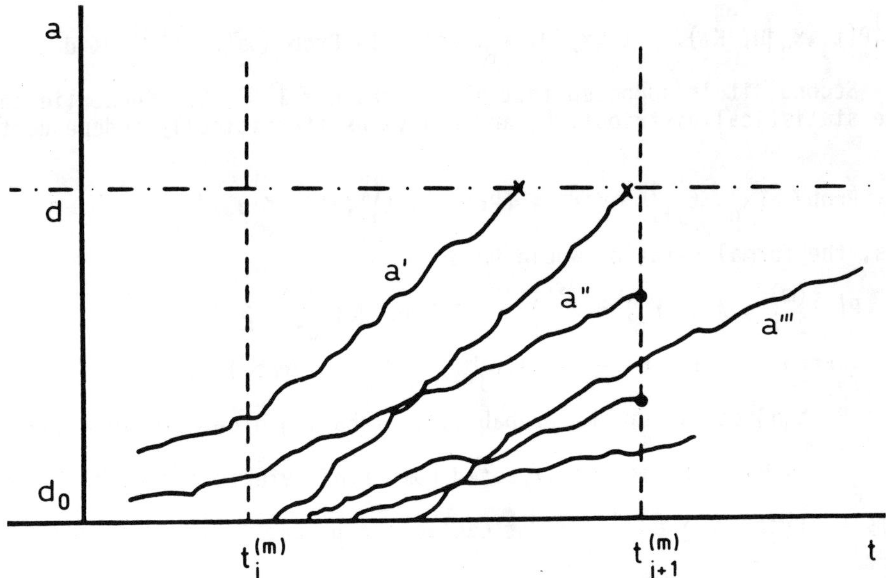

Figure 1. Crack growth as a cumulative stochastic process developing between two maintenance actions occurring at time instants $t_j^{(m)}$ and $t_{j+1}^{(m)}$. Sample functions a' and a''' refer to cases of undetected cracks. a'' shows the case of a revealed and then repaired crack. d_0 is a threshold valued below which cracks can be tolerated and/or cannot be detected (x wall-through defect; • detection event).

where

$$R(\Delta t_j^{(m)}, X | H, K_\beta) = \underset{\text{all } t \in \Delta t_j^{(m)}, n \in N}{\text{Prob}} \{ \tilde{a}(x_n, t) < d \} \qquad (1.6)$$

and

$$T = \sum_j^{0,\ldots,J-1} \Delta t_j^{(m)} .$$

$\tilde{a}(x_n, t) = \{a; t \in \Delta t_j^{(m)}, x = x_n \in X\}$ stays for a stochastic process non-decreasing in time. The stochastic process expresses depth (or another equivalent characteristic connected with geometry and size) of cracks or defects at a discrete number of points $(n = 0, 1, 2, \ldots, N)$. Obviously, a_n depends on material properties, load conditions and stress-corrosion rate. N is a statistical variable that in its turn will depend upon $t_j^{(m)}$.

Let us make two assumptions. First, concern is about the probability that the tank is not in its ruptured conditions at time $t_{j+1}^{(m)}$. Then, owing to the fact that the stochastic process $\tilde{a}_n(x,t)$ is non-decreasing in time it will be [2]

$$\int_{t_j}^{t_{j+1}} R(t',x_n|H, K_\beta) \cdot \lambda(t',x_n|H, K_\beta)\, dt' = 1- \text{Prob } \{ \tilde{a}(x_n,t_{j+1}^{(m)}) < d \}.$$

Second, it is supposed that all cracks $n = 1, 2, 3...$ underlie the same statistical distribution, and behave as statistically independent, i.e.

$$\text{Prob}\{ \tilde{a}(x_n, t_{j+1}^{(m)}) < d\} = (\text{Prob}\{ a(t_{j+1}^{(m)}) < d\})^n .$$

Thus, the formal relation would hold

$$R(t_{j+1}^{(m)}, X|H, K_\beta) = \sum_{n}^{0,...,\infty} (1-\text{Prob } \{a(t_{j+1}^{(m)}) \geq d\})^n .$$

$$. \text{ Prob}\{n\} = 1 - \bar{n} \text{ Prob } \{a(t_{j+1}^{(m)}) \geq d\} + 0(\text{Prob })^2 , \tag{1.7}$$

where Prob $\{n\}$ stays for the probability of having n defects in X and

$$\bar{n} = \sum_{n}^{1...\infty} n \text{ Prob}\{n\}$$ is the expected number of defects per tank. Since terms $0 \,(\text{Prob } \{a \geq d\})^2$ can be neglected, probability of rupture becomes

$$P_R(t_{j+1}, X|H, K_\beta) = \bar{n} \text{ Prob } \{a(t_{j+1}) \geq d\} . \tag{1.8}$$

Upperscript m has been omitted for sake of simplicity.

Cracks belong to two classes, according to their orientation which may be either parallel ($\|$), or perpendicular ($\perp$) with respect to the welding axis or line. It will be $n = n_\| + n_\perp$ with obvious meaning of symbols. Given a crack, one can define probabilities of orientation $p_\|$ and $p_\perp = 1 - p_\|$. Therefore, relation (1.8) would be rewritten as

$$P_R(t_{j+1}, X|H, K_\beta) = \bar{n} \left[p_\| P_\| (t_{j+1}, X|H, K_\beta) + p_\perp P_\perp (t_{j+1},X|H, K_\beta) \right], \tag{1.9}$$

where

$$P_\| (t_{j+1}, X|H, K_\beta) = \text{Prob } \{ a_\| (t_{j+1}) \geq d\} ,$$

$$P_\perp (t_{j+1}, X|H, K_\beta) = \text{Prob } \{ a_\perp (t_{j+1}) \geq d\} .$$

2. THE RUPTURE MODEL H

The problem which has been first afforded consists of the computation of the probability P_α for each identified rupture mode H_α, by the use of the set of data K_β that becomes available during inspection and repair effected at time $\Delta t_1 \approx 5$ yrs.

Tanks under investigation belong to a population of 53 units. All units have the same manufacture, approximate capacity of 31 m^3, inner diameter $D = 2.4$ m, overall (internal) length $2(L+L_o) = 7.48$ m, wall thickness $d = 15.0$ mm. (Fig. 2). Fe 510 Grade D steel plate has been used to produce all of their parts. Yield strength σ_y and ultimate tensile strength σ_u of this material do not show any appreciable

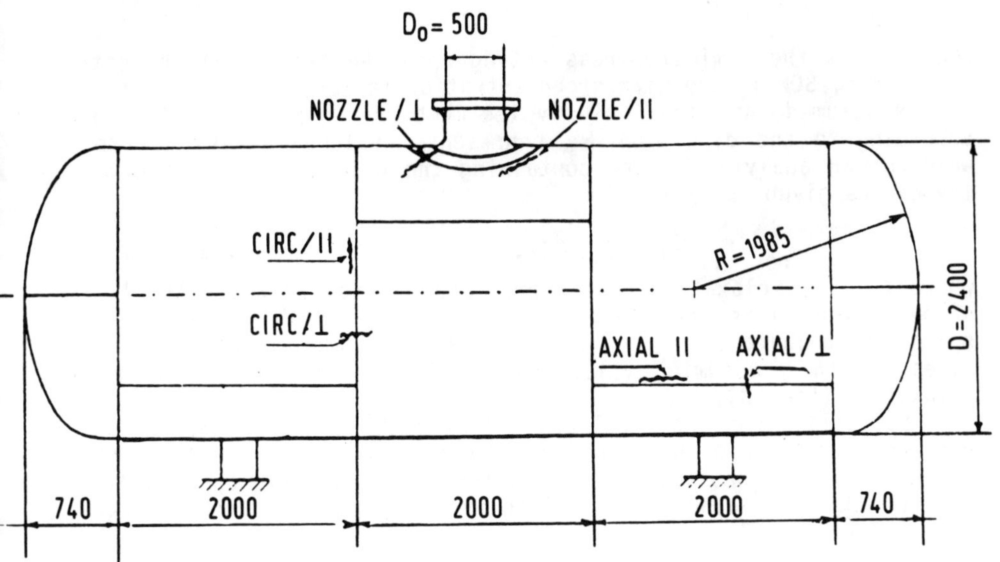

Figure 2. Schematic view of the ammonia storage and transportation tank showing location and direction of the weldings for beltline region, spherical bottoms and nozzle (all lengths expressed in mm).

dependence upon the rolling direction. The dispersion about the expected values $\bar{\sigma}_y = 452.7$ MPa and $\bar{\sigma}_u = 588.1$ MPa is also neglible.

The type of attack which is found in the ammonia tanks is of wall-through type. Cracks or defects localize on weldings and adjacent heat affected zones. These zones indeed are places of high residual stresses that are not eliminated by the subsequent thermal treatment of stress relief. The attention focuses on circumferential and axial weldings of beltline region X_{bl}, axial weldings of spherical bottoms X_{sb}, circumferential and axial weldings of the circular nozzle or opening X_{co}. Other weldings and preferential locations of residual stresses (like fixing shelfs, supporting saddles, reinforcements, etc.) are neglected.

Loads acting upon the tank can be either of internal or external origin. Internal pressure fluctuates about operation pressure $p_m = 1.20$ MPa, because of variations in service conditions and in environmental temperature (design temperature ranges from -20° to + 60°C). Moreover, tanks are installed on cash wagons and subjected to vibrations, and overloads as a consequence of their normal movement and handling, or of accidental bumps. All these loads of different external origin are scarcely known, however, thus reference has been made to design pressure $p_0 = 1.47$ MPa, which sets a rather conservative upper value.

Applied stress is defined as

$$\sigma_{app} = SCF \; \sigma_0 + \sigma_r \; , \tag{2.1}$$

where σ_0 is the nominal stress acting upon the section which contains the defect, SCF is the stress concentration factor. In the analysis, σ_0 is assumed as given only by the membrane component, which varies according to the direction. σ_r represents residual stress due to the welding; an analytical form containing the dependence upon the geometry can be given as [3]

$$\sigma_r = \sigma_{ro} \; f_1 \left(\frac{a}{d}\right) f_2\left(\frac{x}{u}, \frac{z}{d}\right) , \tag{2.2}$$

where f_1 is a relaxation factor produced by the crack itself, f_2 is a shape factor, u is the width of the (welding) bead, x and z are the space co-ordinates. $\sigma_{ro} = g \; \sigma_y$ is the maximum level of residual stress, with g assuming values in [0,1] according to the degree of relaxation (Fig. 3).

Basically, rupture model may follow either a fragile or a ductile pattern, namely H_F or H_D. The first depends upon the size of the defect and material toughness expressed in terms of crack tip opening displacement COD. The second is controlled by the ligament l and flow strength $\sigma_f = (\sigma_y + \sigma_u) / 2$.

Let us refer to the beltline region, unless otherwise stated. Since the two patterns H_F and H_D do not superpose, they are viewed as mutually exclusive. Their range of applicabiilty is expressed in terms of a

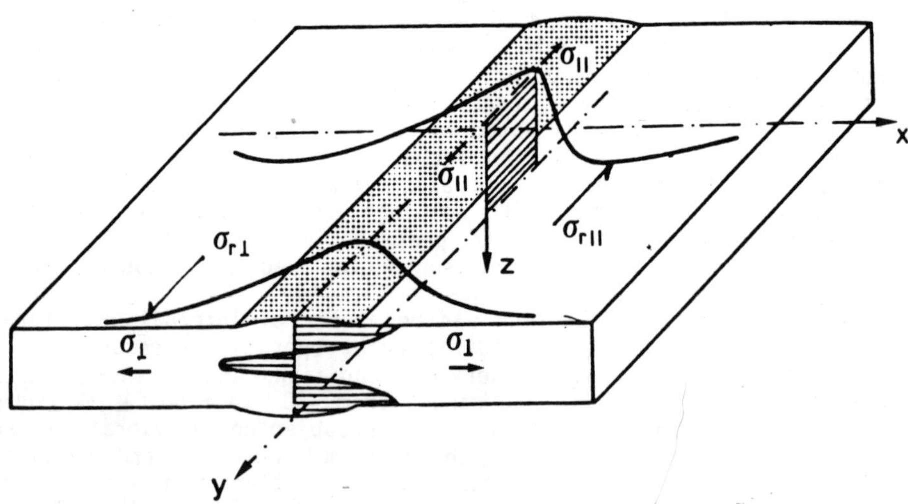

Figure 3. Trend of residual stress along directions x and z perpendicular to welding axis y.

limit value of COD,

$$\delta_{ld} = \frac{K_{IC}^2}{E\,\sigma_y} = \alpha_{ld} \cdot \pi d \cdot \frac{\sigma_y}{E} \cdot \frac{a}{d}\left(1 - \frac{a}{d}\right)^2 \qquad (2.3)$$

Given the critical COD δ_c, if $\delta_c < \delta_{ld}$ or $\delta_c > \delta_{ld}$ the fragile or ductile hypothesis, H_F or H_D, holds respectively. Expression (2.3) is obtained when the critical stress for a surface defect on a plate computed according to linear elastic fracture mechanics (LEFM) as $K_{IC}/\sqrt{\pi a}$, is equated to the stress $\sigma_y(1 - a/d)$ which leads to plastic collapse a notched plate of thickness d. The approximation has been made that in stress and strain computation the beltline region can be treated as a flat plate. K_{IC} is critical stress intensity factor, E is Young modulus and α_{ld} a correcting coefficient. Precisely, α_{ld} takes into account the influence of plastic behavior at crack tip and approximations involved by substituting a sharp to the smooth transition between the two rupture mechanisms. Comparison with experimental data leads to assume $\alpha_{ld} = 2.5$ [4].

The fragile pattern results into a break (before leak) H_{FB} if $\sigma_{app} > \sigma_c$, where σ_c is the critical stress calculated according to LEFM. For the cylindrical geometry it is admitted [5]

$$\sigma_c = \frac{1}{M_p} \sigma_f \arccos\left[\exp\left(-\frac{\pi E \delta_c}{8\,\sigma_f\,c}\right)\right] , \qquad (2.4)$$

$$M_p = \frac{1 - a/d}{1-a/Md} ,$$

with Folias' factor

$$M = \left(1 + 3.22 \frac{c^2}{Dd}\right)^{\frac{1}{2}}.$$

Conversely, in the case of ductile rupture, crack would first pass through the entire thickness of the section if [6]

$$\sigma_{app} > \sigma_{cp} = \sigma_f \left[1 + \frac{ac}{2d(2c + d)}\right]\left[\frac{1- a/d}{1 -a/1.1\,d}\right] . \qquad (2.5)$$

This situation of wall-through defect will have a corresponding limit value of COD,

$$\delta_{ls} = \alpha_{ls}\,\pi\,l\cdot\frac{\sigma_y}{E} \cdot \frac{c}{l}\left(1 - \frac{c}{l}\right)^2 . \qquad (2.6)$$

It can be remarked that ligament l depends upon crack orientation. Indeed it will be $2l \lesssim 2L + \pi D$ and $2l = \pi D$ for the two cases of axial and circumferential weldings. It is assumed $\alpha_{ls} \approx \alpha_{ld} = 2.5$. When $\delta_c < \delta_{ls}$ one has fragile behavior (H_D^f) resulting into a break

(H_{DB}^{f}) or in a leak (H_{DL}^{f}), if

$$\sigma_{app} > \sigma_{c}^{*} = \frac{1}{M}\frac{2}{\pi}\,\sigma_f\;\text{arccos}\left[\exp\left(-\frac{\pi\,E\,\delta_c}{8\,\sigma_f c}\right)\right] \qquad (2.7)$$

or $\sigma_{app} < \sigma_{c}^{*}$, respectively [7]. If $\delta_c > \delta_{1s}$, one has ductile behavior (H_D^{d}) resulting into a break (H_{DB}^{d}) or in a leak (H_{DL}^{d}), when

$$\sigma_{app} > \sigma_{cp}^{*} = \sigma_f\left(1 - \frac{c}{T}\right) \qquad (2.8)$$

or $\sigma_{app} < \sigma_{cp}^{*}$, respectively. This last case means plastic collapse under the hypothesis of perfectly elastic-plastic material.

The complete decision tree and model can be constructed as shown by Fig. 4. The tree has been tested against more than 600 experimental data regarding ruptures of large plates and tubes [4].

3. THE SET OF EXPERIMENTAL DATA K_β

In the reliability computation three input variables are given a statistical treatment, namely COD, crack length 2c and depth a. Mechanical data concerned with the material have been collected by means of ad hoc testing performed on d = 22 mm steel sheet, which has been cut and welded by following a procedure similar to the one adopted for the population of tanks under analysis. COD measurements have been made according to BS 5762, on three-point bending specimens by the use of the so-called linear relation. Two specimens, parallel ($\parallel$) and perpendicular ($\perp$) to the rolling axis, have been used for each of the three metallurgically different conditions of the tank material, i.e. base and welded metal, and heat affected zone (Fig. 5). For the representation of COD values it has been found appropriate to interpret all experimental data by means of a two-parameter Weibull pdf, a result which conforms a recommendation given by the Japan Welding Society [8].

After surfaces are thoroughly polished to remove deposits and brushed, magnetoscopic inspection reveals cracks and gives information about their length 2c (Fig. 6).

Once cracks are discovered, they are brazed. Magnetoscopic inspection, however, does not allow to determine the shape or depth of cracks. It is apparent that crack length and depth are correlated, at least when defects have a small size. It would be appropriate to look for two joint pdf's $f(a_{\parallel}, 2c_{\parallel})$ and $f(a_{\perp}, 2c_{\perp})$, which seems a rather difficult task [9]. Therefore, it has been preferred to start from the crack length marginal cdf and to rely upon the simple exponential form

$$\text{Prob}\{2c > 2C\} = \exp(-2C/2\bar{c}). \qquad (3.1)$$

From the interpolation of experimental data it is found $2\bar{c}_{\parallel}$ = 45.60 mm and $2\bar{c}_{\perp}$ = 13.50 mm for the two directions of cracks with respect to the welding axis. Similarly, an exponential marginal cdf can be used to represent crack depth distribution. To infer expected crack depth $\bar{a}$ it has been supposed that Prob $\{a_{\perp} \geq d\}$ / Prob $\{a_{\parallel} \geq d\}$ = 10 and

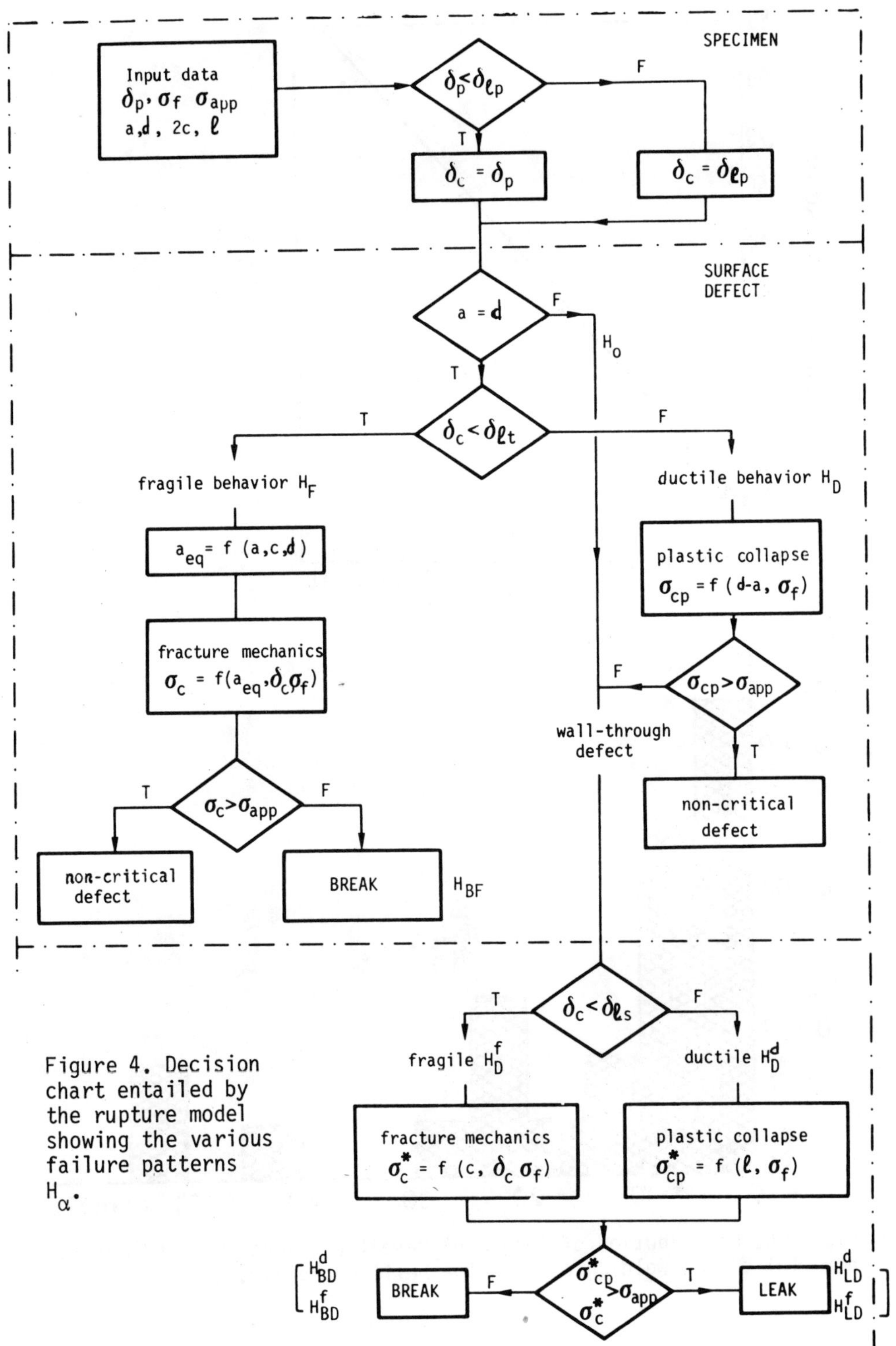

Figure 4. Decision chart entailed by the rupture model showing the various failure patterns H_α.

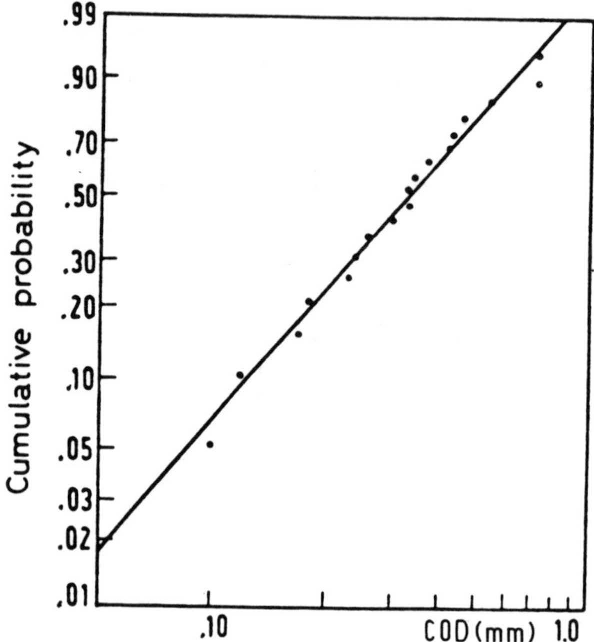

Figure 5. Weibull cdf of COD for the base metal. Case of specimens parallel to the rolling axis on steel sheet.

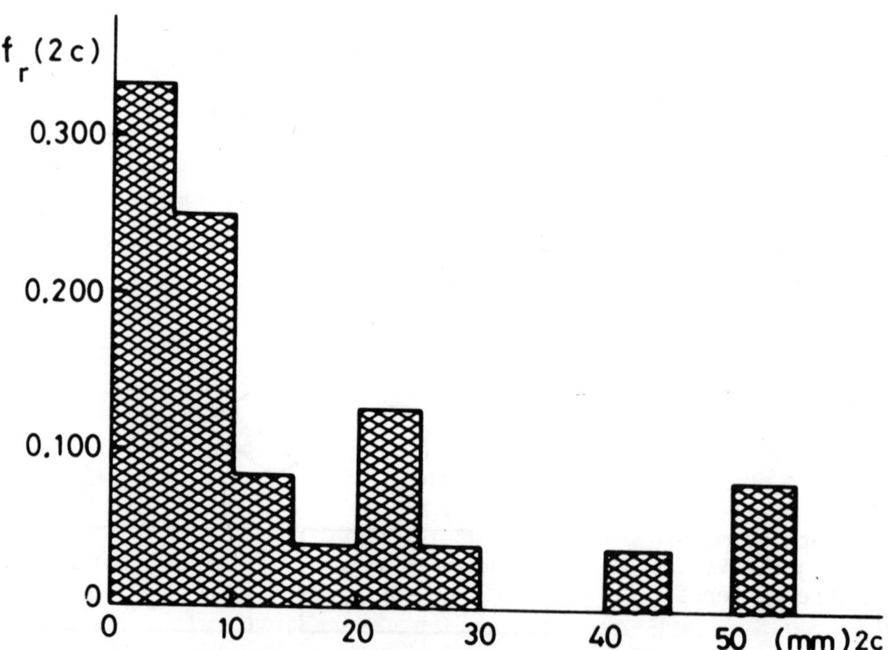

Figure 6. Distribution of length of revealed cracks. Case of cracks parallel to the welding axis in the beltline region.

that a single crack has an inherent probability of spill hazard $p_0 =$ Prob $\{H_0\}$ = Prob $\{a_\parallel \geq d\}$ + Prob $\{a_\perp \geq d\}$ = $1.1.10^{-4}$. It ensues $\bar{a}_\perp = 1.63$ mm and $\bar{a}_\parallel = 1.30$ mm.

Not all existing cracks are detected, however [10]. As for sensitivity of magnetoscopic inspection, cracks are not observed if their length is below threshold values $2c_{0,\perp} \approx 2$ mm and $2c_{0,\parallel} \approx 4$ mm in case of direction perpendicular and parallel with respect to the welding axis, respectively. It is possible to introduce the probability of revealing a generic crack having length 2c in the form [11]

$$r(2c) = \rho\left[1 - \exp(-2\mu c)\right] . \qquad (3.2)$$

Revealed cracks have the exponential pdf $f_r(2c)$ as in (3.1), which allows to obtain the pdf of existing cracks $f_e(2c)$ by means of the relation

$$f_r(2c) = f_e(2c)\, r(2c)\, F_r^{-1} , \qquad (3.3)$$

where F_r represents the fraction of revealed cracks. Clearly, in the computation reference must be made to the pdf of cracks existing before the corrective action $f_e(2c)$ and to the pdf of cracks which remain undiscovered and survive the brazing action,

$$f_b(2c) = f_e(2c)\left[1 - r(2c)\right]\left[1 - F_r\right]^{-1} . \qquad (3.4)$$

On the other hand, for the expected number of cracks the relation holds

$$\bar{n}_b = (1 - F_r)\,\bar{n}_e = \frac{1 - F_r}{F_r}\,\bar{n}_r .$$

It must be noted that ρ and μ depend upon the direction of cracks in the welding. Evaluation of the instrumentation and procedure adopted in magnetoscopic inspection suggests $\rho_\parallel \approx \rho_\perp = 0.99$. Next, by assuming that probability of revealing cracks at their threshold lengths $c_{0,\parallel}$ and $c_{0,\perp}$ is 0.50, it is found $\mu_\parallel = 0.1758$ and $\mu_\perp = 0.3516$. Thus, revealed fractions F_r become 0.8685 and 0.7900, respectively. Furthermore, it can be admitted that the same exponential pdf holds for existing and revealed defects (i.e. $\bar{a}_e = \bar{a}_r$), whereas for remaining defects the exponential pdf would have $\bar{a}_b = \bar{a}_e .\bar{c}_b/\bar{c}_e$. With all these assumptions and data, relevant crack parameters are as in Table 1.

4.   RELIABILITY ASSESSMENT AND OPTIMIZATION

Reliability estimate of the beltline develops along a two step approach. The first step entails the computation of the probability of rupture due to a single crack p_R. Since rupture can be the outcome of a leak or a break, one has $p_R = p_L + p_B$. In its turn, the crack can be either parallel or transverse, therefore $p_L = p_\parallel\, p_{L\parallel} + p_\perp\, p_{L\perp}$, $p_B = p_\parallel\, p_{B\parallel} + p_\perp\, p_{B\perp}$.

Table 1. Parameters characterizing pdf's of revealed, existing and remaining crack population after magnetoscopic inspection and subsequent brazing ($F_{r\parallel} = 0.8685$, $F_{r\perp} = 0.7900$) effected on the ammonia tank at the age $\Delta t_1 = 5$ yrs.

Parameter	Revealed cracks (r)		Esxisting cracks (e)		Remaining cracks (b)	
	$\perp$	$\parallel$	$\perp$	$\parallel$	$\perp$	$\parallel$
$2\bar{c}$ (mm)	13.50	45.60	11.23	40.61	2.695	7.695
$\bar{a}$ (mm)	2.171	1.629	2.171	1.629	0.5210	0.3087
$\bar{n}_\perp$, $\bar{n}_\parallel$	1.770	1.130	2.240	1.301	0.4704	0.1711
$\bar{n} = \bar{n}_\perp + \bar{n}_\parallel$	2.900		3.541		0.6415	

Table 2. Results of the reliability computation for the belt-line region (case $\sigma_{or} = 0.6\,\sigma_y$) obtained through Montecarlo simulation (10^5 runs).

Probability	Existing cracks		Remaining cracks	
	$\perp$	$\parallel$	$\perp$	$\parallel$
$P_F = \mathrm{Prob}\,\{H_F\}$ $p_D = \mathrm{Prob}\,\{H_D\}$ $p_O = \mathrm{Prob}\,\{H_O\}$	$2.76.10^{-3}$ $997.24.10^{-3}$ $0.97.10^{-3}$	$2.11.10^{-3}$ $997.83.10^{-3}$ $0.08.10^{-3}$	$0.55.10^{-3}$ $998.45.10^{-3}$ $< 10^{-5}$	$0.26.10^{-3}$ $999.74.10^{-3}$ $< 10^{-5}$
$p_{BD} = p_{BD}^f + p_{BD}^d$ $p_{LD} = p_{LD}^f + p_{LD}^d$	10^{-5} ~ 1.00	10^{-5} ~ 1.00		
p_{BF}	$1.81.10^{-2}$	$0.47.10^{-2}$	$< 10^{-2}$	$1.15.10^{-1}$
$p_B = p_{BF} \cdot p_F(1-p_o)+$ $\quad + p_{BD}\big[p_D(1-p_o)\mathrm{Prob}\{\sigma_{cp} <$ $\quad < \sigma_{app}\} + p_o\big]$ $p_B = p_B^{\,n}$	5.10^{-5} $1.12.10^{-4}$	1.10^{-5} $0.13.10^{-4}$	$< 10^{-5}$ $\lesssim 10^{-6}$	3.10^{-5} $5.13.10^{-6}$
$p_L = p_{LD}\big[p_D(1-p_o)\mathrm{Prob}$ $\quad \{\sigma_{cp} < \sigma_{app}\} + p_o\big]$ $p'_L = p_L^{\,n}$	123.10^{-5} $27.5.10^{-4}$	16.10^{-5} $2.08.10^{-4}$	$\lesssim 10^{-5}$ $\lesssim 10^{-6}$	$\lesssim 10^{-5}$ $\lesssim 10^{-6}$
$P_B = P_{B\perp} + P_{B\parallel}$ $P_L = P_{L\perp} + P_{L\parallel}$ $P_R = P_B + P_L$	$0.12.10^{-3}$ $2.96.10^{-4}$ $3.08.10^{-3}$		$\sim 5.1.10^{-6}$ $< 10^{-6}$ $\sim 5.1.10^{-6}$	

Admitting that both δ_c and δ_{ld} are statistical variables with their respective pdf's $f(\delta_c)$ and $f(\delta_{lc})$ known, it is possible to compute the probability of fragile behavior

$$p_F = \text{Prob}\{H_F\} = \text{Prob}\{\delta_c < \delta_{ld}\} = \int_0^{\delta_{ld}} f(\delta_{ld}) \int_0^{\delta_{ld}} f(\delta_c) \, d\delta_c \, d\delta_{ld}. \qquad (4.1)$$

and of ductile behavior $p_D = \text{Prob}\{H_D\} = 1 - p_F$.

In the case of fragile behavior, the probability of break before leak is

$$p_{BF} = \text{Prob}\{\sigma_{app} > \sigma_c \mid \delta_c < \delta_{ld}\} = \int_0^{\sigma_{app}} f(\sigma_{app}) \int_0^{\sigma_{app}} f(\sigma_c) \, d\sigma_c \, d\sigma_{app}. \qquad (4.2)$$

Conversely, given the ductile behavior, it is possible to assign probabilities to the other patterns of the tree of Fig. 4, particularly $p_{BD} = p_{BD}^f + p_{BD}^d$, $p_{LD} = p_{LD}^f + p_{LD}^d$.

Thus, the overall probability of break before leak and probability of leak become

$$p_B = p_{BF} \cdot p_F \cdot (1-p_0) + p_{BD}\left[p_D(1-p_0)\text{Prob}\{\sigma_{cp} > \sigma_{app}\} + p_0\right],$$
$$p_L = p_{LD}\left[p_D(1-p_0)\text{Prob}\{\sigma_{cp} < \sigma_{app}\} + p_0\right]. \qquad (4.3)$$

The results of the computation performed by use of Montecarlo method are displayed in Table 2.

The second step consists of the determination of the probability of rupture of the beltline X_{bl}. One has the expected number of cracks before and after the maintenance operation, $\bar{n}_e$ and $\bar{n}_b$ (Table 1). Owing to (1.9) and (4.3), it ensues $P_{Be}(X_{bl}) = 1.12 \cdot 10^{-3}$, $P_{Bb}(X_{bl}) \approx 5.1.10^{-6}$ and $P_{Le}(X_{bl}) = 2.96.10^{-3}$, $P_{Lb}(X_{bl}) \lesssim 10^{-6}$. Corresponding rupture probabilities are $P_{Re}(X_{bl}) = 3.08.10^{-3}$ and $P_{Rb}(X_{bl}) \approx 5.1.10^{-6}$.

To allow a direct, although rough, estimate of the probability of rupture of the ammonia tank it is useful to express these results in terms of probability per unit length of welding $q(X_{bl}) = P/w_{bl}$, where $w_{bl} = 48.1$ m is the total welding length of the beltline. Then a preliminary analysis of stress and strength distribution leads to assume $q(X_{sb}) \approx 2q(X_{bl})$, and $q(X_{co}) \approx 5\,q(X_{bl})$ for the spherical bottoms and circular opening, respectively. Since welding lengths involved are $w_{sb} = 65.0$ m and $w_{co} = 3.54$, it is obtained $P(X) = 1.64\,P(X_{bl})$, a relation that implies $P_{Be}(X) = 0.196.10^{-3}$, $P_{Le}(X) = 4.85.10^{-3}$ and $P_{Bb}(X) = 8.3.10^{-6}$, $P_{Lb}(X) = 1.6.10^{-6}$.

Clearly, the degree of accuracy of the inspection and repair procedure, and the time between maintenance operations have a definite influence upon the probability that the ammonia tank undergoes rupture during the service period. An optimization can be sought, where objective is a certain safety (or reliability) level to be achieved by minimizing the resources involved.

To this purpose, let $E_B = 5.10^6$ \$ and $E_L = 5.10^5$ \$ US be the expected costs suffered in case of break and leak of a single tank, respectively. The difference between the two values allows for the different degree of risk involved. On the other hand, each maintenance operation involves a direct cost e_p which depends upon ρ, $\mu_{\parallel}$, $\mu_{\perp}$ and an indirect cost e_q associated to the outage time. In the range of interest, a reasonable direct cost function is

$$e_p = e_p (F_r') = e_{po} \left[1 - \exp \left(- \frac{e_{p1} F'}{1 - F'}\right)\right] , \qquad (4.4)$$

where $e_{po} = 2.10^4$ \$, $e_{p1} = 2.10^{-2}$ and $F' = \max \{ F_{r\parallel}, F_{r\perp} \}$. In the case under study, (4.4) gives $e_p = 2,500$ \$ as in the premises.

Now, basing on the information collected at age t_1, it is possible to look for the optimum time at which next maintenance operation must be performed. This time interval $\Delta t_2 = t_2 - t_1$ corresponds with

$$\min \left\{ \frac{1}{t_2 - t_1} \left[E_B \int_{t_1}^{t_2}(1 - P_B(u)) \lambda_B(u) du + E_L \int_{t_1}^{t_2}(1 - P_L(u)) \lambda_L(u) du + e_p + e_q \right] \right\}, \qquad (4.5)$$

under the assumed external constraint $P_B(t_2)/P_L(t_2) < 10^{-1}$. To solve the problem, one has first to infer the time dependence of the failure intensity function (1.4). Since the law of propagation of cracks is not known, it has been chosen to rely upon the linear from $\lambda = \lambda_0 \Delta t$, that entails a Weibull pdf of rupture. Then having neglected terms $0(\Delta t)^2$ it ensues for the two failure patterns,

$$\lambda_{Bo} = \frac{2P_{Be}(t_1)}{\Delta t_1^2} , \qquad \lambda_{Lo} = \frac{2P_{Le}(t_1)}{\Delta t_1^2} . \qquad (4.6)$$

Crack inspection and repair procedure, however, leaves a number of defects undected. The situation which results can be assimilated to an equivalent aging of the structure at time t_1

$$\Delta t_B^* = \left(\frac{2 P_{Bb}(t_1)}{\lambda_{Bo}}\right)^{\frac{1}{2}} , \qquad \Delta t_L^* = \left(\frac{2 P_{Lb}(t_1)}{\lambda_{Lo}}\right)^{\frac{1}{2}} . \qquad (4.7)$$

With the help of these relations, it becomes feasilble to rewrite (4.5) in the approximate form,

$$\min \left\{ \frac{1}{\Delta t_2} \left[\frac{\lambda_{Bo} E_B}{2} (\Delta t_2 + \Delta t_B^*)^2 + \frac{\lambda_{Lo} E_L}{2} (\Delta t_2 + \Delta t_L^*)^2 + e_p + e_q \right] \right\} . \qquad (4.8)$$

Owing to the data available regarding rupture probability, it is found $\lambda_{Bo} = 1.57.10^{-5}$ yr^{-2}, $\lambda_{Lo} = 38.8.10^{-5}$ yr^{-2} and $\Delta t_{1B}^* = 1$ yr, $\Delta t_L^* = 0.1$ yr. Finally, if $e_q = 0$, the optimum time interval results $\Delta t_2 = 4.3$ yr with

$P_B(t_2) = 0.22.10^{-3}$, $P_L(t_2) = 3.74.10^{-3}$ and $P_B(t_2)/P_L(t_2) < 10^{-1}$, as requested.

ACKNOWLEDGMENT

The paper has been written in memory of prof. D. Sinigaglia, a friend and colleague of the author. Prof. Sinigaglia had the merit to consider the problem first and to collect some experimental evidence. Mention should be also deserved to the help provided by V. Pistone and S. Venzi from Snam SpA, Milan, especially as for the development of the rupture model and computation.

REFERENCES

[1] GARRIBBA, S., 'Use of Stochastic Processes for Mechanical Reliability Analysis and Examples', Nucl. Engng. and Design, 60, 1980, p. 37.

[2] GERTSBAKH, I.B. and KORDONSKIY, Kh. B., Models of Failure, Springer Verlag, New York, NY, 1969 (tr. of Modeli Otkazov, Sovetskoe Radio, Moscow, 1966).

[3] TERAD, H., 'An Analysis of the Stress Intensity Factor of a Crack Perpendicular to the Welding Bead', Engng. Fracture Mechanics, 8, 1976, p. 441.

[4] VENZI, S., SINIGAGLIA, D. and PISTONE, V., 'Structural Reliability Analysis of Pressure Tanks for Liquefied Gases under Stress-Corrosion Attack', in Structural Safety in Design, Operation and Inspection of Chemical Plants, Clup, Milan, Italy, 1981, p. 149 (in Italian).

[5] KIEFNER, J.F., et al., 'Failure Stress Levels of Flaws in Pressurized Cylinders', in ASTM-STP 536, American Society for Testing and Materials, Philadelphia, Pa., 1973, p. 461.

[6] LARSSON, H., 'Use of Elastic-Plastic Fracture Mechanics in Design', in Proc. 2nd Advanced Seminar on Fracture Mechanics, Ispra Courses, Ispra JRC, Italy, April 2-6, 1979.

[7] HAHN, G.T., SARRATE, M. and ROSENFIELD, A.R., 'Criteria for Crack Extension in Cylindrical Pressure Vessels', Int. J. of Fracture Mechanics, 5, 1969, p. 187.

[8] KANAZAWA, T. et al., 'Outline of JWES Standard for Critical Assessment of Defects with Regard to Brittle Fracture and Some Case Studies', in Proc. Colloquium on Practical Applications of Fracture Mechanics, Bratislawa, Poland, July 10, 1979.

[9] BRUECKNER, A., HAEBERER, R., and MUNZ, D., 'Determination of Crack Size Distributions from Incomplete Data Sets for the Calculation of Failure Probabilities', Reliability Engineering, 11, 1985, p. 191.

[10] RACKWITZ, R. and SCHRUPP, K., 'Quality Control, Proof Testing and Structural Reliability', Structural Safety, 2, 1985, p. 239.

[11] BESUNER, P.M. and TETELMAN, A.S., Probabilistic Fracture Mechanics, EPRI 217-1, Tech. Rep. No. 4, Electric Power Research Institute, Palo Alto, Ca., July 1975.

STRUCTURAL RELIABILITY OF PWR PRESSURE VESSELS

A.C. Lucia
Commission of the European Communities
Joint Research Centre - Ispra Establishment
21020 Ispra (Va) - Italy

ABSTRACT

The subject of this paper is the probabilistic structural reliability
assessment of LWR pressure vessels. Particular attention is paid to
the cumulative damage process due to fatigue crack growth. The results
are reported of an exercise of comparative assessments of failure
probability by two different approaches. NDI requirements to assure
specified levels of reliability are evidenced, together with the
parameters to which the reliability is most sensitive.

INTRODUCTION

The problem of the probabilistic approach to the structural reliability
assessment of pressure vessels has been dealt with in several studies,
starting from the early seventies: Becher /1/, Nilsson /2/, Harris /3/,
Marshall /4,5/, Dufresne-Lucia /6,7/.
As is well known, one great advantage of the probabilistic methods,
in comparison with the statistical one, is constituted by the deeper
knowledge gained on the component itself and by the possibility of
investigating several safety-related structural aspects, allowing:

- application of sensitivity analysis procedures for quantitative
  estimation of the weight of the various parameters on the final
  result;
- propagation of uncertainties from the independent variables to the
  dependent ones;
- comparison of different design and operation solutions;
- ranking of structure zones according to their failure probability;
- analysis of the effect of different inspection strategies and definition of a more rational and effective one.

The obvious disadvantage of the probabilistic approach is its complexity, consequence of:

A. C. Lucia (ed.), Advances in Structural Reliability, 233–254.
© *1987 by ECSC, EEC, EAEC, Brussels and Luxembourg.*

i) need of identifying and correctly modelling damage accumulation
    and failure processes;
ii) need of large amounts of data.

Furthermore, having identified the initial statistical distributions
of the independent variables, it is still necessary to consider any
phenomenon which might later influence these variables and to follow
the evolution of their distributions.

As a consequence, the probabilistic studies are devoting ever
increasing effort in order to cope with the above requirements.
Particular attention has been paid by Dufresne-Lucia /7/ to the col-
lection of large samples of data, to get improved statistical distributions for the main random variables.

NRC and EPRI both have programs underway to achieve the above goal.
The second Marshall Report /5/ underlines the importance of fracture
assessment and comes out with some 57 recommendations on materials,
design, fabrication and inspection.

From all these studies it seems possible to state that the main
factors for the safety of the pressure vessels are:

1) static fracture toughness and, as a consequence, radiation embrittlement and ECC water temperature;
2) position of cracks in the thickness (Lucia /8/);
3) integrity of the cladding (Marshall /5/, Lucia /8/).

On the other hand, some aspects of the probabilistic analysis seem to
need further and deeper attention. The inspection effectiveness is
not clearly known and the influence of in-service inspection on structural reliability has to be better modelled. This will, in turn, need
validation activities for the inspection techniques.

The residual stresses still constitute a potential source of
danger, the knowledge about their value and their use in a FM analysis
being rather vague. Furthermore, residual stresses can be introduced
by in-service repairs without post-weld heat treatment, causing local
stress intensification. Local thermal stresses of relevant intensity
can be generated during ECC water injection by possible quench front
oscillations and local variations in fluid (water/vapour) temperature
and heat transfer coefficients.

There is a need of "data on the fracture behaviour of actual
vessels of well-characterized material and construction and containing
known predetermined flaws" (Marshall /5/), in order to cope with the
complexity and uncertainties of fracture mechanics analysis.

RELIABILITY OF A STRUCTURAL COMPONENT

The deterioration of the integrity of a structural component takes
place continuously under normal conditions of operation, loading and
environment, although emergency and faulty conditions can strongly
speed up the process. Structural reliability can be thought of as the
probability of absence of crossings of a safe level by accumlated
damage processes: the best approach to its assessment is constituted

by the theory of stochastic processes. The general expression of the
unconditional reliability is:

$$R(T,\underline{X}) = \sum_{a=1}^{A} \underset{T,\underline{X}}{P} \{^{a}\underline{f}(^{a}\underline{q}) > {}^{a}\alpha\} \cdot P\{F_{a}\} \tag{1}$$

where $\underset{T,\underline{X}}{P}$ is the conditional reliability with reference to failure criterion F_{a}; $P\{F_{a}\}$ is the probability that criterion F_{a} holds, and the
sum is extended to all the A possible failure modes of the structural
component. The general solution of expression (1) is very difficult.
Simplifying assumptions have to be made.

Let us present now some of the problems tackled and the results
obtained in the course of a study aimed at the estimation of the probability of failure of the pressure vessel of a PWR /7/. This study
led to the development and implementation of the numerical code
COVASTOL /9/. Main simplifying assumptions are: load and resistence
are statistically independent processes; the vessel is subdivided into
regions analysed independently; fatigue and neutron embrittlement are
the only two damaging processes; brittle failure (K_{1C} criterion) is
the only failure mechanism considered.

FATIGUE DAMAGE

The mechanism of fatigue is perhaps the most important example of
degradation of the strength of a structure, caused by the irreversible
accumulation of damage to the internal structure of the material. The
interpretation of damage as the birth and propagation of defects in
the elementary structure of the material, caused by the alternating
stress field acting on imperfections of the crystalline network, on
distortions due to impurities, etc., is generally accepted. Qualita-
tively we can distinguish:

- an initial nucleation stage (defect generation);
- a transition stage (defects coalescence);
- a propagation stage (unstable growth of the largest defects).

In general, being N_F the number of cycles to rupture and N_O, N_T, N_C the
cycles at the end of the nucleation, transition and propagation stages,
one has:

$$N_F = N_O + N_T(N_O) + N_C(N_T) \tag{2}$$

which expresses the fact that each stage is determined by the level
reached in the preceding stage.

The damage accumulation mechanisms are presumably different in
the three stages and in each one they are significantly dependent on
the environmental conditions and the stress field; for this reason the
construction of a unified model of interpretation seems unlikely.

Furthermore, the relative importance (in terms of number of cycles) of the above three fatigue stages depends on the intensity of the alternating load. This rather complex picture can be partially clarified by the consideration that experimentally evident fatigue failures in operating components are mainly caused by fabrication defects, generally introduced during welding procedures. As a consequence, the nucleation and transition stages are relatively more than just a research problem; it is a practical problem of current design, considered by the current standards.

The approaches to fatigue damage modelling can be classified in four broad categories /13/:

i) Criteria for design - Essentially deterministic methods, such as ASME Sect.III standards for design and ASME Sect.XI standards for acceptance criteria of defect propagation rates, they are based on the Palmgren-Miner rule.
This rule assumes a linear accumulation of damage, entailing a time-invariant failure rate which is often in contrast with experimental evidence.

ii) Methods based on the randomization of analytical formulations -
Some hundreds of more or less different relationships can be found in the literature for expressing the fatigue growth of cracks. Some of them are purely theoretical or based on microscopic properties of material, but the most widely employed are semi-empirical and have been developed mainly as interpretative models of experimental results; they allow the prediction of the behaviour of "a" as a function of "N". This prediction is based on the integration of the growth rate, for fixed initial conditions of the defect. The result arrived at is not, however, in general, representative of the real growth situations. This is due both to the fact that the initial conditions have a considerable scatter of values and to the fact that, for the same initial conditions, there is an intrinsic variability in the process of damage by fatigue which leads to a distribution of values "a" at cycle N.
It thus appears natural to consider the relationships for defining the growth rate as stochastic. In this context, therefore, prediction methods can be seen as being based on the integration of the FCG relationships with parameters, initial conditions and loads represented by random variables. These procedures lead to the determination of a distribution of dimensions for the propagated defect, at cycle N, or of a distribution of the number of cycles N for a given propagation from a_o to a_f.
The COVASTOL numerical code belongs to this category.

iii) Methods based on the extreme values theory - The structural reliability, in its most stringent formulation, can be defined as the probability that the largest of the loads envisaged is smaller than the smallest of the resistances hypothesized. This means that what one needs to know is the distributions of the extreme values of the loads and of the resistances, rather than their effective distributions. This observation, together with the

fact that the possible distributions of the extreme values of a
random variable are asymptotically independent of the distribution of the variable itself, leads to the consideration of the
extreme values theory as a fundamental ingredient of structural
reliability.
In these methods the variation of the results is interpreted
on the basis of a probabilistic model of the initial distribution
of the defects, linked to the stress intensity distribution by the
fracture mechanics relationships (Freudenthal, Ang). In this case,
as in the preceding ones, the relationships derived from the mathematical model are rather simple and suitable for current
design use.

iv) Methods based on the simulation of the stochastic process of
    damage accumulation - In general both the loading actions and
    the resistance degradation mechanisms have the characteristics
    of stochastic processes. They can thus be defined as random va-
    riables which are functions of time. The particular load history
    which affects a component is one of the possible realizations of
    the stochastic load process and the same applies for the environ-
    mental condition or for the evolution of the dimensions of a
    defect inside the component.
    The prediction of the component lifetime can, therefore, be based
    on the representation of the stochastic processes acting on the
    component.
    In most cases, the damage accumulation mechanisms can be repre-
    sented by a positive "damage rate" function such that the measure
    of damage is a monotonic increasing function of time.
    Under rather general conditions such accumulation processes can
    be represented by a well-known class of stochastic processes,
    the Markovian class. The simplicity and flexibility of models
    based on Markov or semi-Markov schemes is the reason for their
    frequent appearance in the literature.
    Our numerical code RELIEF /10/ is based on this approach.

MATERIAL PROPERTIES DATA FOR FCG EVALUATION

Let us now go back to the code COVASTOL and to the problems connected
with that type of modelling approach. It is very appropriate for
pointing out the two types of difficulties which have to be overcome
in modellization: dependence of the process on many parameters (consequence: reduction to simpler and less accurate models, or development of complex models and need of very large amounts of experimental
data to get statistical significance in every region of the experimental space) and scattering of data.
     Parameters which might influence the propagation rate are the
following:

- stress parameters: maximum stress, mean stress, frequency and shape
  of the load signal, value of $R = \sigma_{min}/\sigma_{max}$;

- material parameters: chemical composition, heat treatment, presence of cracks;
- environment parameters: corrosion, irradiation, temperature;
- geometric parameters.

The models trying to describe the FCG should attempt to take all these parameters into account. Actually such a complete model does not exist. The different models (empirical, semi-empirical or theoretical) present in the literature can be classified as follows:

i)   phenomenological models;
ii) models based on the dislocation theory (e.g. Bilby-Cottrel-Swinden model);
iii) models based on the material behaviour at crack tip (e.g. McClintock, Pook-Fiort, etc.);
iv) models based on cyclic properties of the material (e.g. Schwalbe).

Actually, theoretical models often bring in parameters which, in practice, are not available and their complexity makes it difficult to use them except for the precise situation they were established for. Hence we have mainly considered phenomenological models, to be validated by comparison with experimental data.

The Paris, Priddle, Forman and Walker relations (this latter one, unlike the preceding three empirical models, relies on the physical concept of effective stress intensity factor) have been selected. Their validation has been made by the use of a population of about 3000 experimental points relative to A533B steel. The comparison of the four models is based on:

- standard deviation
$$\sigma = \frac{1}{n}\left\{\sum_{i=1}^{n}\left[\left(\frac{da}{dN}\right)_i - f(\Delta K_i)\right]^2\right\}^{1/2}$$

where $\left(\frac{da}{dN}\right)_i$ is the experimental value, and $f(\Delta K_i)$ the calculated value;

- determination coefficient r^2:

$$r^2 = \frac{\left|\Sigma(\text{Log}\Delta K_i)\left(\text{Log}\frac{da}{dN_i}\right) - \frac{\Sigma(\text{Log}\Delta K_i)\left(\Sigma\text{Log}\frac{da}{dN_i}\right)^2}{2}\right|}{\left|\Sigma(\text{Log}\Delta K_i)^2 - \frac{(\Sigma\text{Log}\Delta K_i)^2}{n}\right|\left|\Sigma\left(\text{Log}\frac{da}{dN_i}\right) - \frac{\left(\Sigma\text{Log}\frac{da}{dN_i}\right)^2}{n}\right|}$$

measuring the correlation between theoretical curve and experimental points;

- mean of the relative deviations:

$$e = \frac{1}{n} \sum_{i=1}^{n} \left[\frac{f(\Delta K_i) - \left(\frac{da}{dN}\right)_i}{\left(\frac{da}{dN}\right)_i} \right]$$

Table 1 groups together all the results obtained for the different
models. Without entering into detail, only evidence can be given that,
even if the Priddle, Forman or Walker formulas can - in certain cases -
lead to slightly better theoretical forecasts than the Paris one, this
last one was chosen for the entire complex of the experimental conditions considered.

What remains to be done is to choose the C and n coefficients and
their distributions, if any. First of all it is necessary to go back
to the main phenomena and parameters influencing the FCG and to select
the most relevant ones. In order to do so, a series of experiments on
specimens have been made, investigating the influence of the following
parameters:

a) Material:
    base material: laminated A533B Cl.1 steel
                    forged A 508 Cl.3 steel
    weld material
    heat affected zone

b) Stress orientation:
    combined loads on specimens: mode I and II, mode I and III, Mode
    I, II and III

c) Environmental conditions:
    dry environment
    wet environment, with nominal water composition
    wet environment with polluted water (oxygene, chlorine, fluorine
    content above nominal values)

d) Frequency of the loading cycle

e) ΔK range

f) Value of $R = \sigma_{min}/\sigma_{max}$

g) Effect of the 3D stress field when passing from specimen to
    structure

h) Irradiation

Starting from the experimental data collected to study these different
phenomena and according to the real conditions of a PWR vessel, we
have identified a limited number of "classes" or domains defined on
the basis of the most relevant parameters: ΔK, R, environment, frequency.

When dealing with two experimental data distributions, with a
given confidence level, we used the student test in order to decide
whether to mix the data of the two sets or not. At the end we came
out with four classes (see Fig. 1):

TABLE 1 - COMPARISON OF FATIGUE CRACK GROWTH LAWS

ΔK (MPa m)	f (cpm)	Environ-ment	No. of data	PARIS $\frac{da}{dn} = C(\Delta K)^n$				FORMAN $\frac{da}{dn} = \frac{C'(\Delta K)^n}{(1-R)Kc - \Delta K}$				PRIDDLE $\frac{da}{dn} = C''(\Delta K - \Delta K_s)^{n''}$				
				C	n	r^2	e	C'	n'	r^2	e	C''	n''	r^2	e	ΔK_s
$0 < \Delta K < 20$	f	air	720	$3.3 \cdot 10^{-27}$	2.64	0.99	+1.3%	$2.1 \cdot 10^{-19}$	2.68	0.99	+1.10%	$1.5 \cdot 10^{-14}$	0.82	0.86	-9%	$3.8 \cdot 10^6$

ΔK (MPa m)	f (cpm)	Environ-ment	No. of data	PARIS $\frac{da}{dn} = C(\Delta K)^n$				FORMAN $\frac{da}{dn} = \frac{C'(\Delta K)^n}{(1-R)Kc - \Delta K}$				WALKER $C_1(Keff)^{n_1} \exp C_1(LogKeff)^2 + C_3(LcgKeff)^3$					
				C	n	r^2	e	C'	n'	r^2	e	C_1 medium	n_1	C_2	C_3	r^2	e
$20 < \Delta K < 90$	f	air	821	$4.8 \cdot 10^{-25}$	2.32	0.98	-1.5%	$7.2 \cdot 10^{-15}$	2.09	0.98	+0.4%						
$90 < \Delta K$	f	air	98	$7 \cdot 10^{-20}$	1.65	0.73	+1.7%	$9.5 \cdot 10^{-4}$	0.69	0.68	+4.1%	$7 \cdot 10^{-20}$	1.65	$2.8 \cdot 10^{-3}$	$-1.4 \cdot 10^{-4}$	0.69	+2.3%
$\Delta K_t < \Delta K$	$f \leq 10$	water	858	$2.3 \cdot 10^{-14}$	1.04	0.87	-6%	$9.2 \cdot 10^{-2}$	0.48	0.47	-0.4%	$2.3 \cdot 10^{-14}$	1.04	$2.2 \cdot 10^{-3}$	$-9.3 \cdot 10^{-5}$	0.89	+0.9%

$$r^2 = \frac{\left[\Sigma(Log\Delta K_i)\left(Log \frac{da}{dN_i}\right) - \Sigma(Log\Delta K_i)\left(\Sigma Log \frac{da}{dN_i}\right)^2\right]}{\left[\Sigma(Log\Delta K_i)^2 - \frac{(\Sigma Log\Delta K_i)^2}{n}\right]\left[\Sigma\left(Log \frac{da}{dN_i}\right)^2 - \frac{\left(\Sigma Log \frac{da}{dN_i}\right)^2}{n}\right]}$$

$$e = \frac{1}{n} \sum_{o}^{n} \left[\frac{\frac{da}{dN}_{cal} - \frac{da}{dN}_{exp}}{\frac{da}{dN}_{exp}}\right]_i$$

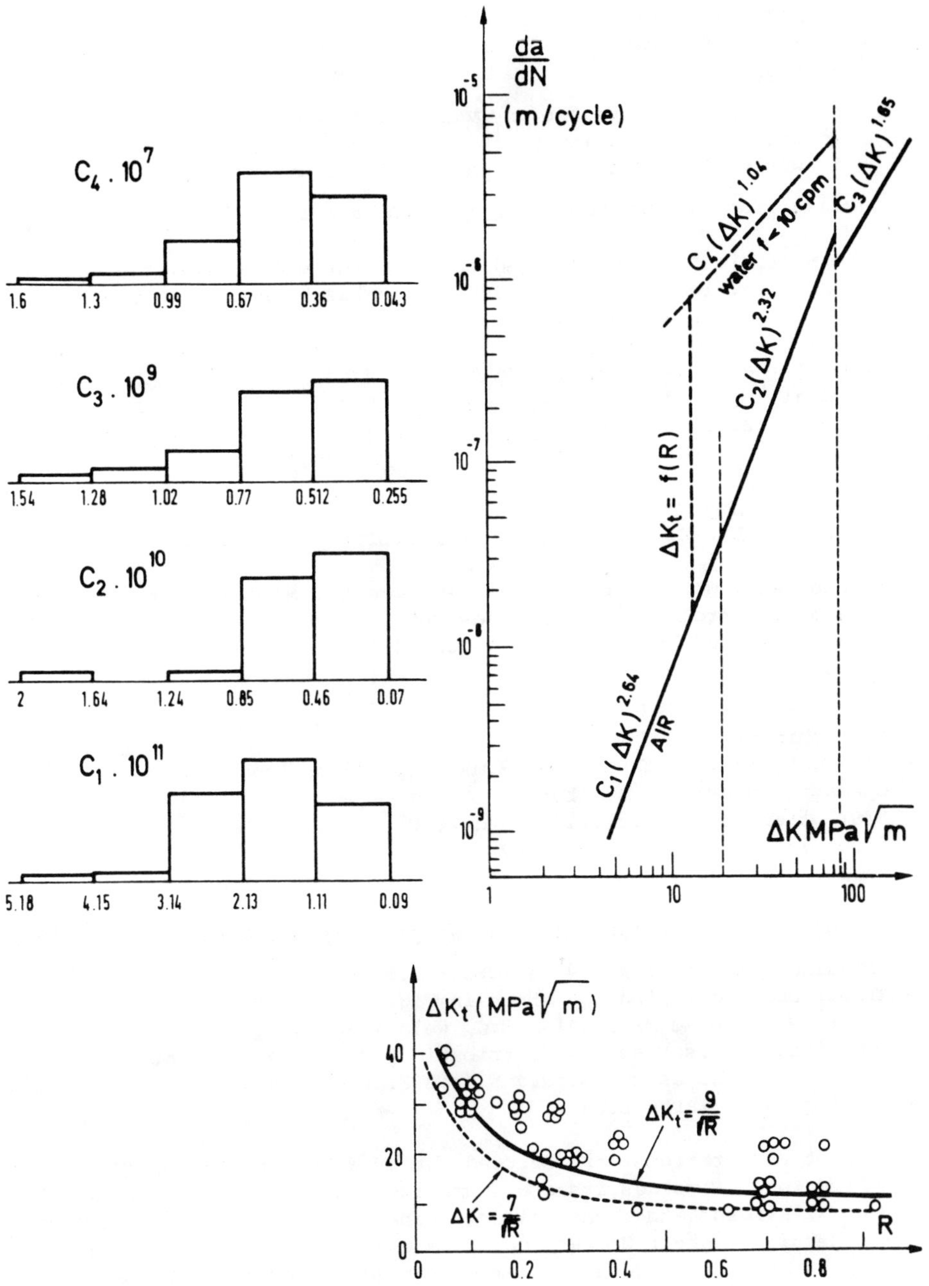

Figure 1. Representation of Paris law in the numerical code COVASTOL.

- 3 classes corresponding to 3 ranges of ΔK:

$$0 \leq \Delta K < 20 \text{ MPa}\sqrt{m}$$
$$20 \leq \Delta K < 90 \text{ MPa}\sqrt{m}$$
$$90 \leq \Delta K < 170 \text{ MPa}\sqrt{m}$$

  in dry environment: for any value of f and R
  in wet environment and f > 10 cpm for any R
  in wet environment, f < 10 cpm and $\Delta K < \Delta K_t$

- 1 class, for wet environment $f \leq 10$ cpm and $\Delta K \geq \Delta K_t$

Furthermore, it has been decided to have, for each class, one histogram of C and one value of n. Their definition has been made on the basis of 4150 experimental points.

The use of different methods of data treatment (graphic method, method of the difference, method of the secant, method of the total polynomial) has led to the conclusion that the main cause of dispersion is constituted by the experimental technique, the data handling methods giving a much lower contribution to the dispersion.

A BENCH-MARK EXERCISE ON FAILURE PROBABILITY COMPUTATION

A simple bench-mark exercise has been made to compare the results obtained by the probabilistic methods developed by Westinghouse (W) from the one part, and by CEC-CEA-Framatome (CEC) from the other.

  Common input data were:

- inside radius: 1994 mm
- wall thickness: 200 mm
- flaw geometry: surface flaw 12 mm deep, 72 mm long
- operating pressure: 157 bar
- RWST water temperature: 10°C and 20°C
- fluence: $5 \cdot 10^{19}$ (n/cm^2,> 1 MeV)
- reactor age: 40 years

  Input data different for the two participants were:

- cladding thickness: W : 4 mm, CEC : 7.5 mm
- RTNDT: CEC : distribution shown in Fig. 2
        W   : base material = 0°C, weld = -20°C
- impurities: W uses normal distributions for P and Cu content
              CEC uses histograms (see Figs. 3 and 4)
- loading case: large LOCA

Different computations were carried out by Westinghouse and by CEA-CEC-Framatome. Some results are summarized in Table 2. Results No. 1 and No. 2 evidence that the failure probability obtained by CEC model are generally higher. Comparison of cases 1 and 7; 6 and 10; 9 and 11 shows that the CEC stress and temperature distributions give a failure probability value 20 to 40 times lower than that obtained by Westinghouse distributions. The effect of K_{1c} and K_{1a} distributions can be

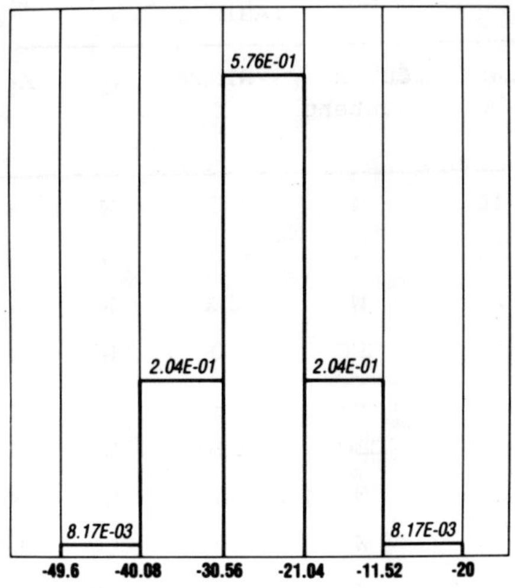

Fig. 2 -   Distribution of RTNDT used in benchmark exercise on
           vessel failure probability.

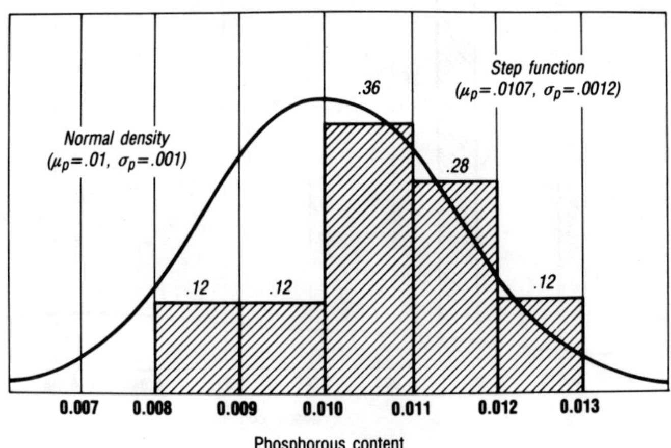

Fig. 3 -   Phosphorus content probability density functions
           used by Westinghouse (normal) and CEC (histogram) in
           the benchmark exercise on vessel failure probability.

TABLE 2

Comp. No.	σ & T data	RWST °C	Cu, P content	RTNDT °C	K_{1c}	K_{1a}	Failure prob. Model W	Failure prob. Model CEC
1	W	10	W	0	W	W	10^{-3}	$8 \cdot 10^{-3}$
2	W	10	W	−20	W	W	$4 \cdot 10^{-3}$	$2 \cdot 10^{-2}$
3	W	10	W	CEC	W	W		$2.3 \cdot 10^{-3}$
4	W	10	CEC	0	W	X		$2 \cdot 10^{-3}$
5	W	10	CEC	−20	W	W	$2 \cdot 10^{-7}$	$< 10^{-7}$
6	W	10	CEC	CEC	W	W		$3.5 \cdot 10^{-4}$
7	CEC	10	W	0	W	W		$1.5 \cdot 10^{-1}$
8	CEC	10	W	0	CEC	CEC		$2.3 \cdot 10^{-3}$
9	W	10	CEC	CEC	CEC	CEC		$5.4 \cdot 10^{-5}$
10	CEC	10	CEC	CEC	W	W		$1.6 \cdot 10^{-2}$
11	CEC	10	CEC	CEC	CEC	CEC		$2.6 \cdot 10^{-3}$

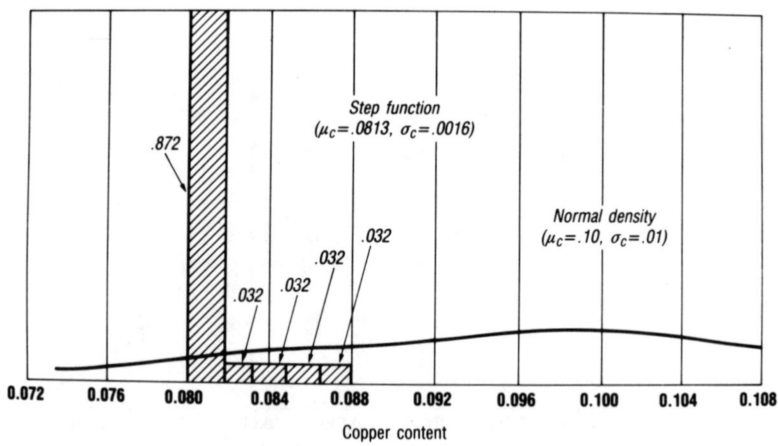

Fig. 4 – Copper content probability density functions used by Westinghouse (normal) and CEC (histogram) in the benchmark exercise on vessel failure probability.

seen comparing cases 6 and 9; 10 and 11: the CEC values give lower
probabilities (factor 20). As far as Cu and P contents are concerned,
comparison of cases 1 and 4; 3 and 6; 2 and 5 shows that the W data
lead to more pessimistic results. Particularly astonishing is the
discrepancy between case 2 and case 5.

As a conclusion, we can say that not only the experimental data
are important in themselves, but also the way chosen to utilize them
can affect the final result. This is particularly true when the result
is determined by the tails of the distributions.

NDI REQUIREMENTS TO ASSURE A GIVEN LEVEL OF STRUCTURAL RELIABILITY

Using the COVASTOL computer code as a basis, a sensitivity analysis
has been performed on all the parameters considered in the code. One
of the inferences from this analysis is the high sensitivity of failure probability to the position of defects in the thickness and in the
various regions of the vessel. This aspect of the sensitivity analysis
is given further attention here and is also discussed elsewhere /8/.

Calculations made with COVASTOL from data compiled on operating
PWRs allow the following conclusions to be drawn:

- Rupture probabilities at the 40th year vary from 10^{-7} to 10^{-12}
  according to the weld location and the safety injection water temperature;

- The evolution of the rupture probability of various welds as a
  function of time is presented in Fig. 5. This figure shows that the
  belt weld is more sensitive to the time than the others. This is due
  to the fact that, for this weld, time has two effects: crack growth
  and steel embrittlement;

- The most dangerous defects are located within the first millimeters
  of the internal layer. If it were possible to remove all the defects
  located at less than 6 mm from the ferritic steel and stainless
  steel coating interface, the rupture probability would be reduced by
  a factor greater than 500 (without taking into account the protective
  effect of the cladding); the effect of a perfect repair procedure
  is indicated in Figs. 6 and 7.

The protective effect of the cladding decreases the fatigue growth
of defects and lowers the overall rupture probability by some orders
of magnitude (Fig. 8). This fact evidences the importance of an early
detection of cladding integrity damage.

Fig. 9 shows the conditional failure probability (assuming a
specified defect and accident) for the belt line weld, at the 40th
year of reactor operation, for a large LOCA, as a function of crack
width (2a) and for different locations (d). The location is expressed
by the distance, d, of the crack centre from cladding-ferritic interface and a is one half of the minor axis of the crack. The crack
length is here always assumed equal to 512 mm. The stress and temperature evolution during LOCA are evaluated assuming an emergency core
coolant supply water temperature of 20°C.

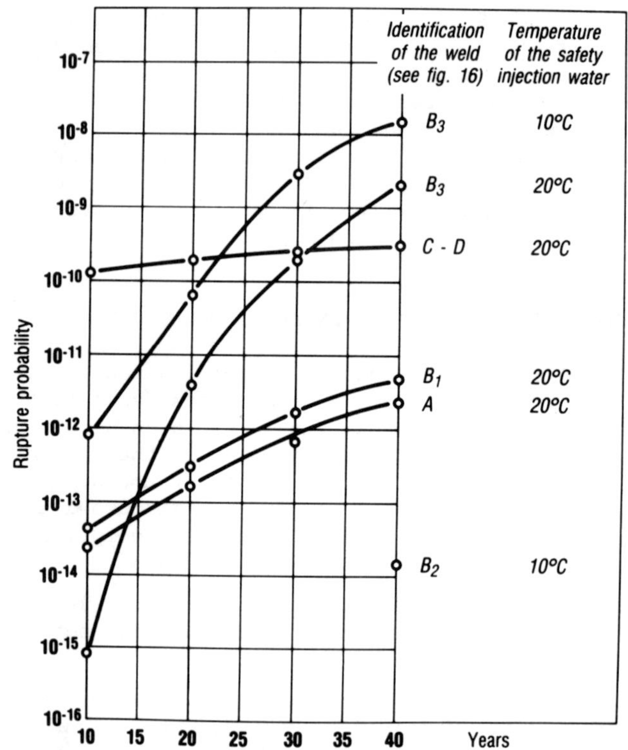

Fig. 5 — Probability of rupture of the welds of a PWR vessel
(see Fig.16) at different temperatures of the safety injection water.

It can be seen that a 12 mm defect leads to a failure probability
lower than 10^{-6} if its inner tip is 12 mm distant from the interface,
while the probability can increase to 10^{-3} if this distance is only
3 mm. Fig. 10 indicates that a surface defect of 3 mm could still
yield a failure probability of 10^{-4} during a large LOCA.

The harmfulness of under cladding defects located in the nozzles
depends on their location. The most harmful defects are located in the
vertical cross section (meridian to vessel) in the vicinity of the
inner corner.

Fig. 10 shows the conditional failure probability caused by an
8 mm undercladding defect (length 64 mm) during an intermediate-size
breach LOCA (40th year of operation). In each zone (A,B,C,D) the
given probability is the average of the values estimated on four
sections (one meridian to vessel, one circumferential to vessel, two
at 45°). The distances of B, C and D from nozzle corner A are about
270, 460 and 790 mm, respectively.

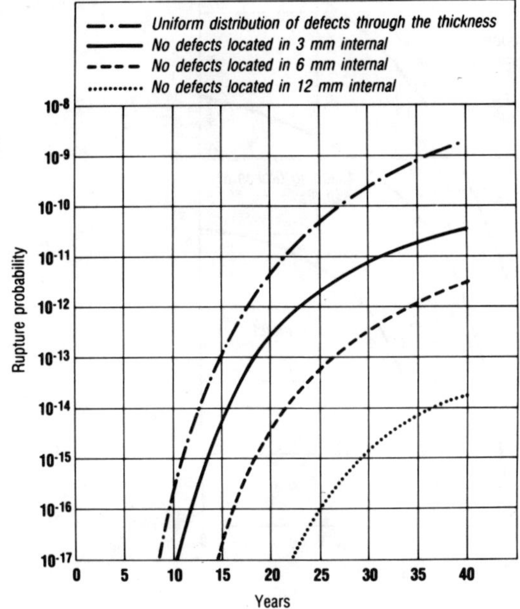

Fig. 6 - Probability of rupture of the beltline weld of a PWR vessel as a function of reactor age and for different crack distributions in the wall thickness.

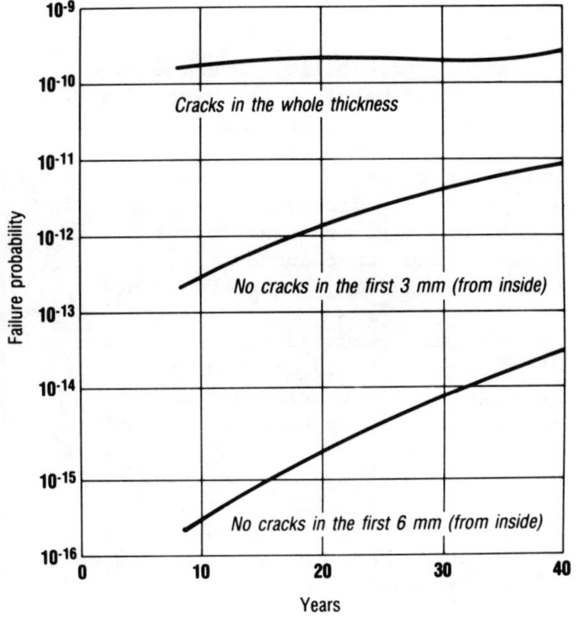

Fig. 7 - Probability of rupture of nozzle attachment weld of a PWR vessel as a function of reactor age and for different crack distributions in the wall thickness.

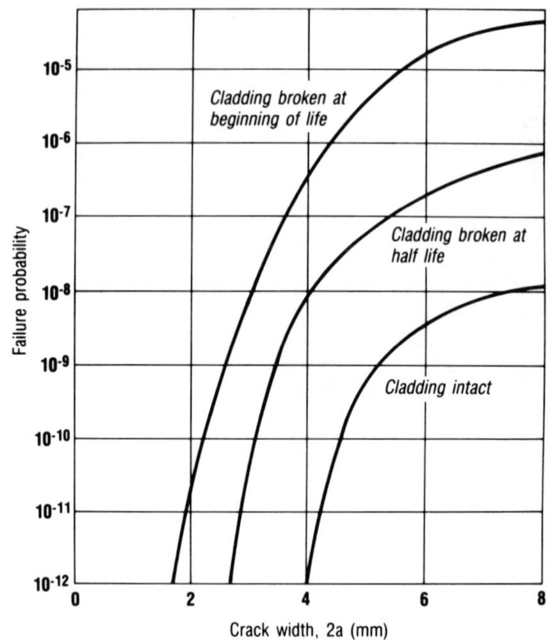

Fig. 8 – Probability of rupture caused by an undercladding
defect in the nozzle region of a PWR vessel.

Fig. 12 shows the conditional failure probability (average values
on four sections) for the nozzle attachment weld, in the case of a
large LOCA (40th year of operation). Defects smaller than 7 mm appear
to make no contribution to the probability of failure.

Of course, calculations carried out on the most stressed section
only give more pessimistic results. This is indicated, for example, in
Fig. 13, where the most stressed section only, meridian to vessel, is
considered for undercladding defects in the inner corner of the
nozzle. The results refer to the case of cladding broken from the
beginning of reactor life.

Figs. 13, 14 and 15 show isoprobability curves in coordinates
representing the crack width (2a) and the distance of crack inner tip
from cladding-ferritic interface. These curves provide a rapid indi-
cation of the degree of resolution that should be reached with NDE,
as a function of crack position, in order to assure ability for ascer-
taining a certain level of structural reliability. For example, in the
case of an intermediate breach LOCA, to assure ability for ascer-
taining a conditional probability of failure lower than 10^{-4} is is
necessary to detect, on the belt-line weld, defects as small as 3 mm in
the first millimetres (inner side) of the wall.

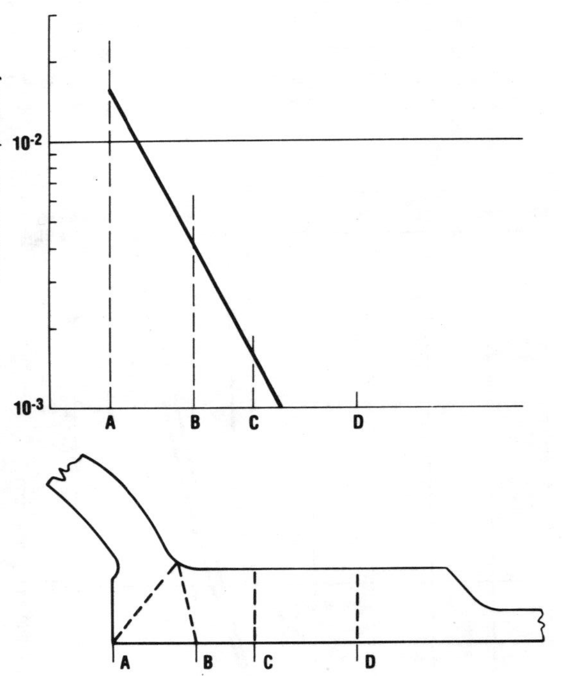

Fig. 9 - Conditional probability of rupture of the beltline weld of a PWR vessel, in case of large LOCA, with ECCS water at 20°C, at the 40th year of operation. d = distance (in mm) of crack centre from cladding-ferritic interface.

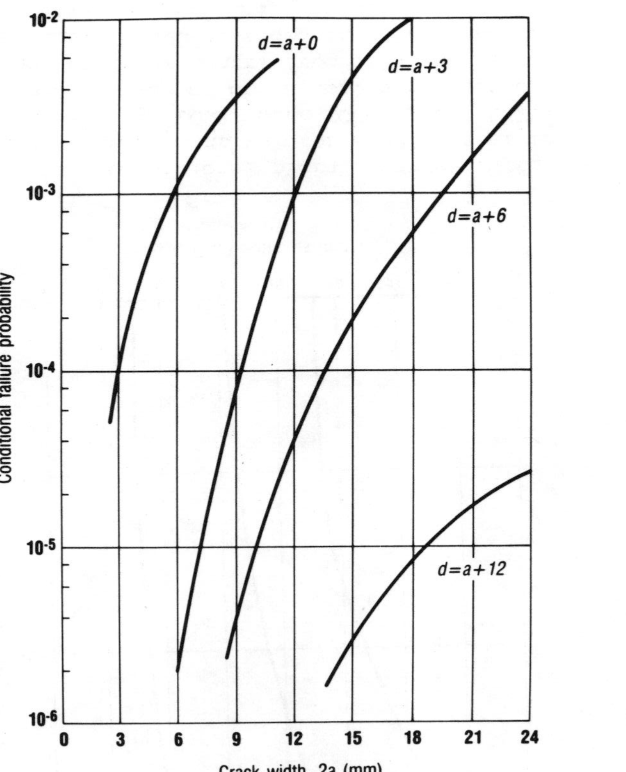

Fig. 10 - Conditional probability of rupture caused by an 8 mm wide undercladding defect in different regions of a PWR vessel nozzle, in case of intermediate LOCA, with ECCS water at 20°C, at the 40th year of operation.

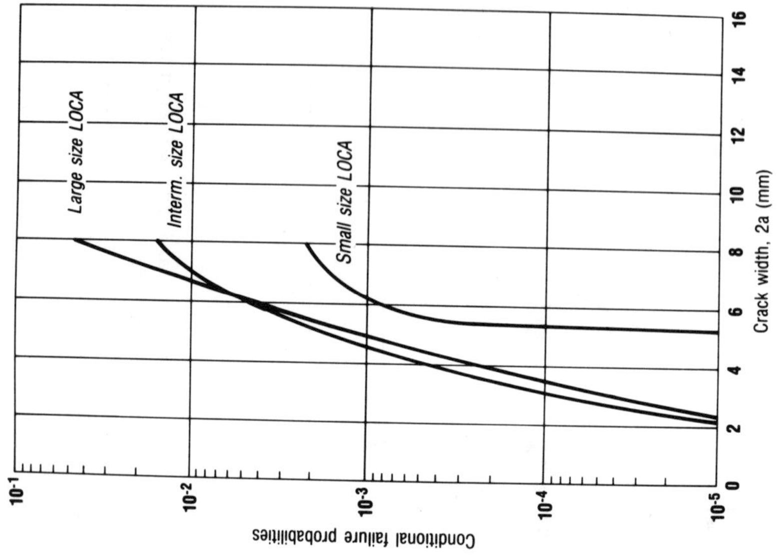

Fig. 11 - Conditional probability of rupture of nozzle attachment weld of a PWR vessel in case of a large LOCA, with ECCS water at 20°C, at the 40th year of operation. d = distance (in mm) of crack centre from cladding ferritic interface. Values averaged on 4 sections (meridian to vessel, circumferential, ± 45°).

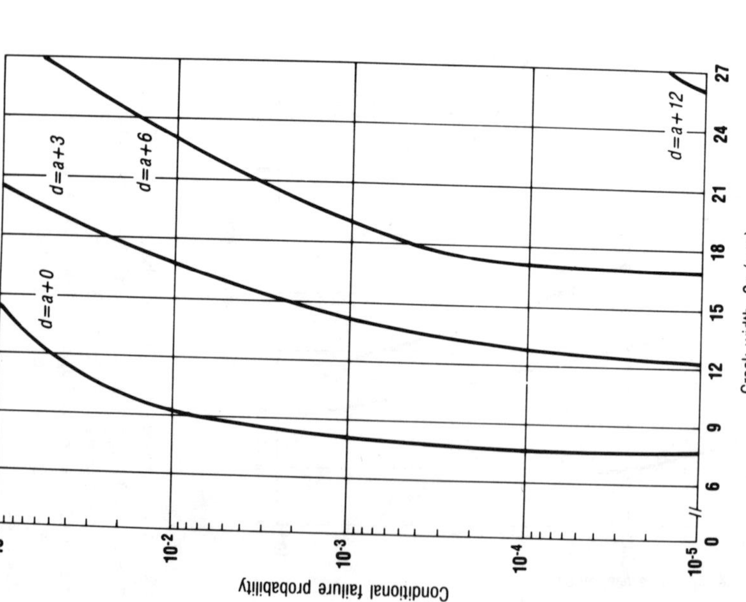

Fig. 12 - Conditional probability of rupture caused by an undercladding defect in the nozzle corner of a PWR vessel for different LOCAs, with ECCS water at 20°C, at the 40th year of operation. Values estimated in the section meridian to the vessel.

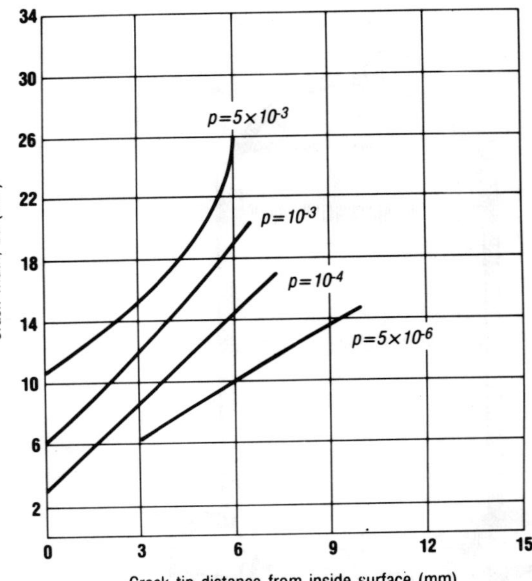

Fig. 13 - Curves of conditional isoprobability of rupture of nozzle attachment weld
caused by large LOCA, with ECCS water at
20°C, in a PWR vessel at the 40th year of
operation. Values averaged on 4 sections
(meridian to vessel, circumferential, ± 45°).

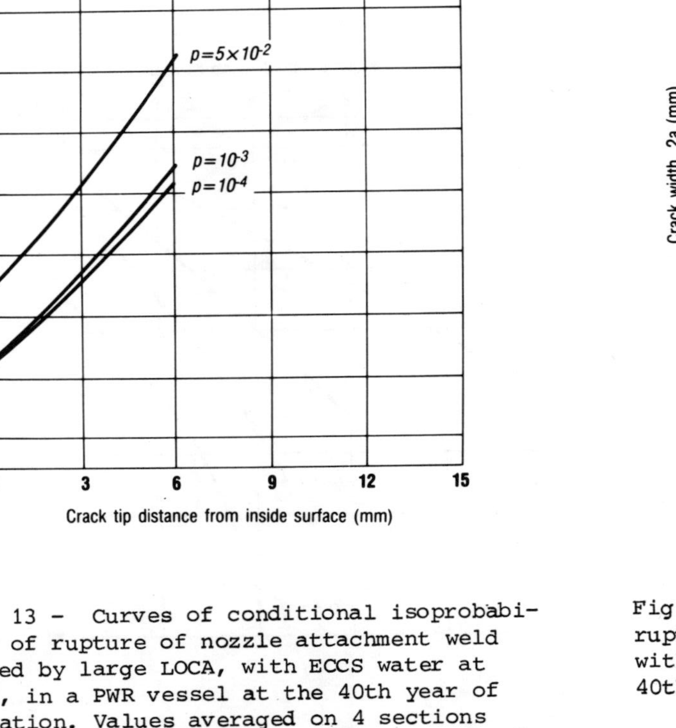

Fig. 14 - Curves of conditional isoprobability of
rupture of the beltline weld caused by a large LOCA,
with ECCS water at 20°C, in a PWR vessel at the
40th year of operation.

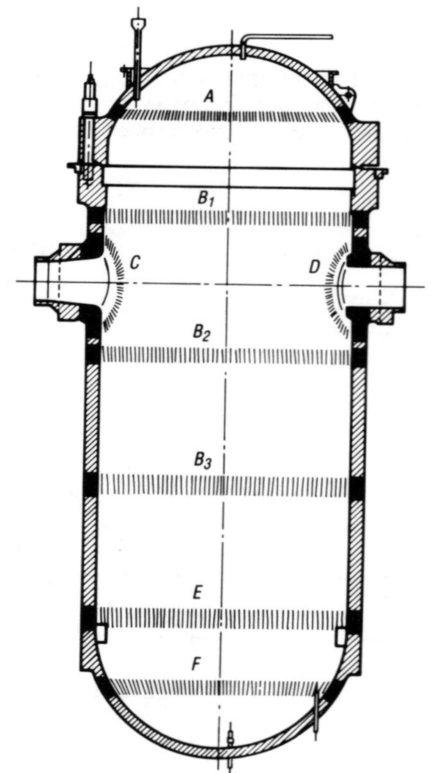

Fig. 15 - Curves of conditional isoprobability of rupture of the beltline weld caused by an intermediate LOCA, with ECCS water at 20°C in a PWR vessel at the 40th year of operation.

Fig. 16 - PWR vessel: weld identi- fication.

CONCLUSION

A few concluding statements can be made:

- The estimation of the structural reliability of a component by a
  probabilistic approach can be made only starting from simplifying
  assumptions on the stochastic processes representing loads and resistance.

- The implications of these assumptions should always receive preliminary investigation.

- The results can be strongly affected by data available on parameters and variables.

- All phenomena considered in this study proceed from normal manufacturing and operating conditions; in particular it has been assumed
  that there has been no serious error during quality control and
  quality assurance operations. But, due to the low failure probability
  obtained, it will be necessary to investigate all types of error,
  the occurrence of which is unlikely and which might result in failures having a higher probability.

- Crack growth in the belt-line is lower than in head weld, but the
  failure probability is higher due to radiation embrittlement.

- Thermal shocks are the most critical loadings. Therefore LOCAs with
  any size of breach are the most dangerous accidents and low temperature of injected water (ECCs) can increase the failure probability
  by a factor 10.

- In the case of the reactor pressure vessel here considered, the inner
  layer (thickness not larger than 50 mm) of the wall determines the
  whole pressure vessel failure behaviour.

- Cladding integrity is of utmost relevance.

- If the final goal of NDE is to assist towards keeping the failure
  probability of a reactor pressure vessel below a predetermined level,
  one should relate requirements on NDE reliability and resolution to
  the specific region to be inspected.

As far as future development is concerned, the probabilistic assessment
of structural reliability has to face two main challenges:

-       estimation of the residual life;
-       decision making.

The first problem, consequence of the existence of many aged
plants, needs deeper investigation (theoretical as well as experimental)
of some aspects of structural behaviour (residual stresses, localized
thermal stresses, etc.) and particular research efforts aimed at the
identification and measurement of significant parameters characterizing structural damage.

From a modelling point of view, the use of more general approaches
based on the use of multidimensional stochastic process theory is

recommended. The decision making, having to take into account also
uncertain and imperfect information could take advantage from the use
of non standard logics, such as fuzzy set and possibility theory,
which can provide rational, formal tools and coherent self-contained
decision framework also in the presence of some uncertainties for
which probability measures seem not adequate.

REFERENCES

1. P. Becker, A. Pederson, 'Application of statistical linear elastic
    fracture mechanics to pressure vessel reliability analysis',
    Nucl. Eng. and Design, 1974.

2. F. Nilson, 'A model for fracture mechanical estimation of the
    failure probability of reactor pressure vessels', 3rd Conf. Int.
    Pres. Ves. Tech., Tokyo, April 1977.

3. D.O. Harris, 'A means of assessing the effects of NDE on the
    reliability of cyclically loaded structures', Materials Evaluation,
    July 1977, pp.57-65.

4. W. Marshall et al., 'An assessment of the integrity of PWR pres-
    sure vessels', UKAEA 1976.

5. W. Marshall et al., 'An assessment of the integrity of PWR pres-
    sure vessels', UKAEA, London, March 1982.

6. J. Dufresne, A.C. Lucia, 'A probabilistic approach to the evalua-
    tion of pressure vessels safety margins', in: Structural Integrity
    of Light Water Reactor Components. Applied Science Pub. (1982).

7. J. Dufresne, A.C. Lucia, J. Grandemange, A. Pellissier-Tanon,
    'Etude probabiliste de la rupture de cuve de réacteurs à eau sous
    pression', Report EUR 8682fr, 1982.

8. A.C. Lucia, G. Volta, J. Elbaz, R. Brunnhuber, 'Requirements for
    NDE reliability as a function of the size and position of defects
    in RPV's', CSNI-IAEA Meeting on Defect Detection and Sizing,
    Ispra, May 1983.

9. A.C. Lucia, R. Brunnhuber, J. Elbaz, 'COVASTOL: a new computer
    code for the evaluation of pressure vessel failure probability',
    5th SMiRT, Berlin, August 1979.

10. G. Armon, D. Basile, A.C. Lucia, 'Estimation of the residual life
    of a structural component under cyclic loading', 8th SMiRT,
    Brussels, August 1985.

11. S. Garribba, A.C. Lucia, G. Volta, 'Reliability assurance on
    design and operation of nuclear reactor structural components.
    A unified approach resulting from experimented techniques and
    cases', 2nd Int. Conf. on Structural Safety and Reliability,
    Münich, September 1977.

12. L.A. Zadeh, 'Fuzzy sets', Memo ERL N.64-44, Univ. of Cal.,
    Berkeley.

13. A.C. Lucia, "Probabilistic structural reliability of PWR pressure
    vessels", Nucl. Enc. and Design, 87, 1985, pp.35-49.

RELIABILITY OF NDE

N.F. Haines

1. Why do we want to know about the reliability of NDE?

Pressure vessel manufacturers and plant operators share a common
responsibility for the safety of pressurised components. The manufacturer
must ensure that his product is fit to enter service and the operator
that it is used and maintained to adequate safety standards. Both rely,
as part of their quality assurance, on the use of NDE techniques to provide technical data on structural integrity. To give confidence in the
use of NDE techniques manufacturers and operators need an assurance on
the reliability of the techniques that are currently available. A number
of countries provide guidance in the form of standards or procedural
codes (see Appendix 1) that recommend methods that may be used. In some
countries these codes are enforced by making their use obligatory by law.
Unfortunately very few of those documents provide any direct guidance on
the reliability that may be achieved using specific techniques.

In this talk I will describe some of the work of the past 10 to 15
years that has been attempting to answer questions on NDE-reliability.
The talk will give a brief description of what we mean by a reliable NDE
technique and how reliability might be measured. The physical parameters
that influence defect detection by ultrasonic methods are discussed
under the headings (1) defect parameters, (2) ultrasonic technique parameters, (3) component parameters, (4) human parameters. The final
sections will deal with the quantitative development of an analytical
model of ultrasonic reliability together with a discussion of its validity and application.

2. What is a reliable NDE technique?

The first task of NDE is to detect flaws: this objective has two
aspects of reliability:
1. A reliable detection method must find all types and sizes of
flaws of concern in the component.
2. The method must be capable of being used by different inspection
teams, all of whom consistently find the defects of concern.
The first objective concerns the physical ability of the method to find
defects whereas the second concern the ease with which the method can be

A. C. Lucia (ed.), Advances in Structural Reliability, 255–301.
© *1987 by ECSC, EEC, EAEC, Brussels and Luxembourg.*

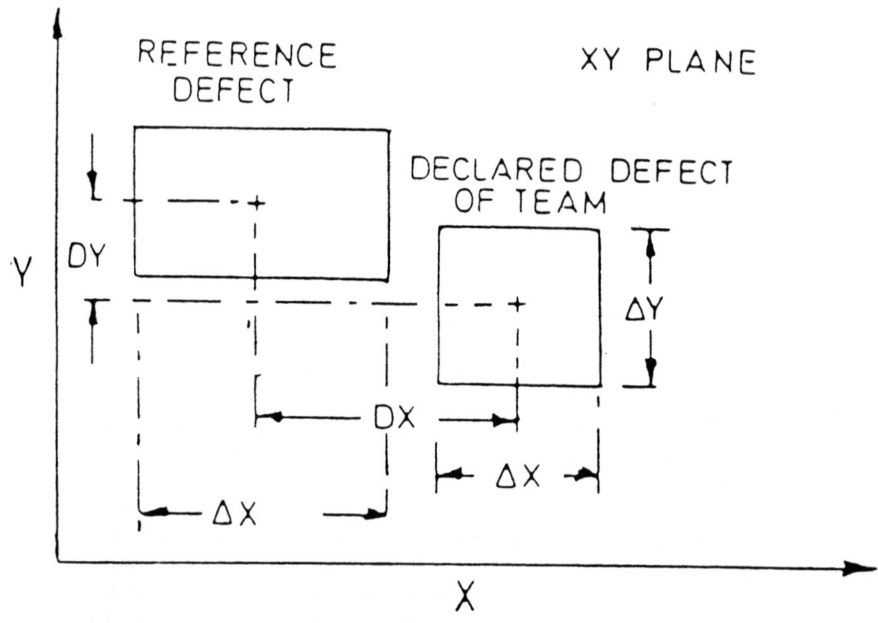

Fig. 1. Definitions of DX, ΔX, DY, ΔY.

used by different operators. Both are essential elements in a <u>reliable</u> NDE crack detection method or procedure.

The second task of NDE is to locate the position of a defect within the component and the third task to measure its size. A <u>reliable</u> technique for location and sizing is thus simply one which can accurately measure position and size and that this can be done consistently by whichever team uses the techniques. (It should be noted that it is quite common in NDE for different techniques to be used for detection and sizing.)

3. How do we quantitatively measure NDE reliability?

The above definition of a reliable NDE detection method requires that a particular defect be consistently found by whichever team uses the technique. Thus a simple quantitative definition of detection reliability is

$$\text{Detection Reliability} = \frac{\text{Number of Teams Detecting Defect}}{\text{Number of Teams Inspecting Component}} \qquad (1)$$

It must be recognized that this applies to the detection of a particular defect and, therefore, must be investigated for all defects of concern in a particular component if we are to fulfill both conditions specified above in Section 3. This would appear to be a difficult task but, as we shall see later, several simplifying factors enable us to reduce the size of the problem significantly.

An alternative definition that could be used for detection reliability in a component is

$$= \frac{\text{Number of defects detected}}{\text{Total number of defects in component}} \qquad (2)$$

however, it is extremely difficult to interpret this definition for any other component unless it contains a similar distribution of types and sizes of defects. The above definition could be highly biased by a few 'difficult' to detect defects which may be present in atypical proportions. (It is the author's view that use of the above definition could be misleading and unless careful consideration is given to the defects included in the component, over optimistic or pessimistic conclusions on reliability of detection could be made.)

Location and sizing accuracy may be defined as shown in Figure 1. The error in location is the distance between the centre of the cuboid enclosing the defect and the centre of the cuboid enclosing the NDE-predicted position of the defect. Thus we may define the NDE position errors as DX, DY and DZ. If the results of N teams are combined to give mean errors we have

$$\overline{DX} = \frac{1}{N} \sum_{i}^{N} DX_{i} \qquad (3)$$

and similarly for $\overline{DY}$ and $\overline{DZ}$.

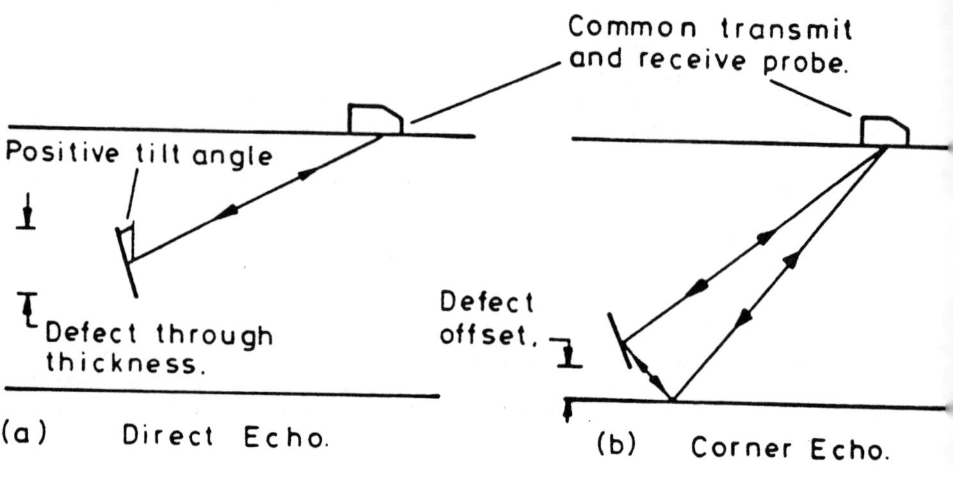

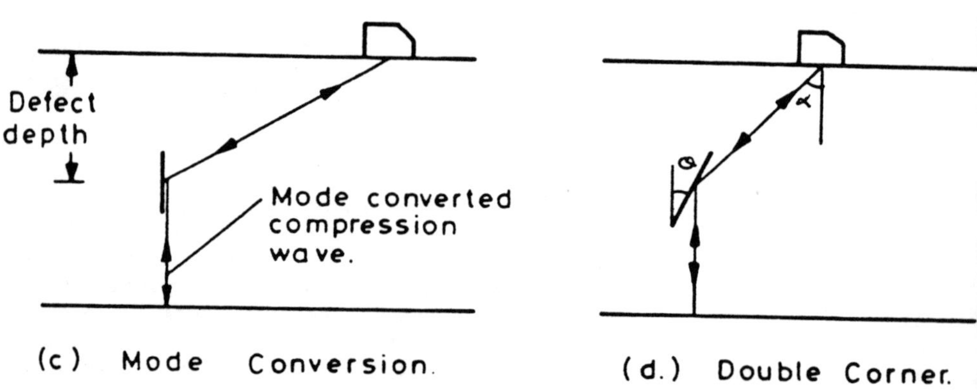

Defect tilt is measured from the vertical direction and
is defined as positive when the top edge of the defect
moves away from the probe.

Fig. 2. Schematic Showing Principal Detection Routes for Pulse-Echo
         Testing.

Sizing may be considered in a similar way by defining the error as

$$ESX = \Delta X - \Delta X_R,\tag{4}$$

where ΔX is the NDE measured X dimension and ΔX_R the real size of the defect. If this is repeated by N teams we may define a mean error as

$$\overline{ESX} = \frac{1}{N} \sum_{i}^{N} (\Delta X_i - \Delta X_R)\tag{5}$$

but it is also necessary to define a root mean square error as

$$\overline{ESX}_{rms} = \pm \sqrt{\sum_{i}^{N} \frac{(\Delta X_i - \Delta X_R)^2}{N}}\tag{6}$$

The latter is obviously necessary if we consider the case of two teams one oversizing by 10 mm and one undersizing by 10 mm. The mean error is calculated by (5) as 0, but the rms error as ± 10 mm which is a more appropriate description of the sizing accuracy.

The above definitions of location and sizing accuracy are defined for a single defect and averaged over the results of a number of teams. It is also possible, as with detection reliability, to define terms averaged over a population of defects for a single team. However, considerable care must be taken in the interpretation of such results when attempting to apply them to other, similar components which may not contain the same distribution of defect types.

4. Which factors affect NDE reliability?

In this section we will briefly review the main groups of parameters that have been found to affect ultrasonic crack detection reliability. It is necessary to restrict attention to ultrasonic methods mainly because research of the past 10 years has concentrated in this area. The general conclusion to some extent may be applied to other NDE methods particularly when discussing 'human' factors. The section is divided into five parts: the first briefly reviews the main ultrasonic detection methods and then four sections covering defect, ultrasonic, component and human factors.

4.1. Ultrasonic crack detection methods.

Figure 2 shows four ways in which an ultrasonic pulse may be transmitted by a probe, reflected from the defect and then returned either directly to the probe or via the opposite component surface. Apart from 'mode conversion' (Figure 2c) all other routes rely for maximum reflected signal on the angles of incidence and reflection at crack and component surfaces being equal. Thus direct echo is most effective if the pulse strikes the defect at 0° and the corner echo if the defect is at the opposite component surface and perpendicular to it. The double corner

Tandem Technique

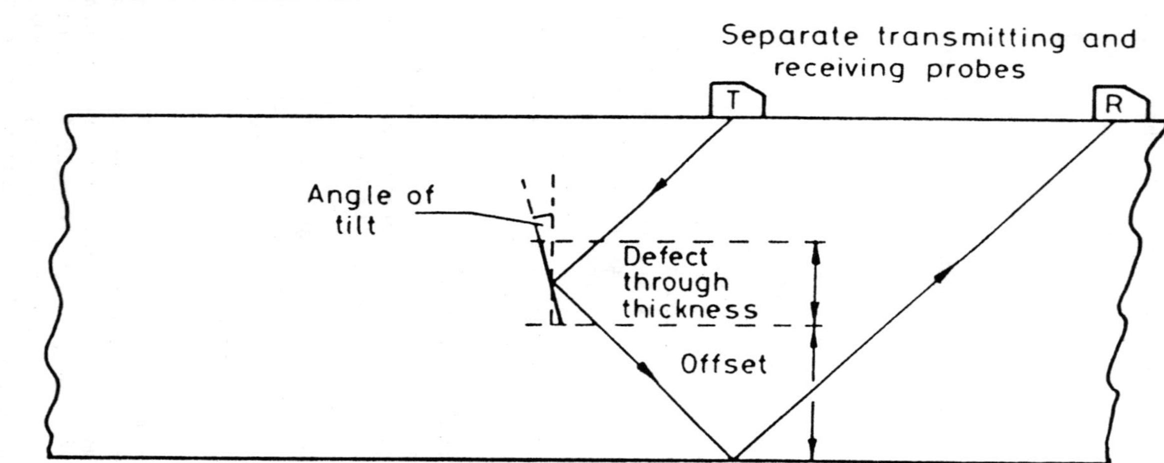

Fig. 3. Schematic Diagram of Tandem Technique.

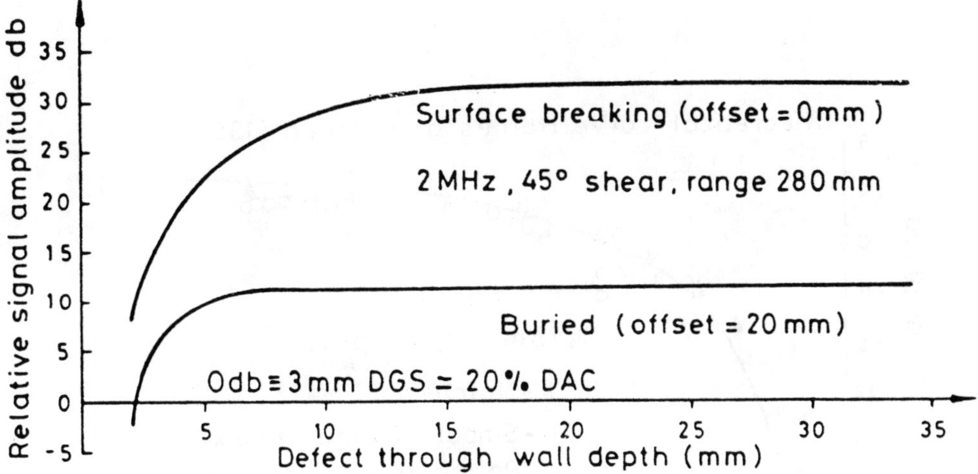

Fig. 4. Variation of Corner Pulse Echo Response with Size for Vertical Defects.

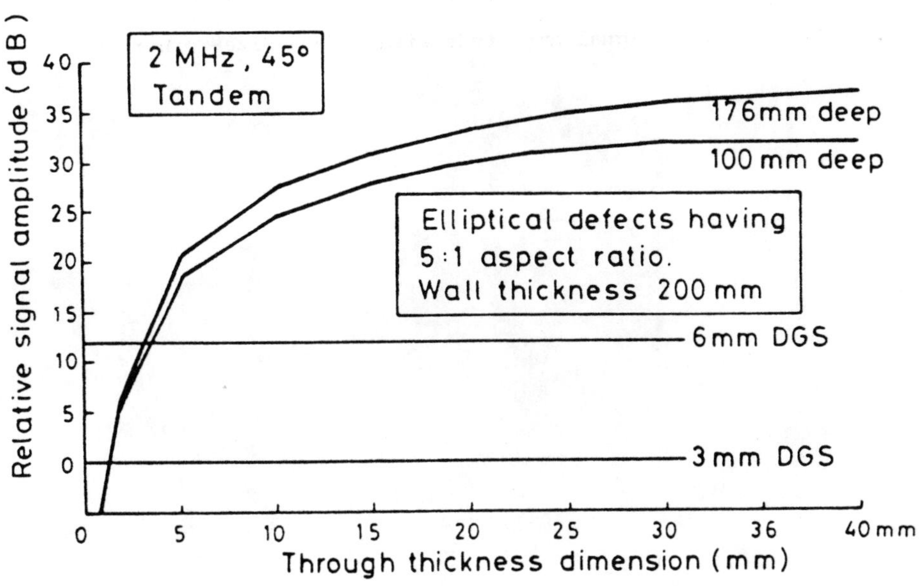

Fig. 5. Variation of Defect Signal with Through Thickness Size for Vertical Defects Detected by the Tandem Technique.

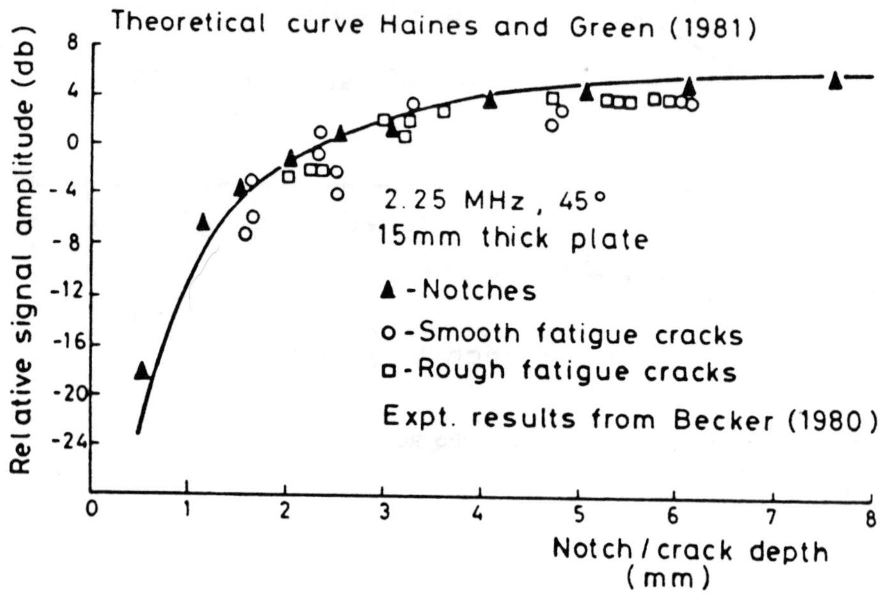

Fig. 6. Variation of Signal Amplitude with Defect Size—Corner Response.

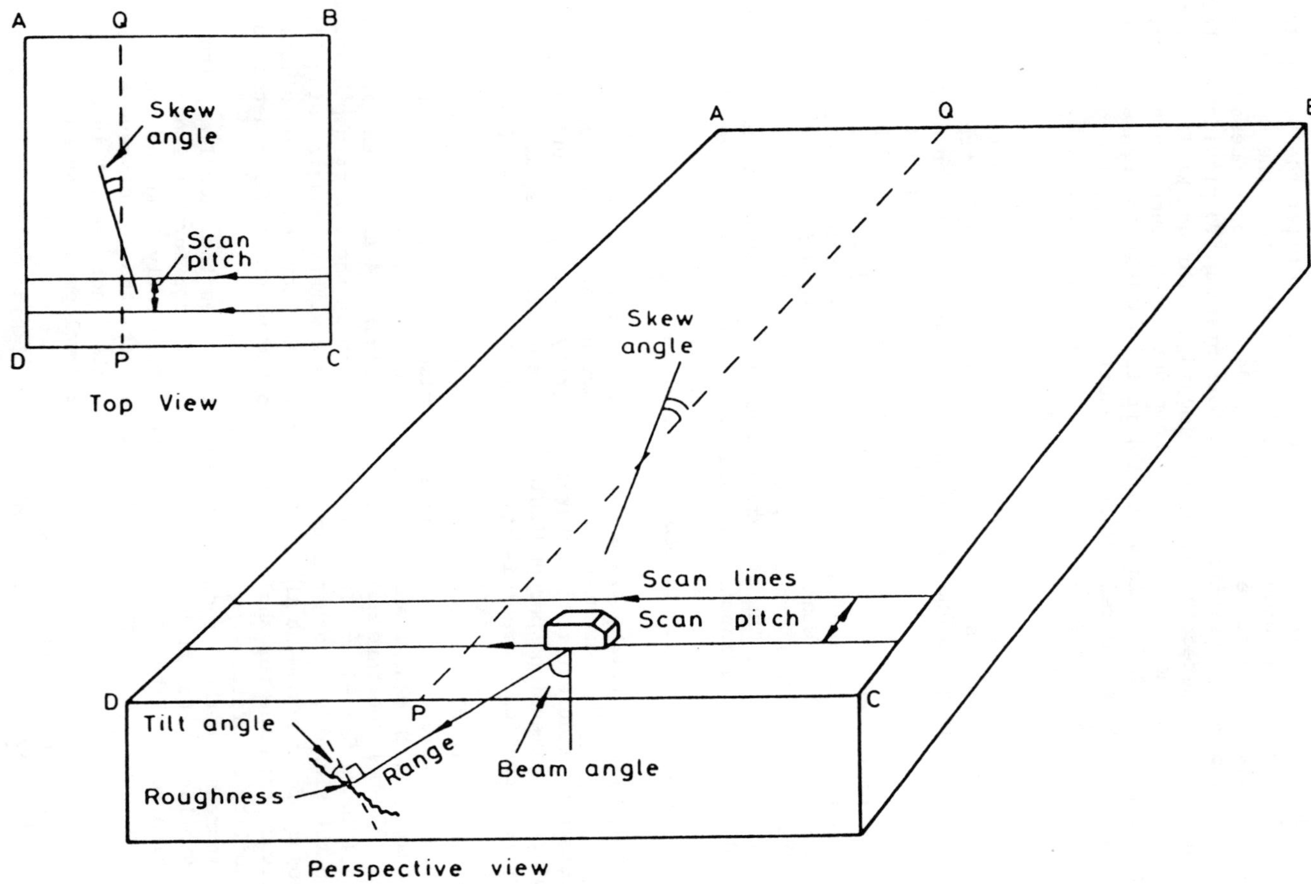

Fig. 7. Simplified Geometry of Ultrasonic Detection System.

(Figure 2d) is most effective if the probe angle (α) is twice the angle
of tilt (θ) of the crack away from vertical.

Mode conversion in Figure 2c would appear to offer the best possibility for detecting through wall cracks. Unfortunately the method is
not particularly effective because the mode conversion process is inefficient and very sensitive to defect orientation. To provide an alternative method of detecting defects lying in the through wall direction
the Tandem technique (Figure 3) has been developed. Once again it may be
seen that, if the defect is vertical, it is possible to ensure that the
angle of incidence and reflection are equal at both defect and inner
wall, thereby obtaining the maximum reflected signal strength.

To complete this brief review of ultrasonic techniques it is important to consider the signal 'strength' or 'amplitude' received when a
defect is detected. It is unnecessary here to consider the full details
of the physical processes involved in the reflection of ultrasonic
pulses, provided it is understood that detection depends on a recognisable signal being displayed or recorded by the receiving electronic
system. We will refer many times, in the remainder of the paper, to
signal strength and its importance in the reliability of ultrasonic
techniques. A simple analogy may be made with receiving a radio signal;
if the transmitter is weak or pointing in the wrong direction, or if the
atmospheric conditions are poor, then detection of the signal is difficult or 'unreliable'. The analogy is useful in emphasising that the important factor in ultrasonics is signal to noise ratio and we will present later experimental results which confirm a direct relationship
between Defect Detection Probability (DDP) and signal to 'noise'.

4.2. Which defect parameters affect detection reliability?

a.   Size. The parameter of most concern when considering the significance of a defect to structural integrity is its size and, in particular the dimensions perpendicular to the local stress field. In
fracture mechanics a crack is typically described in terms of its length
2c) and its through thickness dimensions (a). Both parameters are of
significance in ultrasonic crack detection but the through wall depth
(a) has little effect above a threshold size. The threshold depends on
the component thickness and the technique used but, as a guide, in thick
plate (200 mm) the threshold is 15-20 mm (Figures 4 and 5) and in thin
plate (Figure 6) 3-4 mm. The threshold is simply the defect depth above
which the signal strength changes very little with increasing size. If
we assume for the moment that detection reliability depends on signal
amplitude then the threshold size is also the point at which detection
reliability saturates.

Increase in the defect length (2c) increases the signal amplitude
in a similar way to depth (a), but a second factor further improves
detection reliability. If the defect is sufficiently long that the
repeated scanning pattern (Figure 7) intersects the defect several times,
then the overall probability of detection also increases. This concept
is considered further in Section 5 in the development of an analytical
model of detection reliability.

Further evidence for the small effect of size on signal response
comes from an analysis of the Defect Detection Trials (see list of

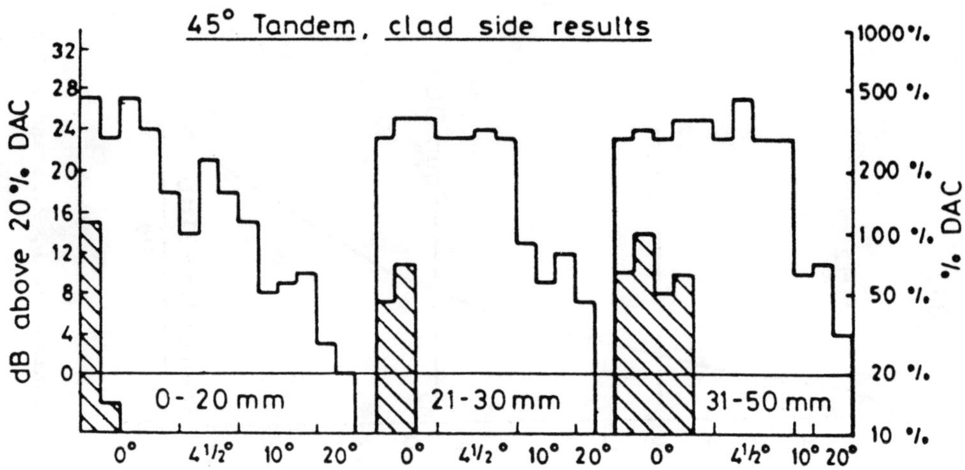

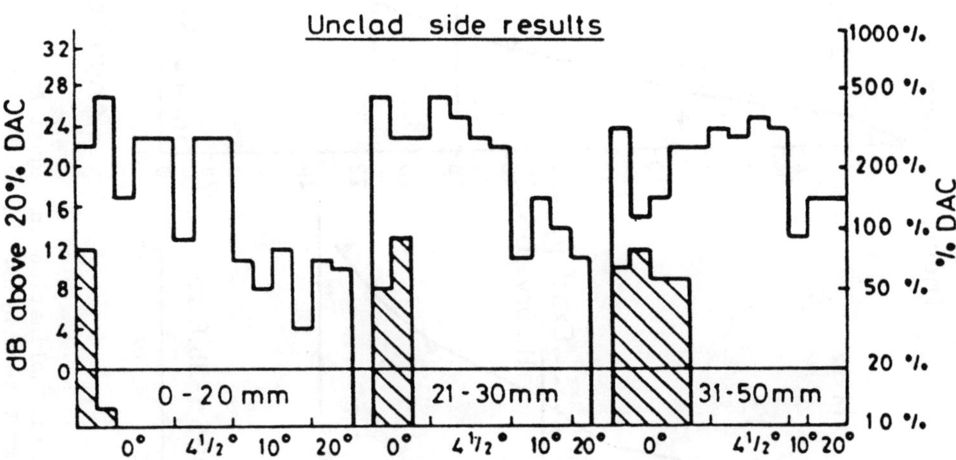

Fig. 8. The Effect of Through-Wall Size on the Tandem Response of all the Defects in Plates 1 and 2.

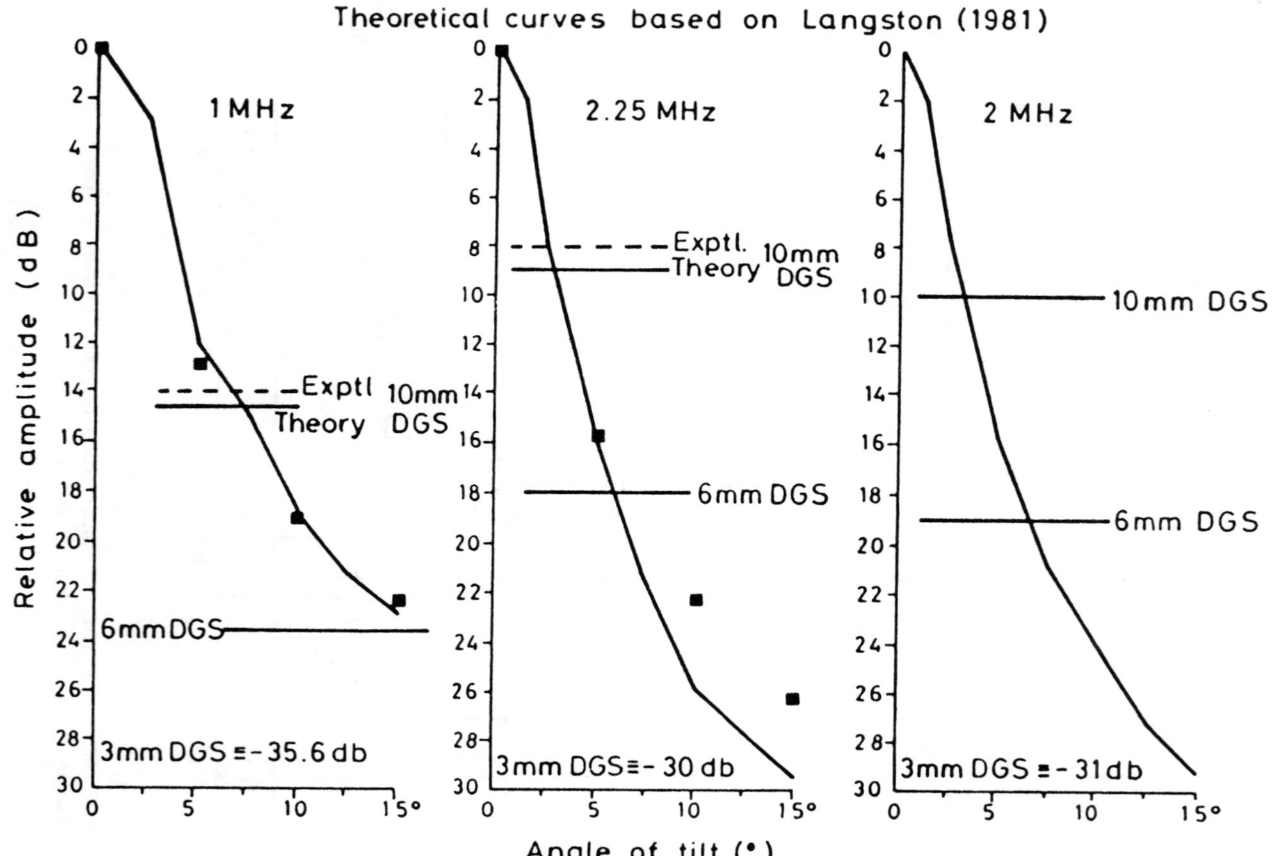

Fig. 9. The Effect of Inclination on Signal Amplitude for the Tandem Technique at 1, 2.25 and 2 MHz (25 × 125 mm Defect Lying 103 mm Through a 176 mm Thick Wall).

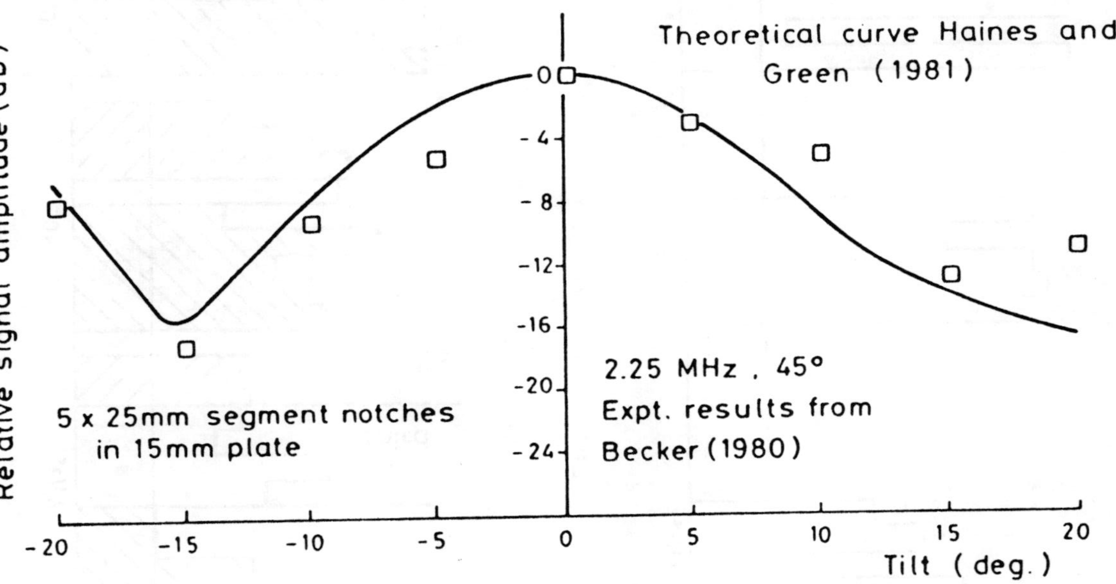

Fig. 10. Variation of Signal Amplitude with Orientation-Corner Response.

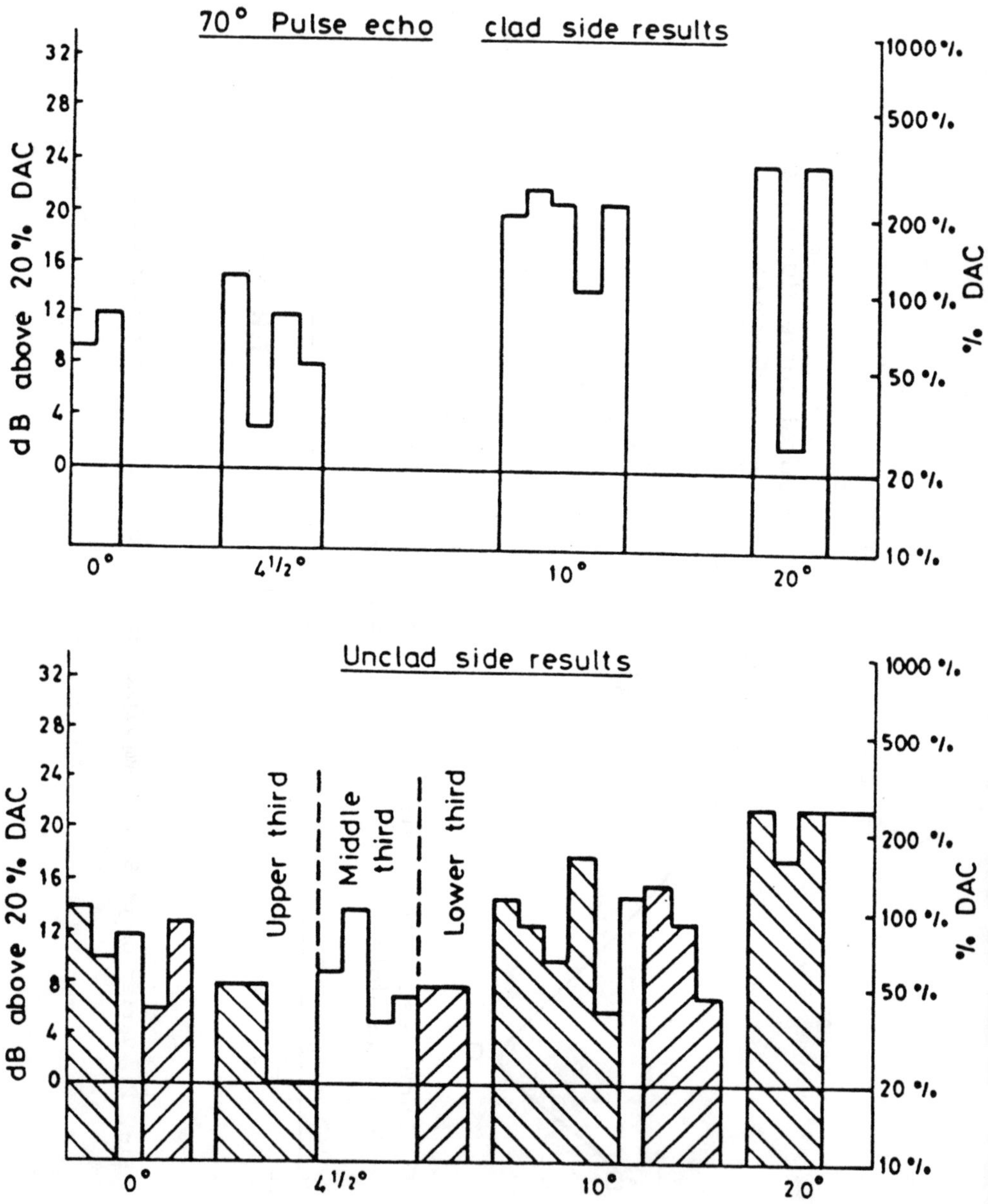

Fig. 11. The Effects of Defect Angle and Through-Wall Position on the
70° Pulse Echo Response of the Plate 1 Defects.

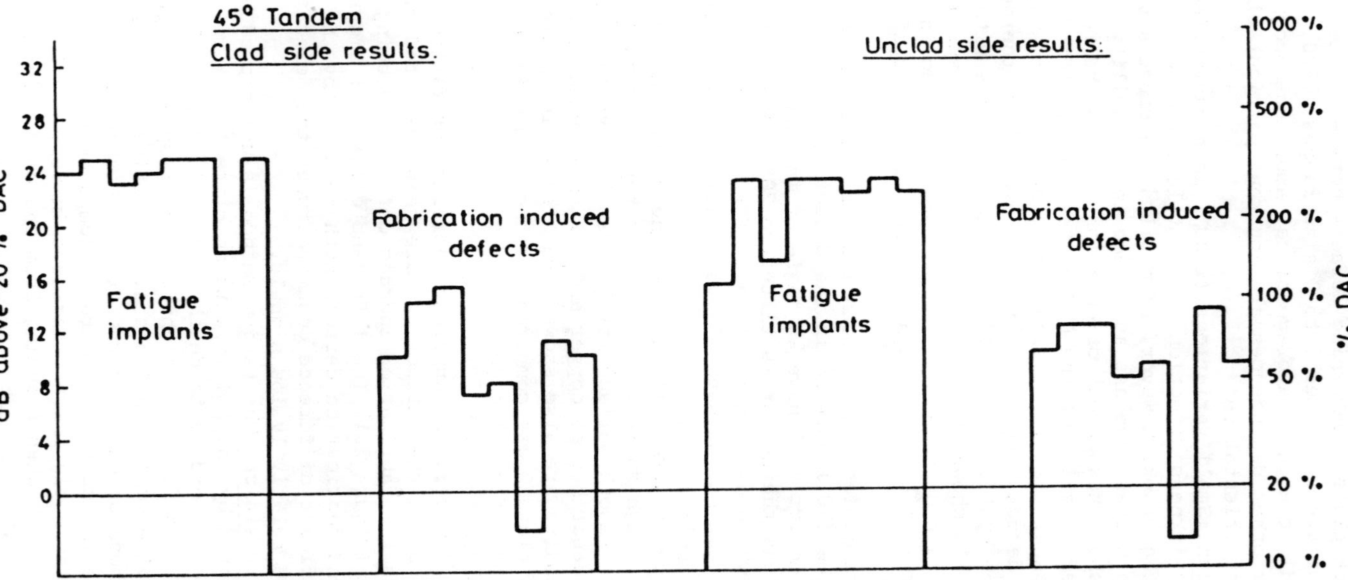

Fig. 12. The Effect of Defect Type on the Tandem Response of the Plate 2 Defects.

reports for further reading). Figure 8 shows the results for some 45 defects in two thick plates containing defects between 10 mm and 50 mm in depth. The groups are separated into 3 size ranges (0-20, 21-30 and 31-50) and little difference of average signal amplitude occurs between the groups. However other dependences are apparent particularly on defect type and defect orientation.

    b.   Orientation. Small variations (± 20°) in defect orientation about the through thickness direction have little effect on the structural significance of a crack. However, such tilts can cause a very significant reduction in signal response and hence in reliability of detection. Figure 8 shows how defect orientation affects the response during the tandem examination of the DDT plates Numbers 1 and 2. For all size groups it is clear that increasing defect orientation away from the vertical decreases defect responses. Figure 9 also shows the decreasing response of a tandem system to 25 × 125 mm planar defects with increasing defect tilt. Figure 10 shows the results for a 5 mm deep surface breaking defect in a 15 mm plate detected by a single probe working in the corner mode (Figure 2b). The signal decreases for both directions of tilt up to ± 15°.

    For defects lying close to the scanning surface it is possible to use high angle probes (70°) working in the direct mode (Figure 2a). Clearly a vertical defect will have an angle of incidence of 20° which will decrease to 0° as the defect is tilted to 20°. Results of this are shown in Figure 11 where defects in plate 1 of the DDT exercise give an increase in signal response between 0° (vertical) and 20° (normal to incident ultrasonic pulse).

    c.   Defect Type. Figure 8 illustrates the significant reduction in signal response for rough surfaced fabrication type defects (hatched areas) compared with smooth surface fatigue cracks. The fabrication induced defects were described as copper and carbon induced cracks within the welds. These results are also shown more clearly in Figure 12 where only the defects of plate 2 are compared for the tandem inspection.

    A similar pattern of results occurs for the inspection of plate 2 by the 45° single probe technique from the unclad side (right hand diagram Figure 13). However, the apparent increase for fatigue cracks is because they are close to the surface and are being detected via the corner route whereas the fabrication induced defects are being detected by the direct route (Figure 2a). The left hand diagram of Figure 13 comprises fatigue and fabrication defects both detected by the direct route. Little significant difference occurs between the defect response and both groups give relatively low signals.

    d.   Defection Position. The tandem method is unaffected by defect position as illustrated by the DDT results on plate 1 (Figure 14). The defects are separated according to depth zone (0-85 mm, 86-170 mm and 171-255 mm) and little difference in the average response of groups can be seen. However, pulse echo techniques are significantly affected by position primarily because of the corner route (Figure 2b) which occurs for defects near the far surface. Figure 15 compares the plate 1 result for 45° probes showing a significant increase in response for defects in the 171-255 mm depth zone when scanning from the clad side. Similarly for the 0-85 mm group when scanning the unclad side.

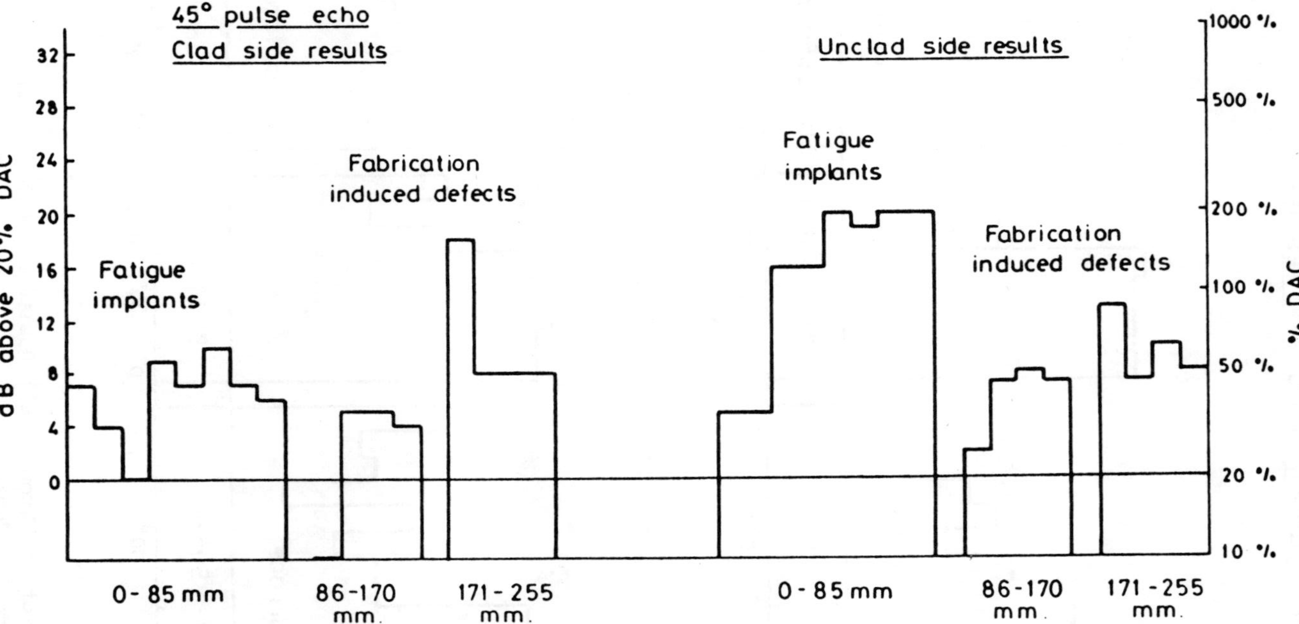

Fig. 13. The Combined Effects of Through-Wall Position and Defect Type on the 45° Pulse Echo Response of the Plate 2 Defects.

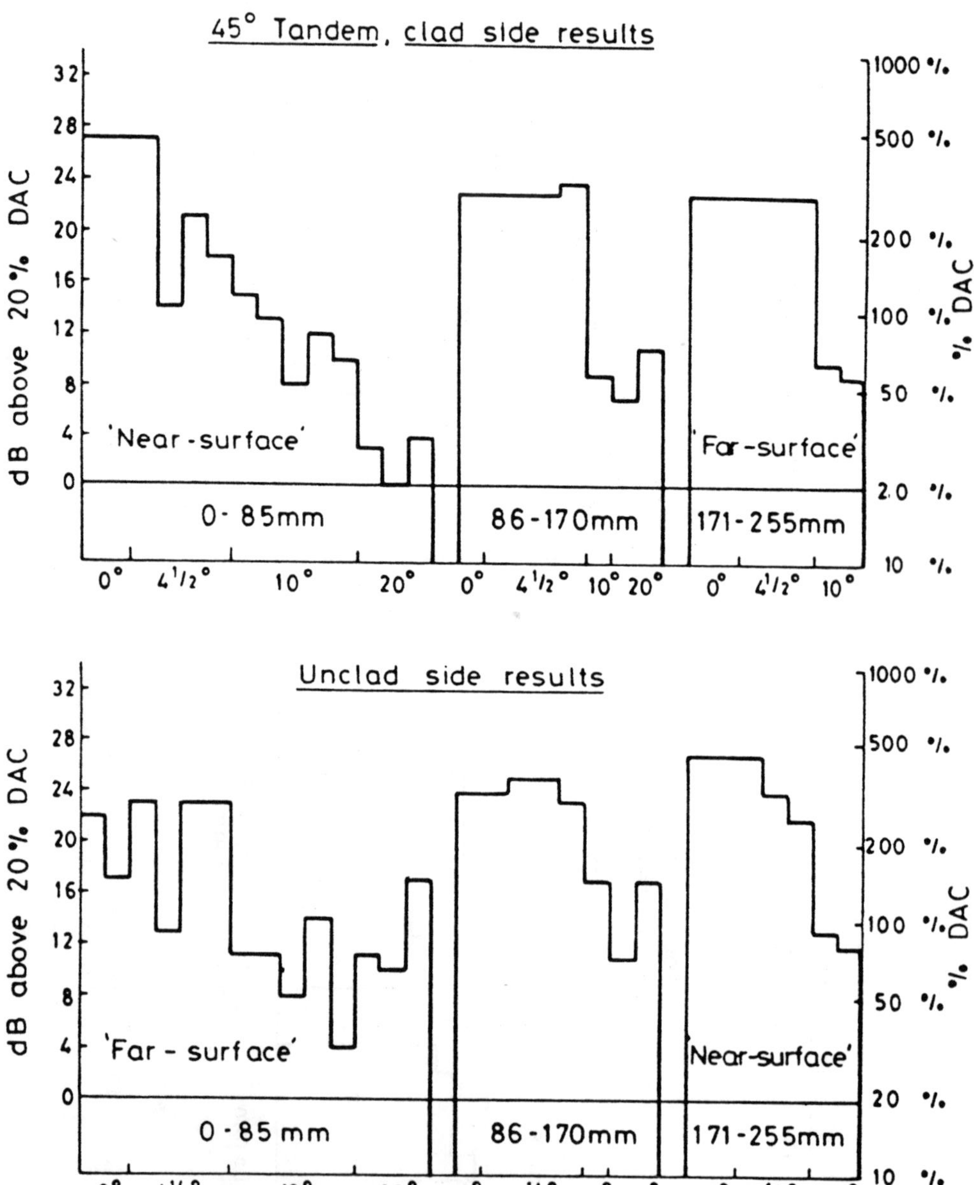

Fig. 14. The Effect of Through-Wall Position on the Tandem Response of
the Plate 1 Defects.

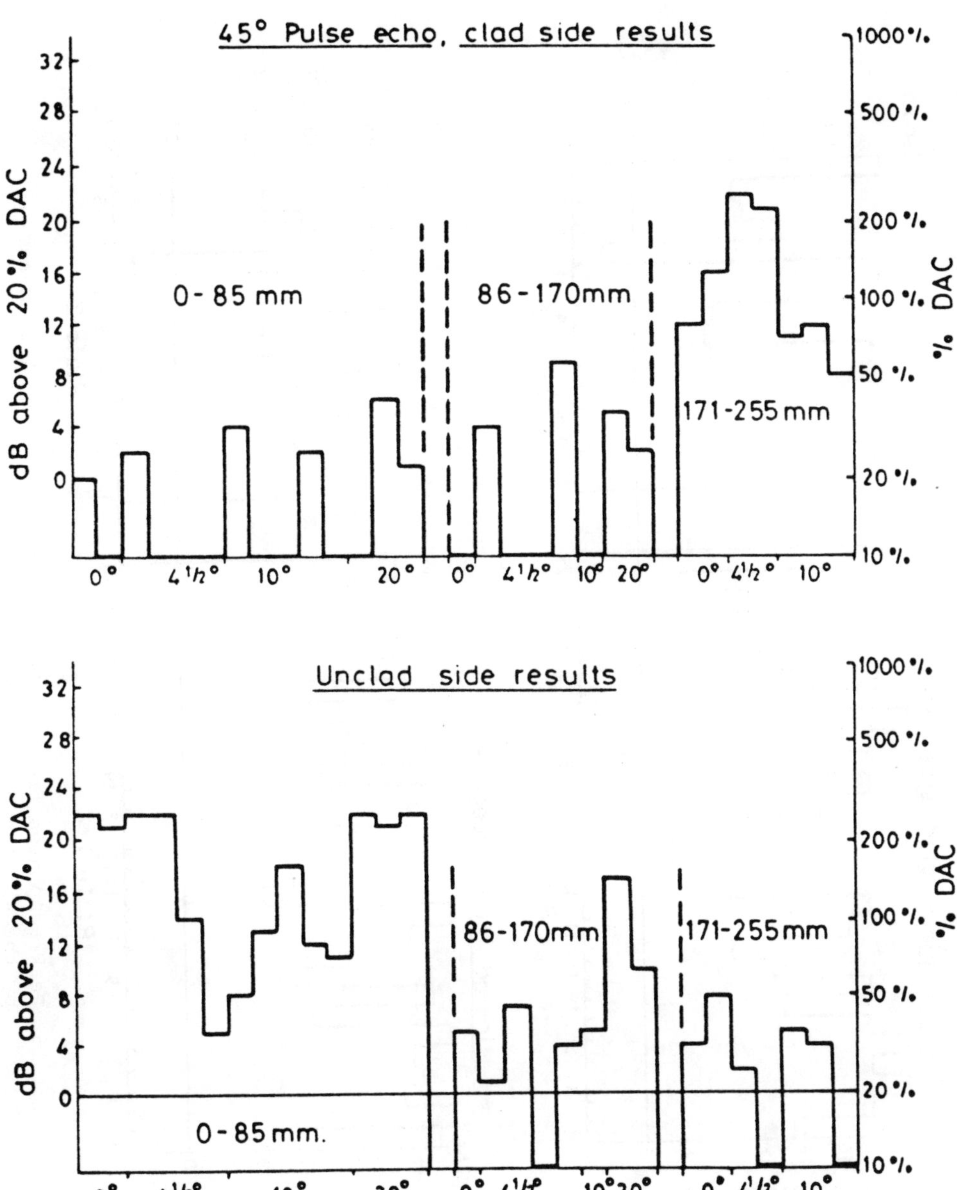

Fig. 15. The Effect of Through-Wall Position on the 45° Pulse Echo Response of the Plate 1 Defects.

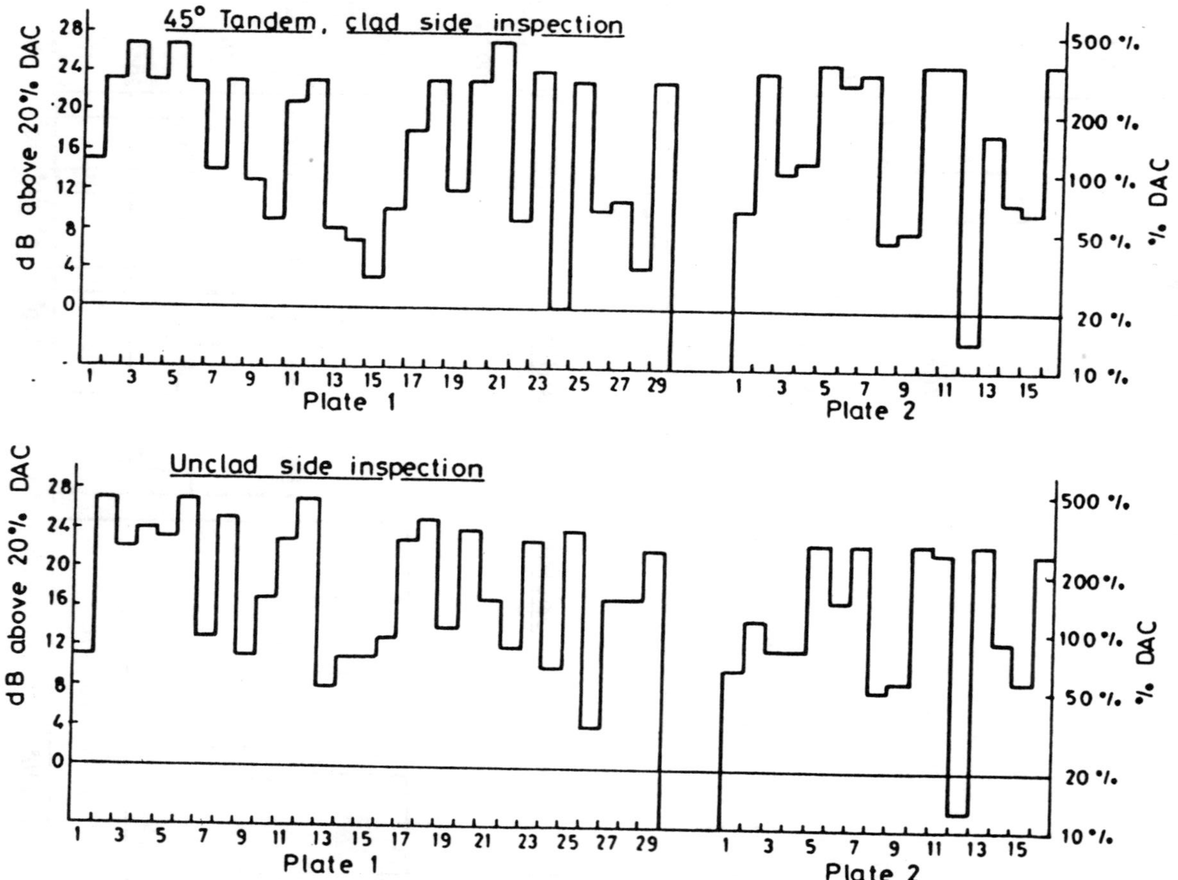

Fig. 16. Peak Responses from the Defects in Plate 1 and 2 as Measured by the 45° Tandem Probes.

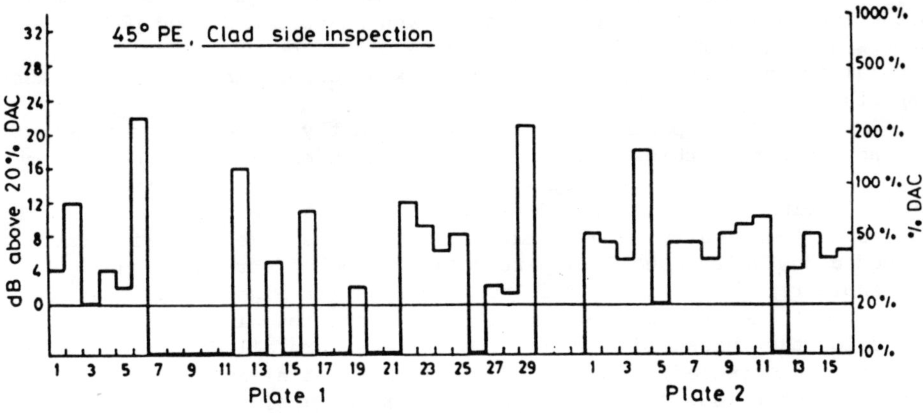

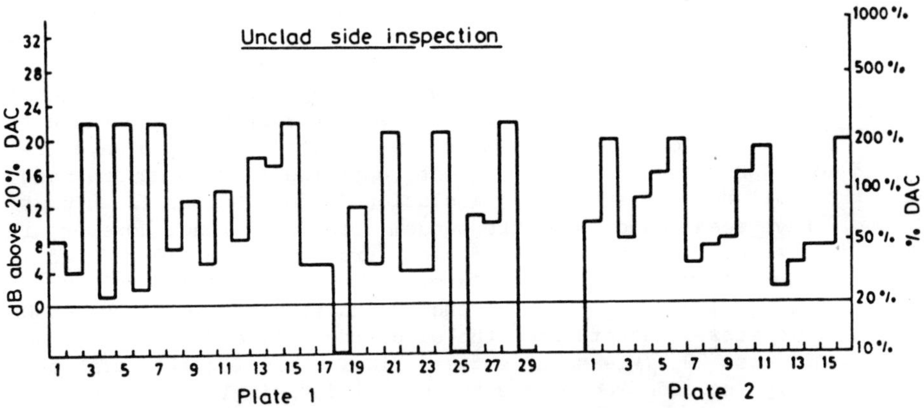

Fig. 17. Peak Responses from the Defects in Plates 1 and 2 as Measured by the 45° Pulse Echo Probes.

4.3. Which ultrasonic parameters affect detection?

Many parameters combine to define the shape of the ultrasonic beam of a transducer, principally the size and shape of the piezoelectric element, the frequency, the angle of the transmitted beam, the range of the defect and the electronic excitation and receiving system. No attempt will be made here to describe the effects of these parameters on defect detection; considerable research has taken place and a good physical understanding of the relative importance of parameters established. Several of the reports listed in Appendix 3 contain information on these aspects.

One important parameter which significantly affects defect detection is the 'reporting threshold'. It was discussed earlier that detection depends on a recognisable signal that can be distinguished from the 'noise' being reflected by a defect. To ensure a consistent approach by different teams carrying out similar inspections some countries have issued codes which specify a particular method of establishing a threshold reporting level. The level must be set high enough that the majority of the 'noise' from small discontinuities within the component are below the threshold, but low enough to ensure all defects of significance are detected. Because defect size is not always the most important parameter (see Section 5.2.) in determining signal response, it is sometimes difficult to define a reporting threshold that fulfills both needs. To illustrate this point let us look at the results from the DDT exercise. Figure 16 shows all the results from plates 1 and 2 using the tandem method. The vertical axis gives the maximum signal strength observed for each defect on a scale in dB on the left and % DAC on the right. 0 dB is equivalent to 20% DAC which is approximately the signal amplitude from a 3 mm diameter circular reflector at the same range. Using Figures 16 and 17 we can assess the results that would have occurred in the DDT exercise if different reporting thresholds are used. At 20% DAC all but one small slag inclusion (defect Number 12 in plate 2) would have been found using the tandem technique alone. However, if the 45° pulse echo method were used alone from the clad side of the plates (Figure 17 upper) at least 13 out of the 45 defects would be missed at the 20% DAC reporting threshold. Raising the level to 50% DAC tandem would miss 6 defects from the clad side only and the 45° (clad side) would miss 29. (Please note these figures are for illustration purposes only, no code recommends the use of 45° single probe methods alone.)

The introduction of a 'reporting threshold' removes the need for us to consider the 'noise' levels in ultrasonic detection, hence we may consider signal to 'threshold' rather than signal to 'noise' ratio in our analysis of detection reliability.

4.4. Which component factors affect detection reliability?

a. Component Geometry. The shape of a component can have a major influence on ultrasonic techniques. The simplest geometries are flat plate or cylindrical pressure vessels, providing the radius of curvature is not too small. Components with their opposite surfaces parallel provide the best opportunity for high reliability inspection, because of all the techniques of Figures 2 and 3 may be used. However, for non-

FROM D. DAVIS
I. MECH. E. CONF. 1982
PAPER C134/82
UNDERCUT
VESSEL FLANGE
VESSEL HEAD
SEAL SURFACE
KEEP INSULATION, LUGS, INST. LEADS CLEAR OF THIS AREA.
2T+3"
SHOULD BE LEFT CLEAR OF HEAD SUPPORT
FLOOR DRAIN
1
24
UNDESIRABLE HEAD-TO-FLANGE CONFIGURATION
PREFERRED HEAD-TO-FLANGE CONFIGURATION

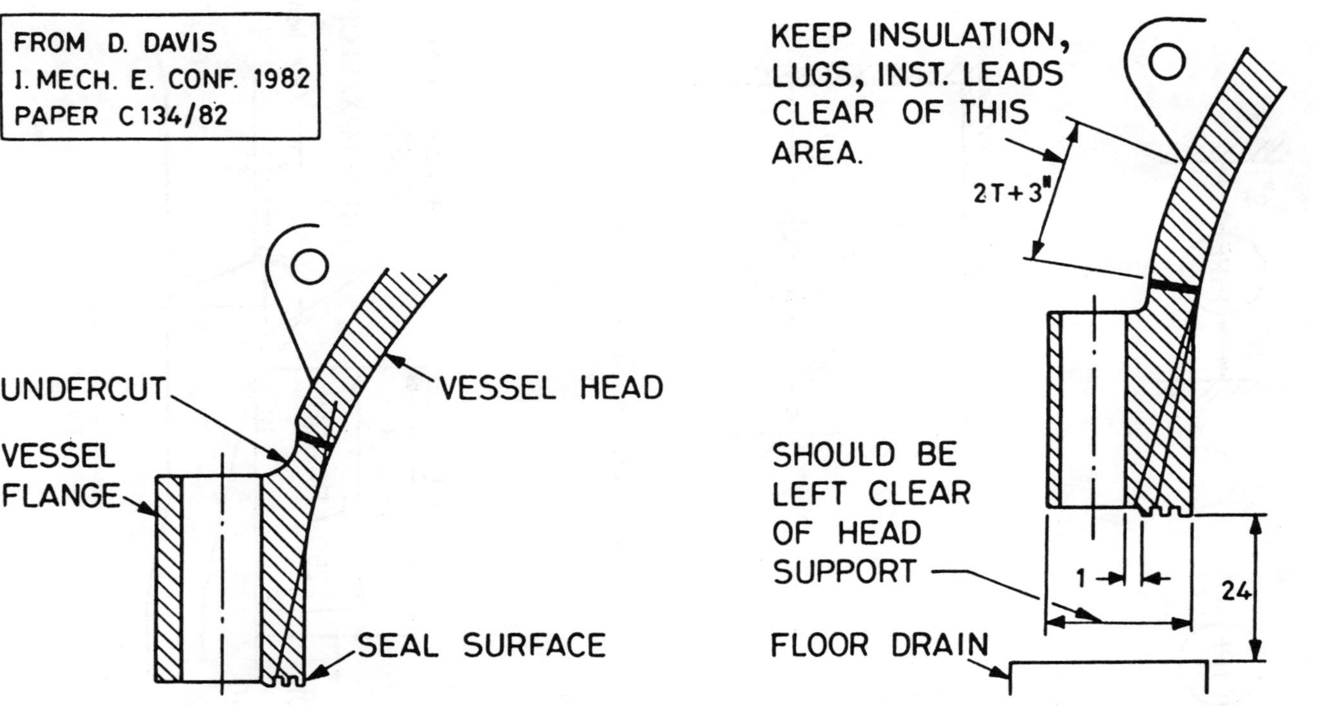

Fig. 18.

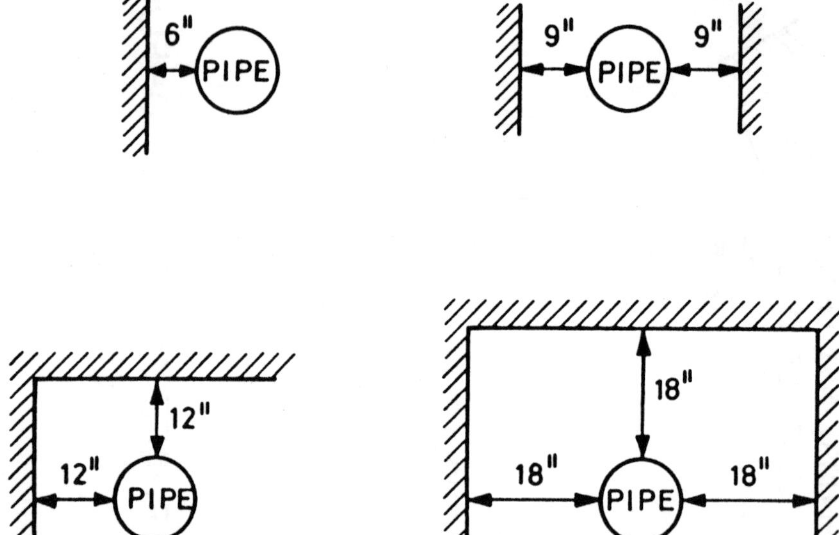

Fig. 19. Radial Clearance Requirements for Piping Surfaces.

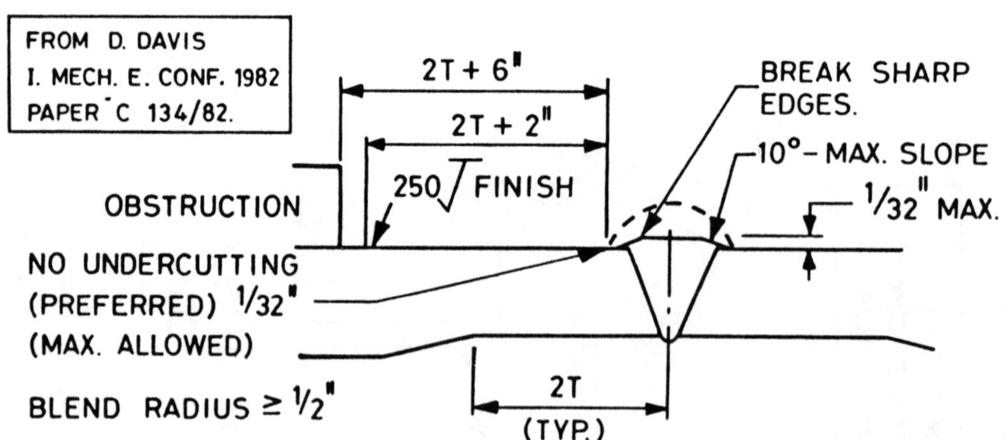

Fig. 20. Pipe to Pipe Welds.

parallel geometries (e.g., pressure vessel nozzles) tandem and corner
mode techniques are difficult to use both in application and interpretation. Some idea of the difficulty is illustrated by Figure 17 where
(as discussed above) inspection from one side of the DDT plates only
using the single probe 45° method 13 defects were missed at the 20% DAC
level. However, some of these defects would have been detected via corner
(Figure 2b) or double corner (Figure 2c) routes. Figure 15 (upper) shows
that in plate 1 only 10 defects out of 22 were detected in the upper
depth zones where corner mode detection would be impossible. These results are the equivalent of inspecting a non-parallel sided component.

For defects close to the inspecting surface it is possible to use
the 70° single probe technique and Figure 11 (upper) shows that all 14
defects lying within 85 mm of the scanning (clad) surface were detected
at the 20% DAC level.

Component design can also affect inspection reliability by either
restricting access (Figures 18 and 19) or by presenting geometric discontinuities, unseen by the inspector, which appear ultrasonically as
defects (Figure 20). Careful consideration at the design stage of plant
can significantly enhance reliability of in-service inspection.

b.   Component Material. If the material contains a large number of
inclusions (slag, porosity) or has a grain structure which causes attenuation and scattering of the ultrasonic pulses, a significant 'noise'
level will be received. The noise level will clearly limit the setting
of a reporting threshold level and attenuation will weaken the signals
from real defects. The net result is a reduction of the signal to noise
(or reporting level) ratio with a corresponding reduction in detection
reliability.

Ferritic steels used in pressure vessel construction are generally
good for ultrasonic inspection. Significant improvements have taken
place in steel manufacturing in the last 15 years and this is why
threshold levels may be set as low as 10 to 20% DAC. Steels produced
before 1970 occasionally contained higher densities of slag and porosity
which sometimes limited threshold levels to nearer 50% DAC.

Cast iron and cast austenitic materials present the most difficult
problems for ultrasonic inspection and in many cases are impossible to
test. Austenitic steels are usually inspectable but some weldments are
difficult to inspect because of a very inhomogeneous grain structure
which both redirects and scatters the ultrasonic beams. (Appendix 3.)

c.   Component Surface. In most ultrasonic applications the probe
is 'coupled' to the component surface by a thin layer of fluid (oil,
grease, water). The maintenance of a uniform parallel layer is important,
since small changes in either thickness or angle between probe and
component surface can significantly affect the amplitude and direction
of the transmitted signal. The problem can be used either by the component surface not being smooth or, in manual inspection, simply by
variations in the pressure applied by the operator. Coupling variations
thus can cause both a systematic attenuation and random fluctuations in
received signal amplitudes. In Section 5 variability of this type is
built into our analytical model of ultrasonic reliability.

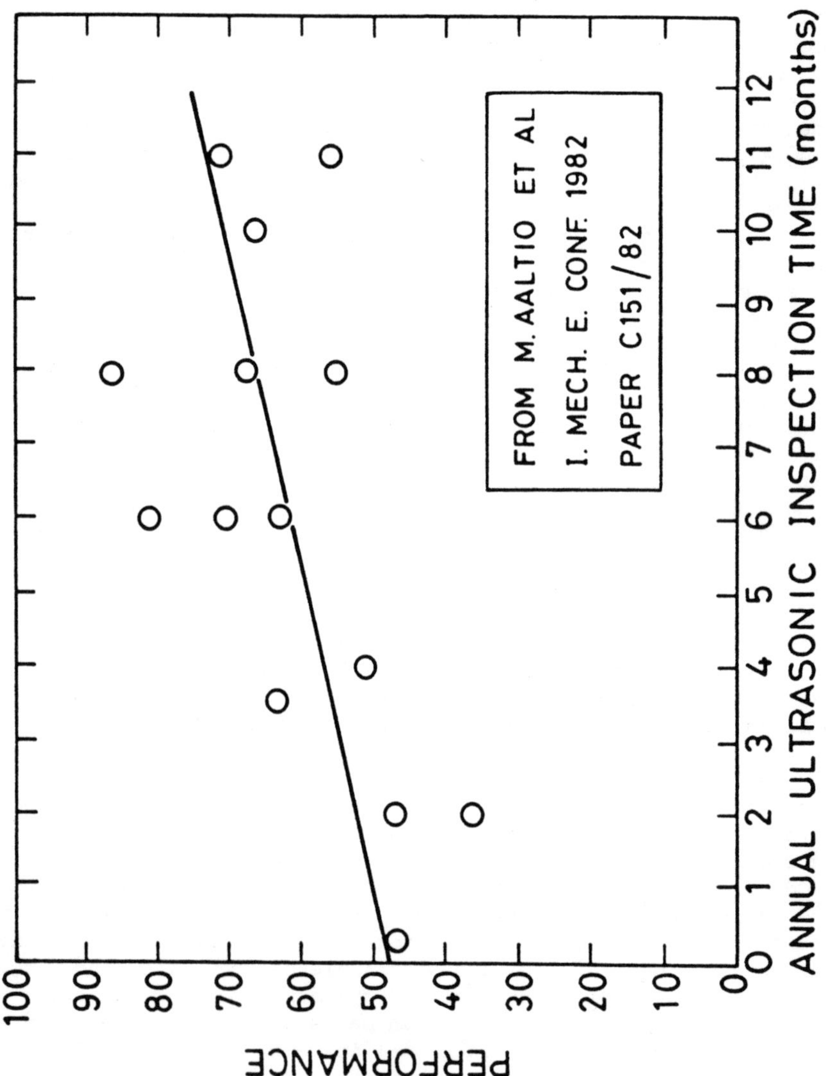

Fig. 21. Inspectors Experience Improves the Performance of the Ultrasonic Inspection.

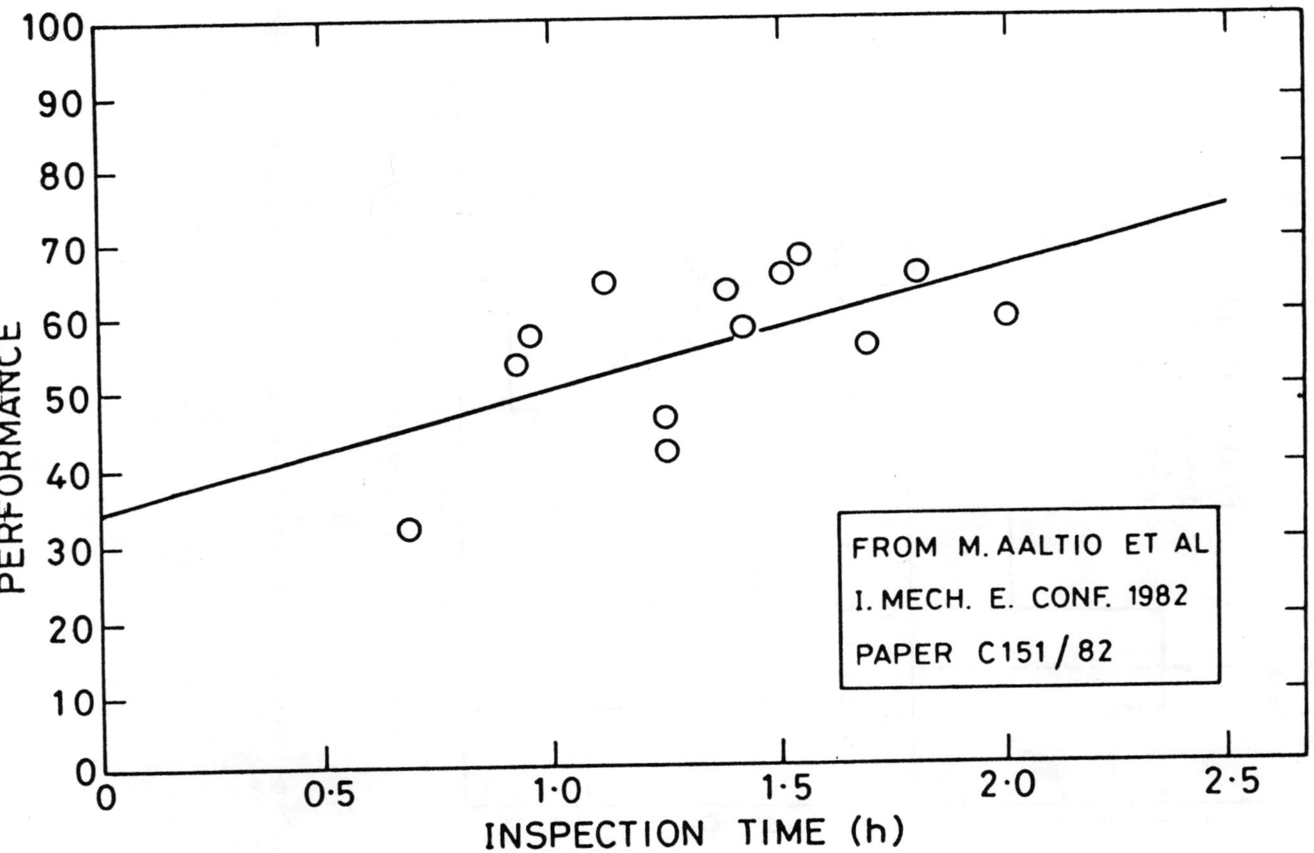

Fig. 22. Increased Inspection Time Improves the Performance of the Ultrasonic Inspections.

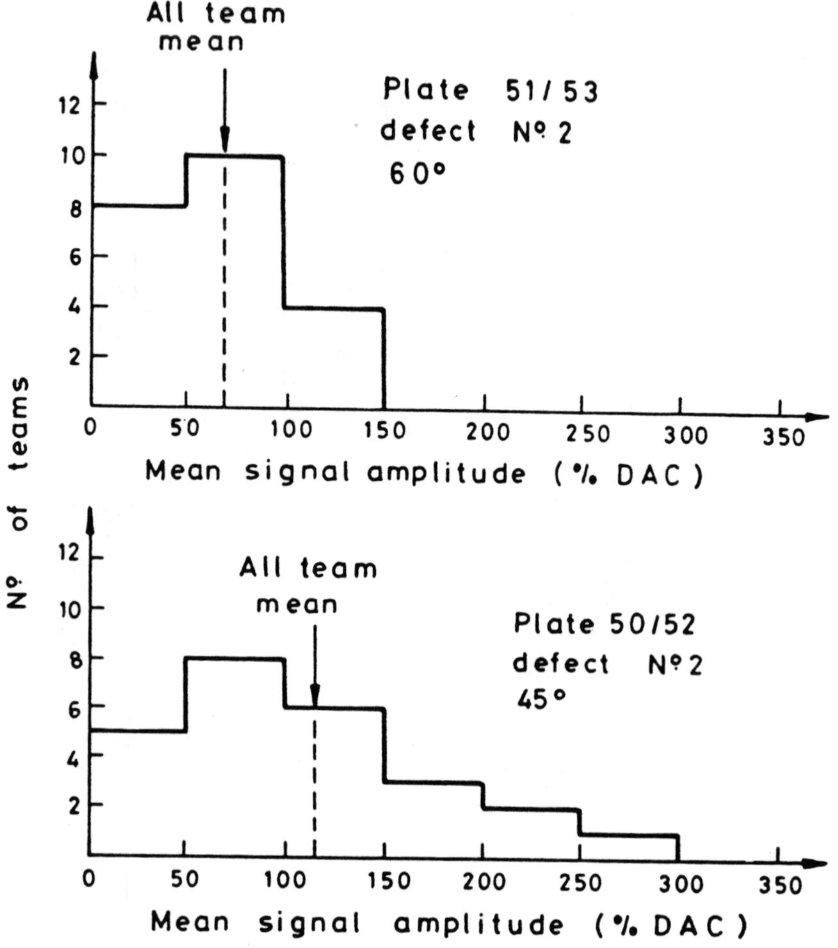

Fig. 23. Comparison of Distributions of Signal Amplitudes from Two PISC Defects.

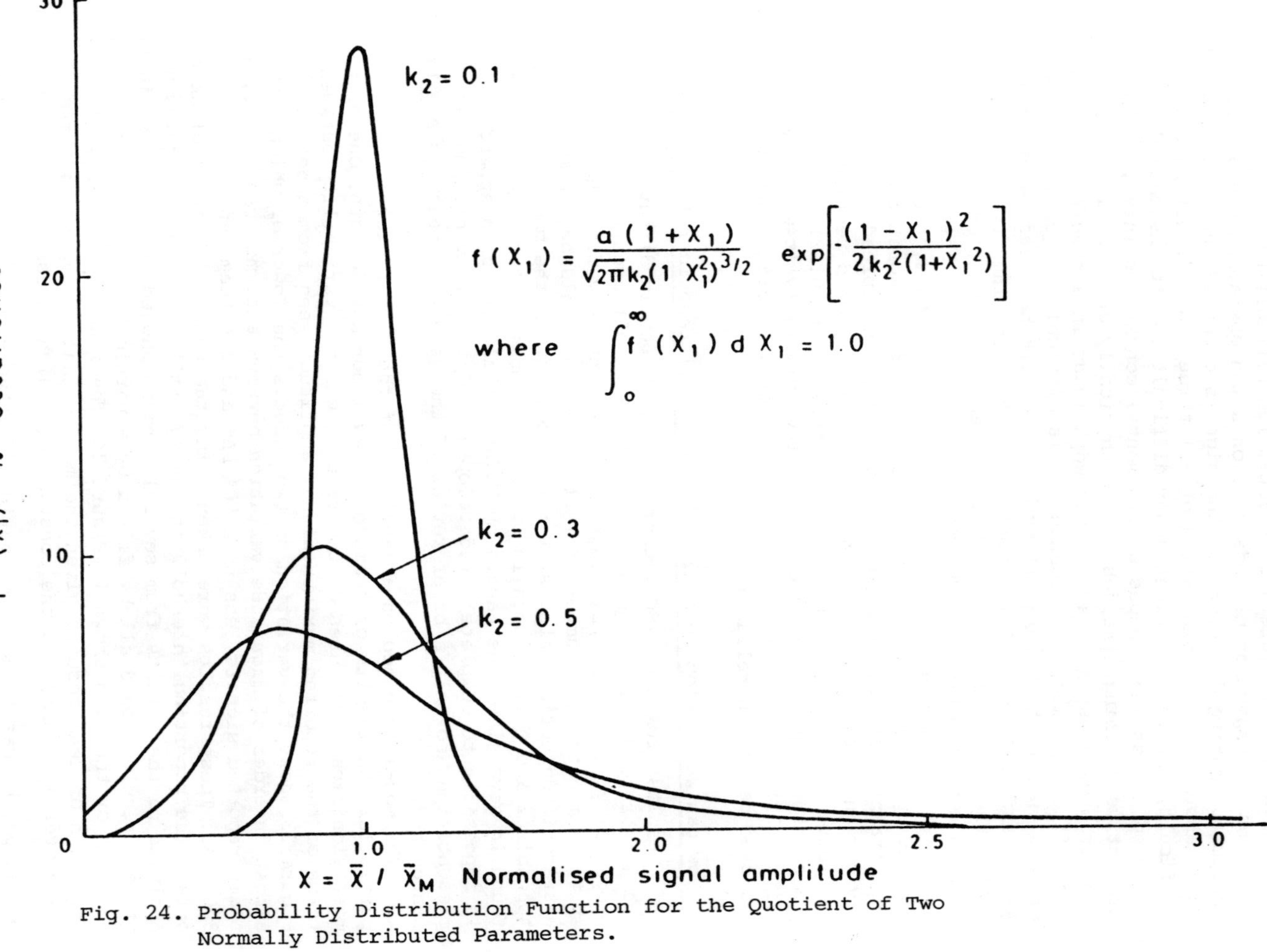

$$f(X_1) = \frac{a(1+X_1)}{\sqrt{2\pi}\,k_2(1\ X_1^2)^{3/2}} \ \exp\left[-\frac{(1-X_1)^2}{2k_2^2(1+X_1^2)}\right]$$

where $\displaystyle\int_0^\infty f(X_1)\,dX_1 = 1.0$

Fig. 24. Probability Distribution Function for the Quotient of Two Normally Distributed Parameters.

4.5. Which human factors affect detection reliability?

Manual inspection methods rely on a man operating the ultrasonic equipment, making decisions on the signals observed and recording data. In a laboratory the combination of all three tasks is not particularly difficult, however, on plant under difficult conditions caused by restricted access and arduous environmental conditions (heat, humidity, protective clothing) the job can be physically and mentally exhausting. Further pressures, such as alloted work loads in a specific time can also affect operator performance. It is difficult to quantify these effects on detection reliability but some work has been reported by Aaltio and co-workers (see Appendix 3). They examined two specific aspects, the effect of operator training and the effect of restricting time on a particular job. The results show that the best untrained operators were as good as the best experienced operators, but the worst untrained operators were far below the level of the worst trained operator. Thus training and experience tends to reduce the spread of operator behaviour and thereby improve overall reliability of detection. The graph in Figure 21 confirms that using operators who regularly perform ultrasonic inspection can improve reliability. Similarly it was shown (Figure 22) that allowing a longer time for a specific inspection task improved detection reliability.

5. How can we develop an analytic model of ultrasonic reliability?

Results from the PISC 1 exercise (see Appendix 3) have been used by the author to develop a quantitative model of ultrasonic defect detection probability. The method, although developed entirely from the PISC 1 results, demonstrate a number of features which should be common to all ultrasonic detection methods. A formal outline of the model is shown in Appendix 2 and further details of the development of the model may be found in several of the papers by Haines et al listed in Appendix 3. It is impossible to completely redevelop all of the arguments in this presentation, so only a brief review of the essential features are presented here.

The model is based on an analysis of the PISC 1 exercise and in particular the results of teams on a few important defects. One of the most significant features of the results was the variability between teams of the reported maximum signal amplitude seen from a defect. Figure 23 shows the variation on two defects the lower of which shows almost an order of magnitude variation between extreme teams, i.e., 5 teams reported signals less than 50% DAC and one team reported up to 300% DAC. These results were taken using the same calibration block to obtain the reporting threshold and using similar ultrasonic equipment. Analysis of the results from several defects showed that the distribution of team results could all be fitted by a particular probability density function of the form shown in Figure 24. The function predicts the distribution of team results given the mean signal to reporting threshold ratio over all teams and the constant k_2, which is essentially a standard deviation. k_2 was found by comparison for the PISC results to be 0.5.

A second observation from the PISC results was that on long defects

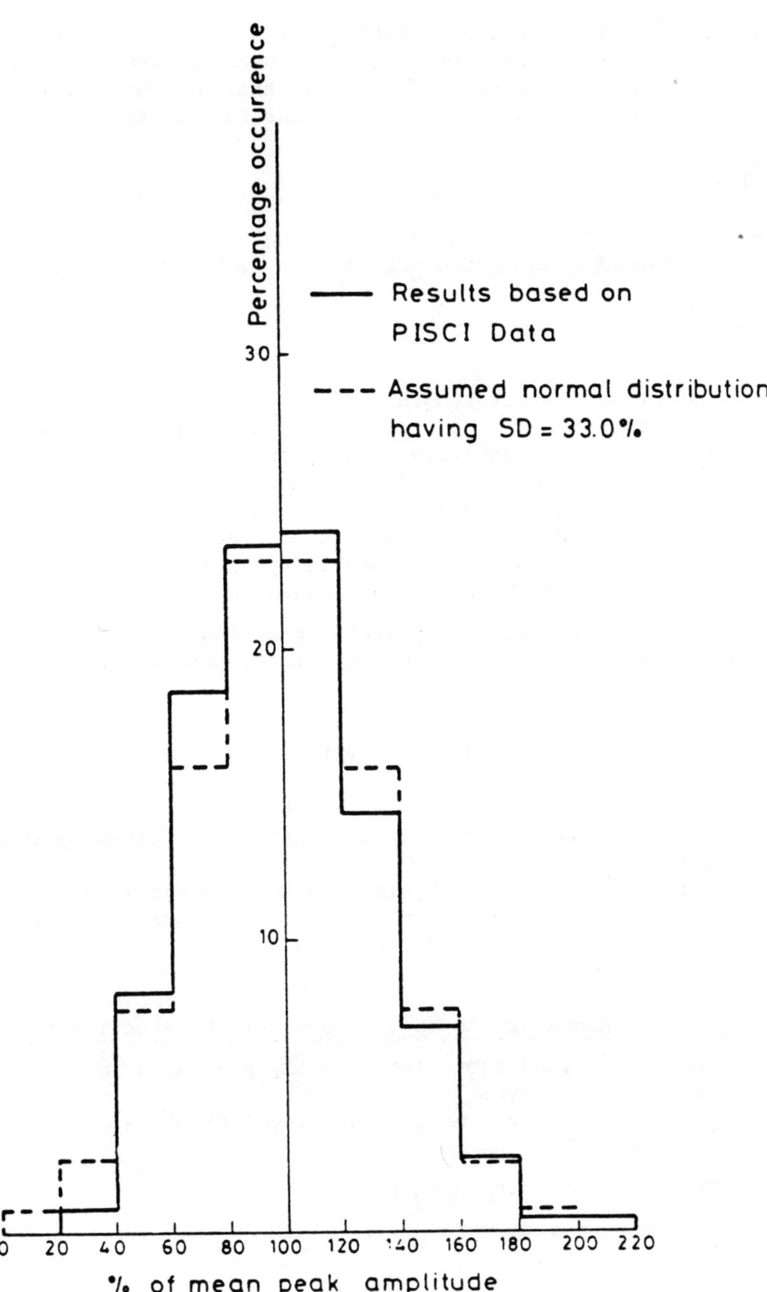

Fig. 25. Distribution of Normalised Signal Amplitudes.

(large 2c) Defect Detection Probability was slightly greater. The result, as suggested in Section 5.2.a., is because several scan lines intersected the same defect. If there is the same probability of detection (P) on each scan then over N scans the DDP becomes

$$1 - (1-P)^N,$$ (7)

where N must be related to the defect length 2c and the scan pitch. For PISC the value of N was found empirically to be given by

$$N = \frac{2c}{25},$$

where c is the semi crack length in mm.

It was also possible to analyse data from separate teams on long defects to deduce a probability density function for the signal amplitude variation between separate scan lines over the same defect. Perhaps not surprisingly the data was found to be well approximated by a normal distribution as shown in Figure 25. The probability of detection on any single scan line is simply the probability that the measured signal amplitude (χ) exceeds the reporting threshold (χ_T). If the probability density function for signal amplitude on a scan is $Q(\chi, \bar{\chi})$, where $\bar{\chi}$ is the mean signal amplitude over N scan lines, then the probability of detection on a single scan is

$$P(\chi > \chi_T \mid \bar{\chi}) = \int_{\chi_T}^{\infty} Q(\chi, \bar{\chi}) \, d\chi = P(\bar{\chi}, \chi_T).$$ (8)

The probability of detection over N scans is then given by substituting P in Equation (7) above.

The remainder of the analysis to develop a model for Defect Detection Probability (= Detection Reliability as defined in Equation (1)) is given in Appendix 2. Curves developed from Equation (3) of Appendix 2 are shown in Figure 26.

6. What data is necessary to apply the model of detection reliability?

The model outlined above and formally presented in Appendix 2 requires 6 input parameters:

1. An estimate of the mean signal level from the defect being considered ($\bar{\chi}_M$).

2. The reporting threshold (χ_T).

3. The between team variability constant (similar to a standard deviation) k_2.

4. The between scan standard deviation k_1.

5. The length (2c) of the defect.

6. The minimum distance between scan lines over which the signals from the defect are uncorrelated (d).

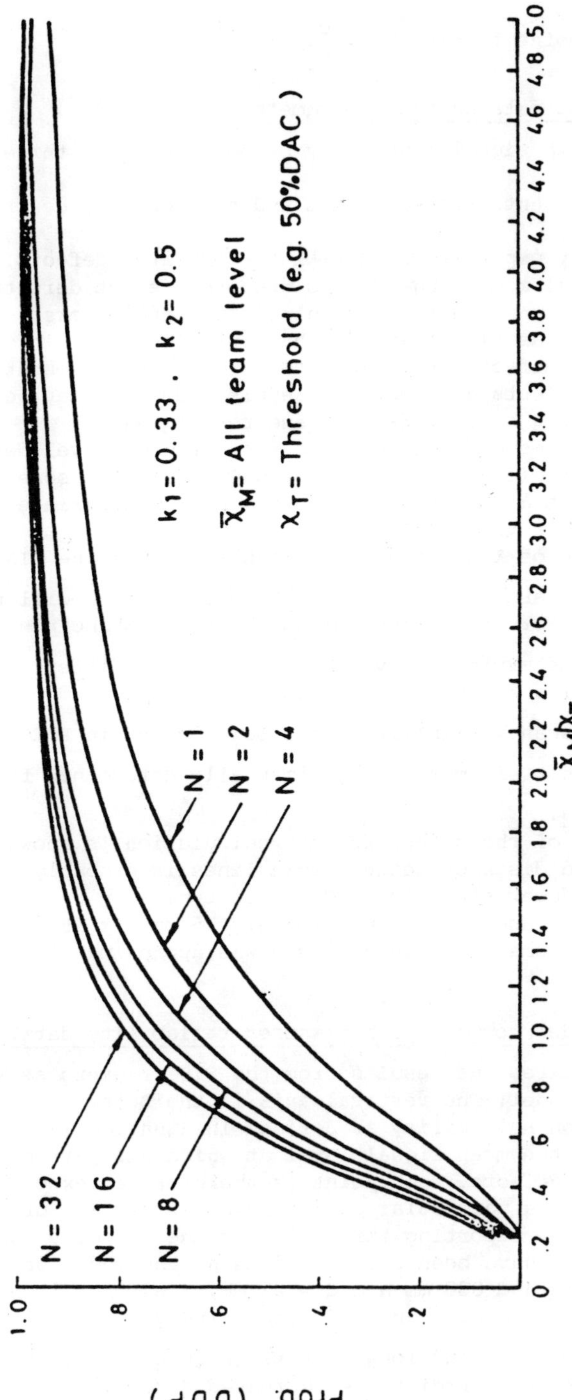

Fig. 26. DDP versus Ratio of All Team Mean to Threshold Level.

(Items 5 and 6 combine to give $N = \dfrac{2c}{d}$.)

7. Where can we get data on these parameters?

1. Estimates of mean signal levels may be obtained from three main sources:
   (a) Real components tested by NDE and subsequently destructively examined.
   (b) Laboratory experiments on real or simulated defects.
   (c) Theoretical modelling methods used to predict defect response as a function of defect and ultrasonic parameters.

Exercises such as PISC 1 and DDT have provided much information on signal levels from defects. Figures 8 to 17 of this presentation give data on specific defects and response levels and these may be reasonably used when considering similar defects and techniques.

2. The reporting threshold is usually a fixed signal level defined in a particular procedure (e.g., 20% DAC, 50% DAC etc.). (In some cases it is necessary to convert a reporting level to a more appropriate unit, e.g., 3 mm flat bottom hole to X% DAC.)

3. The only measure of k_2 currently available is from the PISC 1 exercise where it was measured as 0.5. In the future PISC 2 will provide new information on k_2. For other inspections it is, in principle, possible to experimentally determine the variability between different inspectors and hence determine a k_2 specifically for that work.

4. k_1 the scan to scan variability was also measured in PISC 1 and found to be 0.33. As with k_2 it may be experimentally determined for other inspection conditions.

5. The length (2c) of the defect for any calculation is known.

6. The uncorrelated distance between scan lines is probably closely related to the transducer size. In PISC probes typically had elements of 20-25 mm and d was found to be approximately 25 mm. It is a reasonable assumption that for smaller transducers d was approximately equal to the transducer size.

8. How does the model compare with measured reliability data?

Figure 27 compares the results from the PISC 1 exercise with the prediction of the model. The vertical axis is the Defect Detection Probability (= Detection Reliability as defined in Equation (1)) and the horizontal axis is the mean signal level at which all participating teams detected the defects. Each point is therefore an experimentally observed result from a particular defect using either 45° or 60° single probe techniques. The reporting threshold for the exercise was $\chi = 50\%$ DAC. The solid lines have been calculated using the model described above for 2c = 25 mm and 800 mm and d = 25 mm, i.e., N = 1 and 16. The values of k_1 and k_2 were 0.33 and 0.5 respectively. If we further separate the results for short and long defects (Figure 26), then the model can be seen to correctly predict the increased detection probability.

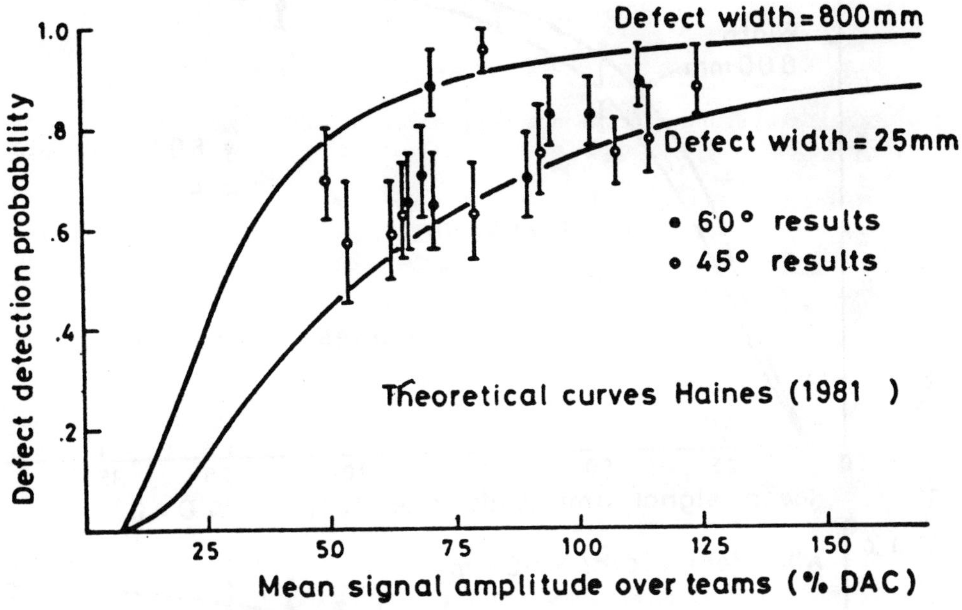

Fig. 27. Relationship Between DDP and Mean Signal Level for Defects in PISC 1 Exercise.

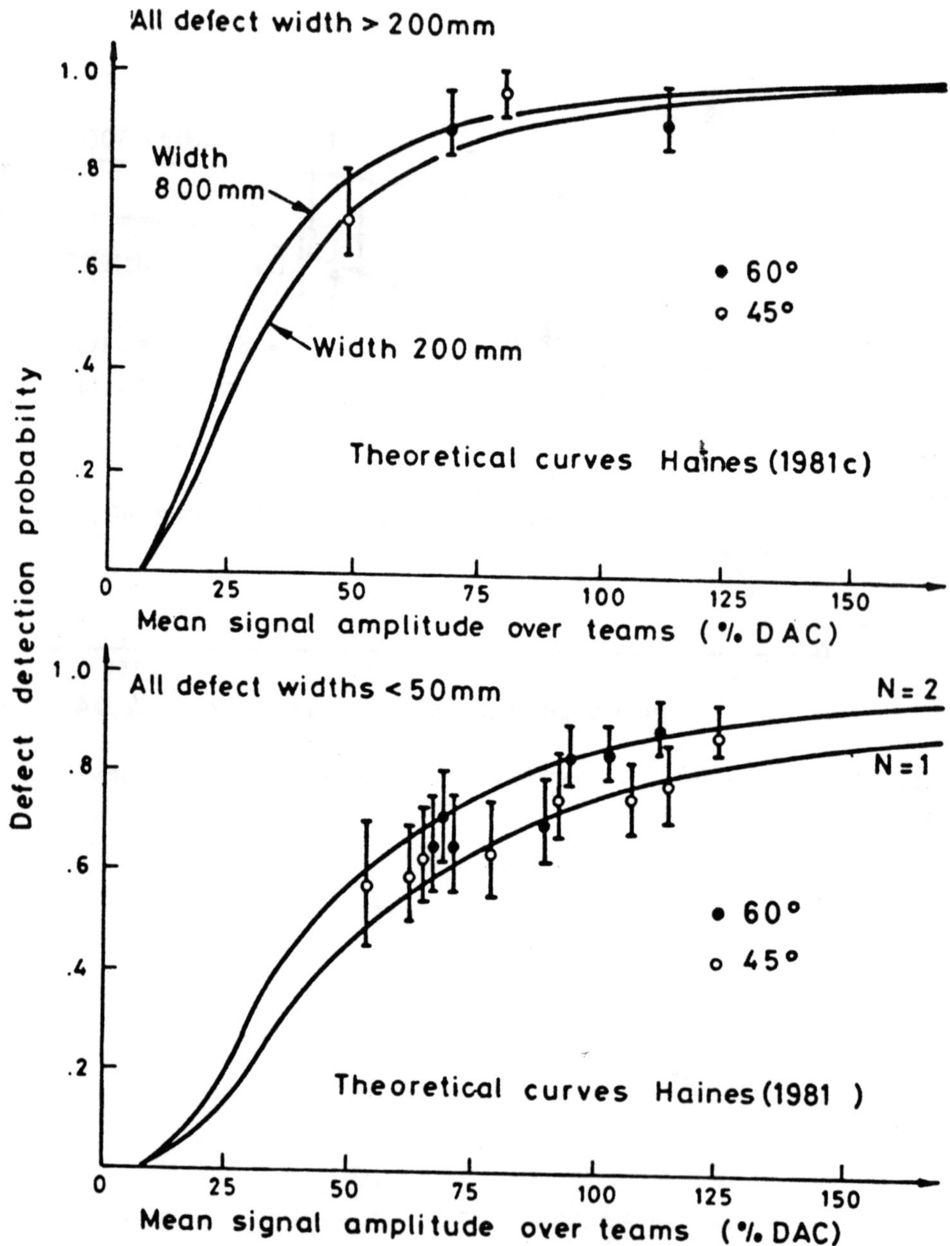

Fig. 28. Comparison of DDPs for Narrow and Wide Defects. All Data from PISC 1 Exercise.

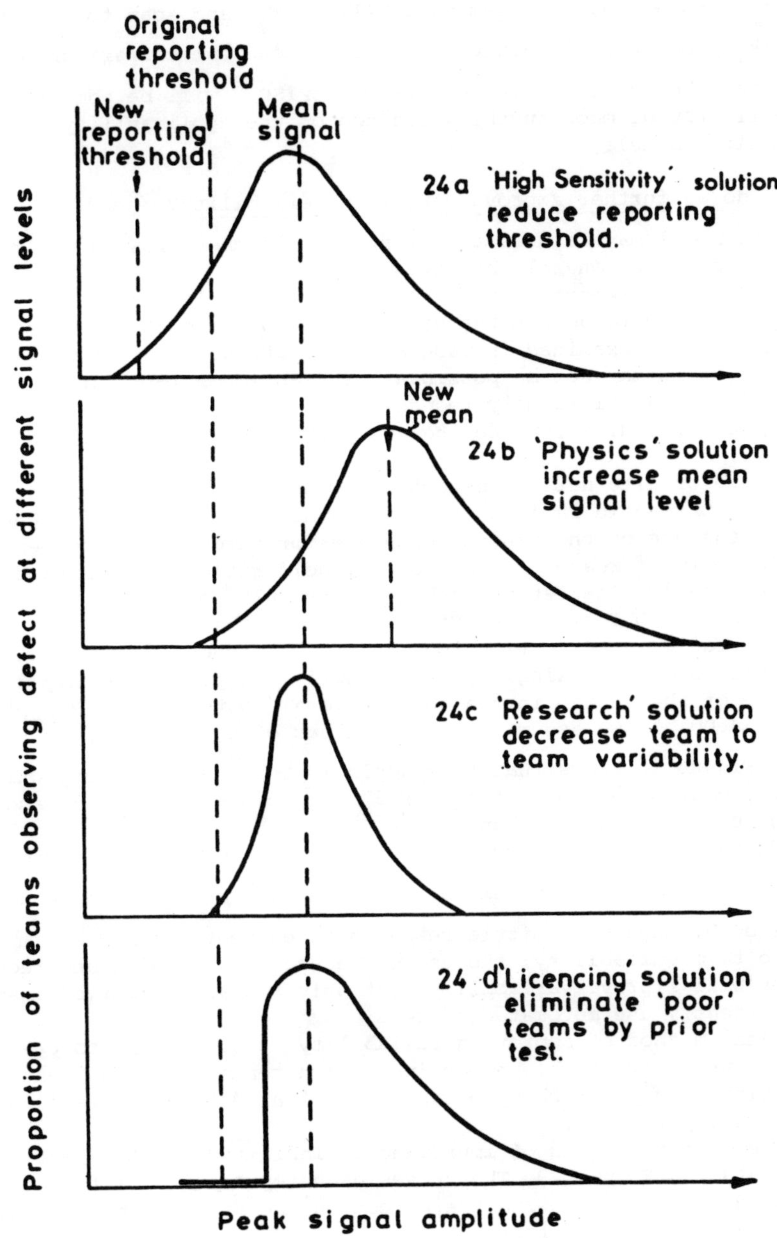

Fig. 29. Diagrammatic Representation of Various 'Solutions' to the Problem of NDE Reliability.

PISC 1 is the only exercise that has provided sufficient data on such factors as team to team variability (k_2) and scan to scan variability (k_1), hence it is not possible to extend the comparison of the model to other experimental programmes. PISC 2 will be completed in 1985 and will provide much further evidence on the applicability of the reliability model.

9. How do we further improve detection reliability?

The model may be used to illustrate a number of ways of improving ultrasonic detection reliability:

1. Reducing the reporting threshold (χ) clearly increases the proportion of teams likely to detect the same defect. Unfortunately, unless the component being examined is made of a low attenuation - low noise material, this may not be possible. (It was impossible during the PISC 1 exercise to significantly reduce the threshold below 50% DAC because the number of indications coming above the reporting threshold increased rapidly.)

2. The use of alternative inspection techniques which give a higher signal to threshold ratio is an obvious method of increasing DDP. For example the use of the tandem technique for through wall buried defects significantly increases signal to threshold ratios for planar defects. As a particular case let us look at the results of defect number 27 on plate 1 of the DDT exercise. Figure 16 shows that if we were restricted to clad side only inspection the tandem technique gave a signal of 80% DAC, whereas the 45° single probe (Figure 17 upper) gave only 25% DAC. Thus use of the tandem increased the signal to threshold (20% DAC) from 1.25 to 4.00. If we assume the same values of k_1 and k_2 apply as for PISC 1 Figure 26 shows that this would increase DDP from approximately 0.70 to about 0.95. (Defect number 27 has 2c = 51 mm, a = 55 and tilt = 20°, assuming d = 25 mm then N = 2).

3. Reducing the variability between teams (k_2) would make a significant improvement in detection reliability as illustrated by Figure 24 for k_2 values of 0.3 and 0.1. Little research is currently aimed at improving team to team variability; the use of mechanical scanning and recording systems will improve variability, but only a small number of inspection jobs currently use automated systems.

4. A final method of improving reliability is to attempt to pre-select by some type of test the better inspection teams. Various countries are attempting this approach by establishing 'validation centres' for NDT methods and personnel.

These four methods of improving reliability are diagramatically illustrated in Figure 29. The curves each represent a different 'solution' discussed above, by showing the effect on the proportion of teams liable to miss a particular defect.

10. What information is available on accuracy of defect location and sizing?

Much less effort has been put into understanding the parameters

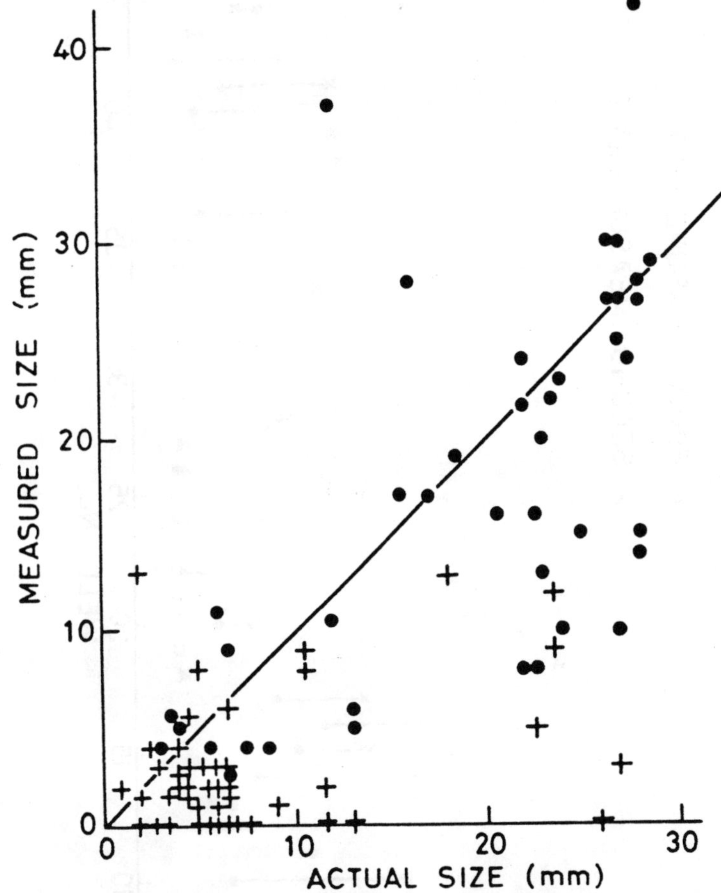

Fig. 30. Graph Comparing Ultrasonic Measurements of the Through Thickness Sizes of Cracks With the Actual Sizes Obtained by Destructive Examination. Results for One Operator Only. (Results Courtesy of the Welding Institute.)

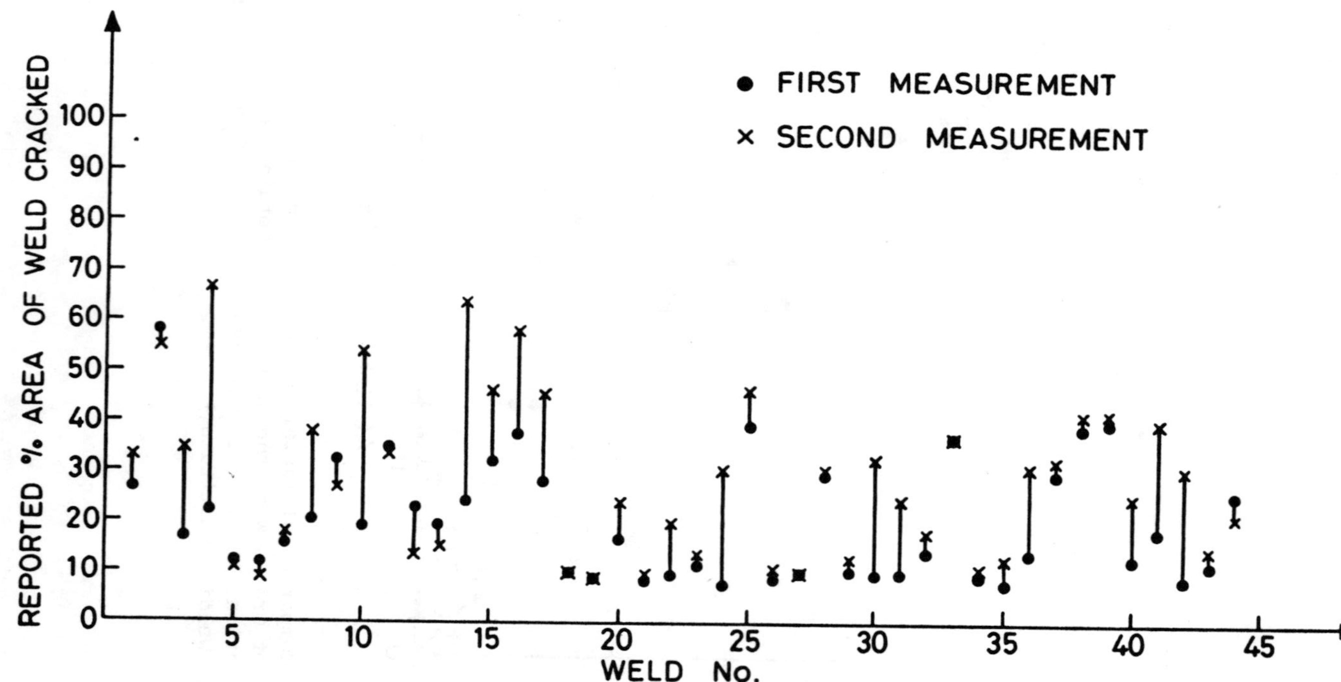

Fig. 31. Ultrasonic Measurements of Defective Welds.

which affect defect location and sizing than into defect detection. The
PISC 1 exercise concluded, in line with earlier studies, that small defects could be more often oversized and large defects undersized. However, the methods used relied entirely on so called 'amplitude drop'
techniques as recommended in the ASME codes (see Appendix 1). In the DDT
exercise much more precise sizing and location was achieved by all participating teams due partly to the use of more advanced ultrasonic
methods and partly to the nature of the defects.

A major exercise was run by the UK welding institute on sizing in
which the defects were separated into planar and non-planar types.
Taking the latter first the report concluded that 'conventional' ultrasonic testing was generally reliable for positioning of non-planar defects and estimating their length (2c). However, defect through thickness
sizing is unreliable with a tendency to undersize. The report on planar
defects concludes:
1. 'The measurement of planar defect cross sectional size is subject to
errors, largely due to difficulties in estimating the true position of
defect extremeties arising from the complex interaction of ultrasound
with unknown flaws. Measurement of defect position and length are subject
to errors of a similar magnitude for the same reasons. Poor surface
finish and access are likely to increase errors.
2. Typical errors on planar defect cross-section measurement for a manual
operator using the 20 dB drop method were +5.2 mm, −7.2 mm of the true
value, while errors incurred by the time-of-flight technique, the most
accurate, were +4.1 mm, −3.1 mm, both at the 95% confidence level. This
represents a drop in performance with respect to tests performed on
non-planar defects, errors being +2.0 mm, −3.6 mm and +1.7 mm, −2.3 mm
for the two techniques respectively, again at the 95% level.

The mean errors for both the Maximum Amplitude and dB Drop sizing
methods are negative, showing a tendency to undersize. There are also
less pronounced trends which indicate that through-thickness sizing is
more accurate with lower angles and higher frequencies.

Figure 30 shows a graph of measured versus actual defect size for
many of the defects in the Welding Institute tests. The majority of the
defects were undersized as may be seen from the greater number of points
below the solid line.

Repeatability of sizing of defects using different teams also cause
problems. Figure 31 shows the results on 44 defects located in a set of
similar reactor structures. The solid dots were the first set of measurements taken and the crosses a set taken later, but not after further
reactor operation. Although relatively close agreement occurred on many,
significant increases occurred on at least 15 defects. Unfortunately no
information is available on the actual defect sizes present.

11. Summary.

In this lecture I have proposed that NDE reliability comprises two
factors:
   (1) A reliable detection method must find all types and sizes of
   flaws of concern in the component;
   (2) The method must be capable of being used by different inspection
   teams, all of whom can consistently find the defects of concern.

The first requirement is a physical limitation of the method chosen and the results have been presented specifically on parameters that affect ultrasonic techniques. It has been proposed that signal to 'noise' on signal to reporting threshold level is the most important factor in achieving reliability.

The second requirement of reliability is the consistency with which different teams use the same methods. It has been demonstrated that results from the PISC 1 exercise showed a large variation in team behaviour.

An analytic model of Defect Detection Probability has been described and shown to fit well the results of the PISC 1 exercise. The model requires 6 input parameters each of which may be determined by reference to similar tests, experimental studies or theoretical analysis. The required parameters are:

(1) The mean signal level at which teams detect, or are expected to detect, a defect.
(2) The reporting threshold level.
(3) The approximate spread of team results.
(4) The variation between scans on a single defect.
(5) The length of the defect (2c).
(6) The scan pitch over which responses from a defect are uncorrelated.

Using the model a number of suggestions are made as to ways in which reliability may be further improved.

Finally a brief review is given of the results of exercises to measure location and sizing accuracy. Further work, however, is required before it will be possible to analyse sizing results to give details for specific methods on particular defects. Work so far has only separated defects into Planar and Non Planar types.

APPENDIX 1.

A summary of British Standard, UK Electricity Supply Industry Standards and CEGB NDT Engineering documents on inspection methods.
(A short list of other national standards on NDT of nuclear plant is also included.)

British Standards - NDT
(u = ultrasonic, r = radiography, mp = magnetic particle, ec = eddy current, p = penetrant, v = visual)

1. General
1. Glossary of Terms used in NDT, 5 parts (includes, u, r, mp, ec, p) 1963-65, BS 3683.
2. Methods for NDT of Pipes and Tubes. 5 parts, 1965 - 66, BS 3889 (includes u, mp, ec, p).
3. Methods for NDT of Steel Castings. 1966, BS 4080 (includes u, r, mp, p).
4. NDT of Steel Forgings. 3 parts 1967 - 68, BS 4124 (includes u, mp, p).

5. Recommendations for NDT of concrete. 5 parts, 1969 - 74, BS 4498.

2. Ultrasonic Methods
6. Calibration Blocks for use in Ultrasonic Flaw Detection 1978, BS 2704.
7. Methods for Ultrasonic Examination of Welds. 3 parts, 1972 - 78, BS
   3923.
8. Methods for Assessing the Performance Characterisations of Ultrasonic.
9. Flaw Detection Equipment. 3 parts, 1972 - 78, BS 4331.
10. Ultrasonic Testing and Specifying Quality Grades of Ferritic Steel
    Plate. 1980, BS 5996.

3. Radiography Methods
11. Radiographic Examination of Fusion Welded Butt Joints in Steel. 2
    parts. 1973, BS 2600.
12. Terminology of Internal Defects in Castings as Recorded by Radio-
    graphy. 1956, BS 2737.
13. Methods for Radiographic Examination of Fusion Welded Circumferential
    Butt Joints in Steel Pipes. 1973, BS 2910. Apparatus for Gamma Radio-
    graphy. 1978, BS 5650.

4. Magnetic Particle Methods
14. Magnetic Flaw Detection Inks and Powders. 1982, BS 4069.
15. Methods for Magnetic Particle Testing of Welds. 1969, BS 4397.
16. Methods for Assessing Block Light used in NDT. 1969, BS 4489.
17. Contrast Aid Paints used in Magnetic Particle Flaw Detection. 1973,
    BS 5044.
18. Magnetic Particle Flaw Inspection of Finished and Machined.
19. Solid Forged and Drop Stamped Crankshafts. 1974, BS 5138.
20. Methods for Magnetic Particle Flaw Detection. 1981, BS 6072.

5. Penetrant Testing Methods
21. Penetrant Testing of Welded or Brazed Joints in Metals. 1969, BS 4416.

6. Visual Methods
22. Guide to the Selection of Low Power Magnifiers Used for Visual In-
    spection. 1974, BS 5165.

Electricity Supply Industry Standards NDT
1. Ultrasonic Probes: Medium Frequency, Miniature Shear Wave.
2. Angle Probes, Issue 1, December 1979. E.S.I. Standard 98-2.
3. Registration of Non-destructive Testing Personnel, Issue 1, December
   1981. E.S.I. Standard 98-3.
4. Ultrasonic Testing of Ferritic Steel Castings, Issue 1, February
   1982. E.S.I. Standard 98-4.
5. Automatic Ultrasonic Testing of Steel Tubes. Part 1, Seamless and
   Welded Tubes, Outside Diameters of 12 mm up to but not including 160
   mm. Issue 1, November 1981. E.S.I. Standard 98-5.
6. Ultrasonic Testing of Ferritic Steel Plate, Issue 1, July 1981. E.S.I.
   Standard 98-6.
7. Ultrasonic Probes: Normal (0°) Compression Wave Probes for Contact

Testing, Issue 1, February 1982. E.S.I. Standard 98-7.
8. Ultrasonic Probes: Low Frequency Single Crystal Shear Wave, Angle
   Probes, Issue 1, January 1982. E.S.I. Standard 98-8.
9. Manual Ultrasonic Testing of Welds in Ferritic Steel Sections. Part
   I - Butt Welds in Ferritic Steel Sections Greater than 10 mm Thick,
   Issue 1, September 1981. Part 3 - Butt Welds in Tubes of Outside Dia-
   meters between 37.5 mm and 100 mm inclusive, Issue 1, November 1981.
   E.S.I. Standard 98-10.

<u>C.E.G.B. Engineering Documents (NDT) Section 702</u>

1. Eddy Current Crack Detection Apparaturs, Magnatest Ed. 400, Issue 1,
   September 1970. Ref. SR/PITB/NNDTU/1.
2. The Non-destructive Examination of Butt Welds in Chromium-Molybdenum -
   Vanadium Material, Issue 1, March 1972. Ref. SR/ES/NNDTU/2.
3. Guidance on the Training of Personnel Engaged in NDT, Issue 2, March
   1977. Ref. ES/PITB/NNDTU/3.
4. Guidance on the Instruments used in the Identification of Materials,
   Issue 1, May 1972. Ref. ES/PITB/NNDTU/5.
5. Practical Limitations of Radiographic and Ultrasonic Examination in
   Measuring Imperfections in Welds, Issue i, July 1976. Ref. ES/QA/
   /NNDTU/7.
6. Surface Crack Detection, Magnetic Particle and Penetrant Methods,
   Issue 1, November 1977. Ref. ES/QA/NNDTU/9.
7. Recommended Ultrasonic Methods of Examination of Pipe Welds in High
   Pressure and Hot Reheat Pipelines. December 1978.

<u>Foreign Standards and Codes (NDT)</u>
<u>U.S.A.</u>

1. ASME Boiler and Pressure Vessel Code - Section V 'Non-Destructive
   Examination', 1980 Edition. All Addenda to December 1979.
2. ASME Boiler and Pressure Vessel Code - Section XI 'Rules for Inservice
   Inspection of Nuclear Power Plant Components', 1980 Edition.
3. U.S. Nuclear Regulatory Commission Guide 1.150 Ultrasonic Testing of
   Reactor Vessel Welds during Pre-service and In-service Examinations.
   June 1981.

<u>West Germany</u>

1. Nuclear Engineering Committee Safety Regulations KTA 3201.3, Components of Primary Circuit of Light Water Reactors - manufacture.
   October 1979.

<u>France</u>

1. RCC-M Design and Construction Rules for Mechanical Components. France
   1980.

APPENDIX 2.

Development of a model of ultrasonic reliability.

<u>Probabilistic Model of Variability in Ultrasonic Testing</u>

   Suppose that on a flat plate specimen a series of parallel scan lines
each intersect a defect and a set of N peak signals are observed (χ_1,

$\chi_2 \ldots \chi_N$). Assuming the scan lines to be sufficiently separated as to
be independent then we would expect the set of χ values to have some
distribution density function $Q(\chi, \bar{\chi})$ where $\bar{\chi}$ is the mean of the N scan
values. (This will only be true for a defect where each scan line inter-
sects regions which have similar parameters e.g. approximate same
position in depth, size, orientation.)

$$\text{Prob } (\chi > \chi_T | \bar{\chi}) = \int_{\chi_T}^{\infty} Q(\chi, \bar{\chi}) \, d\chi = P(\bar{\chi}, \chi_T). \tag{1}$$

If N scans intersect a nominally similar region of a defect then the
probability that χ is greater than χ_T on at least one line is given by

$$\text{Prob } (\chi > \chi_T | \bar{\chi}, N) = 1 - [1 - P(\bar{\chi}, \chi_T)]^N = P_N(\bar{\chi}, \chi_T). \tag{2}$$

The second part of the problem is to consider the variation in the
mean signal level observed within a group of teams using nominally the
same procedure and equipment on a defect. Suppose that each team measures
after N scans a mean $\bar{\chi}$ then we have for n teams a set of means ($\bar{\chi}_1$,
$\bar{\chi}_2 \ldots \bar{\chi}_n$). Assuming that no sub-groups exist over which a systematic
change in $\bar{\chi}$ occurs then we may treat the set of all $\bar{\chi}$ as random variations having a distribution about some mean $\bar{\chi}_M$ where

$$\bar{\chi}_M = \frac{\bar{\chi}_1 + \bar{\chi}_2 + \ldots \bar{\chi}_n}{n}.$$

Let us define the probability density function for team means as
$f(\bar{\chi}_M, \bar{\chi})$.

We may now evaluate the DDP for a defect on which the all team average
is $\bar{\chi}_M$ and the threshold for detection is χ_T by the following analysis.
The fraction of all teams observing the defect at a mean signal level
between $\bar{\chi}$ and $\bar{\chi} + \delta\bar{\chi}$ is $f(\bar{\chi}_M, \bar{\chi})\delta\bar{\chi}$. The probability that a team observing
the defect at a mean value of $\bar{\chi}$ will detect over N scans at least once
at the threshold χ_T is $P_N(\bar{\chi}, \chi_T)$. Thus the fraction of teams detecting
the defect is the product

$$\delta(\text{DDP}) = f(\bar{\chi}_M, \bar{\chi})\delta\bar{\chi} \cdot P_N(\bar{\chi}, \chi_T).$$

To consider all the teams observing the defect at all possible mean
values we must sum over all values of $\bar{\chi}$ giving

$$\text{DDP} = \int_0^{\infty} f(\bar{\chi}_M, \bar{\chi}) P_N(\bar{\chi}, \chi_T) \, d\bar{\chi}. \tag{3}$$

Equation (3) is used in Section 5 to generate a set of curves that give DDP as a function of $\bar{\chi}_M$ or more specifically it is shown that, because of the forms of $f(\bar{\chi}_M, \bar{\chi})$ and $P_N(\bar{\chi}, \chi_T)$, it is possible to plot a single set of curves of DDP against $\bar{\chi}_M/\chi_T$ for different values of N. The specific forms of both $f(\bar{\chi}_M, \bar{\chi})$ and $Q(\chi, \bar{\chi})$ have been derived from the analysis of PISC data and shown to fit functions of the form:

$$Q(\chi, \bar{\chi}) = \frac{1}{\sqrt{2\pi}\ k_1\bar{\chi}}\ \exp\left[-\frac{(\chi-\bar{\chi})^2}{2k_1^2\ \bar{\chi}^2}\right] ,$$

where $k_1 = 0.33$ and

$$f(\bar{\chi}_M, \bar{\chi}) = \frac{\bar{\chi}_M\bar{\chi} + \bar{\chi}_M^2}{k_2\sqrt{2\pi}\ (\bar{\chi}^2 + \bar{\chi}_M^2)^{3/2}}\ \exp\left[-\frac{(\bar{\chi} - \bar{\chi}_M)^2}{2k_2(\bar{\chi}^2 + \bar{\chi}_M^2)}\right] ,$$

where $k_2 = 0.5$. It will be readily recognised that the scan to scan distribution $Q(\chi, \bar{\chi})$ is a normal distribution having a standard deviation proportional to the mean value. The team to team distribution function is the distribution of the ratio of two parameters where each is normally distributed and having a standard deviation proportional to the mean.

APPENDIX 3.

A selection of reports on NDE reliability for further reading.

General Work on NDT Reliability
1. OECD, Plate inspection programme, PISC, EUR 6371EN Vol. 1-6, 1979.
2. Status report on the UKAEA Defect Detection Trials - March 1982.
   B. Watkins, R.W. Ervine and K.J. Cowburn. ND-R.801(R). June 1982.
3. CEGB Inspection of Plates 1 and 2 in UKAEA Defect Detection Trials.
   K.J. Bowker et al., Dec. 1982. CEGB Report NW/SSD/82/0159/R.
4. Progress towards an understanding of reliability in NDE. N.F. Haines.
   Specialised Meeting Ispra, May 1983. EUR 90661 EN.
5. Assessment of the reliability of Ultrasonic Inspection Methods,
   N.F. Haines, D.B. Langston, A.J. Green, R. Wilson, I. Mech. E.
   Conference, London, Oct. 1982, paper C153/82.
6. The reliability of ultrasonic inspection for thick section welds:
   some views and model calculations, J.M. Coffey, I. Mech. E. Conference, London, Oct. 1982, paper C159/82.
7. Reliability and Defect Sizing, M. Aaltio and K.P. Kauppinen, I. Mech.
   E. Conference, London, Oct. 1982, paper C151/82.
8. Effectiveness and reliability of US in-service inspection techniques.
   S.R. Doctor, F.L. Becker and G.P. Selby, I. Mech. E. Conference,

London, Oct. 1982, paper C150/82.

9. Size Measurement and Characterisation of Weld Defects by Ultrasonic Testing. Part 1. Non-planar defects in ferritic steels. Welding Institute, U.K. reference 3527/4/77.

10. Size Measurement and Characterisation of Weld Defects by Ultrasonic Testing. Part 2. Planar defects in ferritic steel. Welding Institute, U.K. reference 3527/10/80. Nov. 1980.

11. A method of predicting the reliability of ultrasonic crack detection from an analysis of PISC 1 data, N.F. Haines, CEGB report RD/B/ /5209N81.

12. UKAEA Defect Detection Trials: Application of some results to model validation and assessment of technique capability. D.B. Langston and R. Wilson, CEGB Report to be issued September 1984.

Effects of Defect and Ultrasonic Parameters on Detection

1. Relationship between ultrasonic indications and flaw size in pressure vessels welds, H.J. Meyer, B+W-TUV Conference on NDE, Washington DC, November 1976.

2. A practical model of ultrasonic reflection from planar surfaces. N.F. Haines, CEGB Report RD/B/N4995, January 1981.

3. A practical model of the ultrasonic tandem technique, D.B. Langston, CEGB Report RD/B/5074N81, July 1981.

4. Description and Application of Elastic Scattering Theory for Smooth Flat Cracks to the Quantitative Prediction of Ultrasonic Defect Detection and Sizing, J.M. Coffey and R.K. Chapman, Specialist Meeting, Ispra, May 1983. EUR 9066 1 EN.

5. Practical model of ultrasonic reflection from surface breaking planar defects. N.F. Haines and A.J. Green, CEGB Report RD/B/5073N81, August 1981.

Reporting Thresholds Calibration

1. Properties of cylindrical boreholes as reference defects in ultrasonic inspection. H. Wustenberg and E. Mundry. Journal of NDT, August 1971.

2. A practical method of relating flat bottom to side drilled hole calibration reflectors for NDT. N.F. Haines, CEGB Report RD/B/5064N81, April 1981.

Subject Index